Integral equations and applications

Integral equations and applications

C. CORDUNEANU

The University of Texas at Arlington

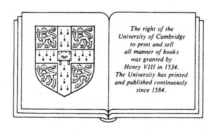

The right of the
University of Cambridge
to print and sell
all manner of books
was granted by
Henry VIII in 1534.
The University has printed
and published continuously
since 1584.

CAMBRIDGE UNIVERSITY PRESS

Cambridge

New York   Port Chester

Melbourne   Sydney

CAMBRIDGE UNIVERSITY PRESS
Cambridge, New York, Melbourne, Madrid, Cape Town, Singapore, São Paulo, Delhi

Cambridge University Press
The Edinburgh Building, Cambridge CB2 8RU, UK

Published in the United States of America by Cambridge University Press, New York

www.cambridge.org
Information on this title: www.cambridge.org/9780521340502

© Cambridge University Press 1991

This publication is in copyright. Subject to statutory exception
and to the provisions of relevant collective licensing agreements,
no reproduction of any part may take place without the written
permission of Cambridge University Press.

First published 1991
This digitally printed version 2008

A catalogue record for this publication is available from the British Library

Library of Congress Cataloguing in Publication data

Corduneanu, C.
Integral equations and applications / C. Corduneanu.
p. cm.
Includes bibliographical references and index.
ISBN 0 521 34050 0
1. Integral equations. I. Title.
QA431.C596 1991
515'.45 – dc20 90-40858 CIP

ISBN 978-0-521-34050-2 hardback
ISBN 978-0-521-09190-9 paperback

Contents

Preface

In the author's view, this book has at least three objectives. First, the book aims to serve as a (graduate) textbook of integral equations. The first chapter introduces the reader to the subject, in the third chapter several basic facts are included on Volterra type equations (both classical and abstract), while the remaining chapters cover a variety of topics to be selected to suit the particular interest of the instructor and students. Second, the book is aimed to serve as a reference in the field of integral equations and some of their applications. Of course, I cannot claim to provide comprehensive coverage of this fast-developing area of research, but I hope that the topics featured in the book will convince the reader that integral equations constitute a very useful and successful tool in contemporary research, unifying many particular results available for other classes of functional equations (differential, integrodifferential, delayed argument). Third, the book provides a good number of results, and describes methods, in the field of integral equations, a feature that will help the young researcher to become acquainted with this field and continue the investigation of the topics whose presentation in the book suggests further development.

Most of the material included in the book is accessible to any reader with a reasonable background in real analysis, and some acquaintance with the introductory concepts of functional analysis. There are several sections which require more sophisticated knowledge of functional analysis (both linear and nonlinear). In such cases, some direction is given, and adequate references are provided. These references are usually to books and monographs dealing with such topics in more depth; only in a very few (perhaps half a dozen) cases is the reader referred to journal papers containing such results. Of course, there are many references to journal papers, but those are aimed at indicating the origination of the results, or contain supplementary material directly related to the text.

Since a rather detailed description of the contents is given in the introduction, together with some historical considerations related to integral equations, I do not find it necessary to dwell upon this aspect here.

I think it is appropriate to make precise the relationship between this volume and the volumes I have published in the past on this subject (see C. Corduneanu [4] and [6] in the References). The material in [6] is covered in the first chapter of this book, but under different assumptions. For instance, the basic Fredholm theory is presented in the L^2-space, instead of the space of continuous functions. Unlike in [6], the idea of approximating general kernels by means of finite-dimensional kernels is used. With regard to [4], in practical terms, there are no repetitions. Some topics, particularly the admissibility results, have been dealt with again but under totally new conditions. More precisely, new developments related to some topics in [4] have found a place in the present book. The same is true with regard to the stability problems for nuclear reactors. Although more than 60% of the contents of [4] is devoted to convolution equations, there are relatively few convolution results in this book.

With regard to the list of references at the end of this volume I would like to mention that I have tried to include any book on the theory of integral equations, old or new, including books dedicated to applications or numerical treatment of these equations. I have left aside titles dealing with the theory of singular integral equations, which have many applications in continuum mechanics and other fields. I have also avoided titles on stochastic integral equations, despite the growing interest in this area. The reason for these omissions is that it is impossible to cover adequately in a single volume such a diversity of topics.

It is generally understood that carrying out a project like this necessitates a good deal of cooperation and interaction. I am particularly indebted for various forms of help in connection with this project to the following colleagues: S. Aizicovici, from Ohio University, Athens, Ohio; T. A. Burton, from Southern Illinois University, Carbondale; D. A. Carlson, from Southern Illinois University; M. Kwapisz, from the University of Gdansk, Poland; A. Korzeniowski, from the University of Texas at Arlington; N. H. Pavel, from Ohio University, Athens, Ohio; J. J. Schaffer, from Carnegie-Mellon University in Pittsburgh; O. Staffans, from Helsinki University of Technology, Espoo, Finland. Professor Staffans read the whole manuscript and suggested many corrections and improvements. It is my duty and my pleasure to express my sincere thanks to all the colleagues mentioned above. Several copies of the manuscript have circulated since the summer of 1988, and numerous remarks and comments have reached me in time for me to take them into account in the final version of the manuscript.

The historical sketch of the theory of integral equations in the introduction was written in conjunction with G. Bantaş from University of Iaşi, Romania. I thank him for his cooperation.

In the technical preparation of the manuscript, constant help has been provided by two of the secretaries in the Department of Mathematics at the

University of Texas, Arlington: Marjorie Nutt and Sue Sholar. I gratefully acknowledge their support during the project.

Also, I would like to take this opportunity to acknowledge the following sources of support for the completion of this project (in the form of a Summer research salary, or as travel grants, or both): The Graduate School of the University of Texas at Arlington; the Oberwolfach Mathematisches Forschungsinstitut; the Office of the Chief of Naval Research; the National Research Council of Italy (CNR): the US Army Research Office (Durham, NC). This support has certainly helped to bring the project into being in a shorter time.

Finally, my gratitude is directed to Cambridge University Press whose interest in this project has provided me from the start with strong motivation and encouragement.

C. Corduneanu

Introduction

It is legitimate to consider J. Fourier (1768–1830) as the initiator of the theory of integral equations, owing to the fact that he obtained the inversion formula for what we now call the 'Fourier transform': under adequate restrictions on the functions f and g, from $g(x) = (2\pi)^{-1/2}\int\exp(ixy)f(y)\,dy$ one derives $f(y) = (2\pi)^{-1/2}\int\exp(-ixy)g(x)\,dx$. Of course, one can interpret the inversion formula as providing the inverse operator (an integral operator!) of the Fourier integral operator. This interpretation was adopted towards the end of the last century by V. Volterra, who identified the problem of solving integral equations with the problem of finding inverses of certain integral operators.

Abel [1], [2] dealt with the integral equation known as 'Abel's equation', namely

$$\int_0^x (x-t)^{-\alpha}u(t)\,dt = F(x), \quad 0 < \alpha < 1,$$

where $u(t)$ stands for the unknown, while $F(x)$ is an assigned function. The solution of Abel's equation was provided by the formula

$$u(t) = -\pi^{-1}\sin\alpha\pi\frac{d}{dt}\int_0^t (t-x)^{\alpha-1}F(x)\,dx.$$

There are two interesting features related to Abel's equation. First, the so-called kernel $k(t) = t^{-\alpha}$ has a singularity at the origin, and second, Abel's equation is a convolution type equation. Both features have had a significant impact on the development of the theory of integral equations. It is interesting to point out that Abel found his integral equation in the case $\alpha = \frac{1}{2}$, starting from the following mechanical problem: find a curve C lying in a vertical plane xOy, such that a material point sliding without friction on C, and starting with zero initial velocity at a given point P_0 on C, reaches the origin O on C in an interval of time which is a given function of the y-coordinate of P_0. The method of solution used by Abel was neither simple nor rigorous, as pointed out by Schlömilch [1]. Schlömilch proposed a more rigorous approach, based on the reduction of multiple integrals to simpler ones.

Abel's work on integral equations exerted a tremendous influence during the following century. In 1860, Rouché [1], in 1884 Sonine [1], in 1888 du Bois-Reymond [1] were dedicating their efforts to the solution of various integral equations, particularly Abel's equation. du Bois-Reymond is credited with the introduction of the term 'integral equation'. In 1895, Levi-Civita [1] generalized the results obtained earlier by N. Sonine, still being concerned with Abel's type equations. Goursat [1] dealt with the solution of Abel's type equations, regarded as a problem of inversion of an integral, and in 1909 Myller [1] dealt with some problems in mechanics by using Abel's type equations as well as Fredholm's type equations (see Section 1.1). One century after Abel introduced and studied the integral equations now known as Abel's equations, mathematicians like Tamarkin [1] and Tonelli [1] were dedicating their attention to this subject.

But the year 1895 marked a new beginning in the theory of integral equations, due mainly to Volterra [2]. Unlike most of his predecessors, who were aiming at finding the solution of the equation by means of formulas or who dealt with special cases of what we now call 'Volterra equations' (the term was introduced by T. Lalesco who wrote his thesis on this topic with E. Picard), Volterra considered more general equations such as

$$\int_0^x k(x,t)u(t)\,dt = F(x),$$

or

$$u(x) + \int_0^x k(x,t)u(t)\,dt = F(x),$$

in which u stands for the unknown function. Volterra, who is also one of the founders of functional analysis, regarded the problem from a functional analytic standpoint, his main concern being the proof of the existence of the inverse of an integral operator. Volterra himself was not at first concerned with the various physical applications of the concept of an integral equation, the preoccupations which played a preponderant role later in his activities. In the same year 1895 when Volterra began his fundamental contribution to the theory of integral equations, Le Roux [1] published a notable paper in which integral equations appear as a powerful tool in investigating partial differential equations.

It is interesting to note that more than a half century before Volterra produced his contributions, Liouville [3] discovered the fact that the differential equation $y'' + [\rho^2 - a(x)]y = 0$, under initial conditions $y(0) = 1$, $y'(0) = 0$, is equivalent to the integral equation (of Volterra type and second kind)

$$y(x) - \rho^{-1} \int_0^x a(t) \sin \rho(x - t) y(t) \, dt = \cos \rho x.$$

Also, apparently unaware of Abel's investigations concerning integral equations, Liouville [1], [2] dealt with the singular integral equation

$$\int_x^\infty (s - x)^{-1} y(s) \, ds = f(x),$$

for which the solution is given by the formula

$$y(x) = -\pi^{-1} \frac{d}{dx} \int_x^\infty (s - x)^{-1/2} f(s) \, ds.$$

Five years after Volterra made his first famous contribution to the theory of integral equations, Fredholm [1], [2] built up a new theory for integral equations containing a parameter of the form

$$y(x) + \lambda \int_a^b k(x, t) y(t) \, dt = f(x).$$

Fredholm's theory, basically representing the extension of solvability of a linear finite-dimensional system $x + \lambda Ax = b$ to the infinite-dimensional case, has constituted one of the most valuable sources for the foundation of what is now known as linear functional analysis. It is instructive to notice that the fundamental paper by Banach [1] provides in its title the motivation for the necessity of new mathematical structures (Banach spaces), while Riesz (see Riesz and Sz.-Nagy [1]) proceeds to the generalization of Fredholm's theory to the case of abstract operators in linear spaces. In the years following the publication of Fredholm's results, illustrious mathematicians of the beginning of this century made their contributions to this subject: Poincaré [1], [2], Hilbert [1], Picard [1], Weyl [1], Fréchet (see Heywood and Fréchet [1]) and Schmidt [1]. The theory was considerably enriched and developed, and its connections to other branches of science were emphasized. In particular, mainly owing to Hilbert and Schmidt, the theory of equations with symmetric kernel was substantially developed, and important results about the orthogonal series were obtained.

Owing to the tremendous success that Fredholm's theory enjoyed during the first decades of this century, the theory of Volterra equations remained somewhat in its shadow. Most books on integral equations written during that period, as well as in more recent years, dismiss Volterra's equation by noticing that the corresponding integral operator has only the eigenvalue zero, and is therefore uninteresting. This approach is not very consistent because: first, if we deal with Volterra equations on infinite intervals, then the difference between them and Fredholm's equations is no longer so striking (with respect to the spectral properties); second, for Volterra equations it is possible to develop a theory of local existence, making these equations very useful in modelling various phenomena from applied fields of science; third, with the advance of nonlinear functional analysis during the past three decades, it has been possible to develop the study of Volterra nonlinear equations to a level which has not yet been reached by the theory of Fredholm nonlinear equations (both in terms of the number of results obtained by the researchers involved in this field, and the variety of the applications). We hope that this feature is adequately emphasized in this book.

Volterra was deeply preoccupied with the possible applications of the theory of integral equations in other fields of science. Among the most interesting applications found for the Volterra equations, we should first mention the so-called 'hereditary mechanics', also known as the 'mechanics of materials with memory'. The roots of this theory can be traced back to Boltzmann [1]. But Volterra advanced a rather sophisticated theory in the 1920s (see Volterra [2]), a theory which underwent some modification during the last three decades and which we can consider as still developing. In his studies during the 1920s on hereditary mechanics, Volterra was led to equations with infinite retardation (delay). Realizing that the mathematical apparatus was not yet developed at that time for undertaking successfully such investigations, Volterra 'cut' the delay, transforming his equations into equations with finite delay. Unfortunately, at that time, the theory of equations with finite delay was practically nonexistent, so that real progress had to be postponed.

Another field in which the theory of integral equations of Volterra type has found significant applications, beginning in the 1920s, is population dynamics. In Volterra and d'Ancona [1] one finds a synthesis of the first generation research pertaining to this field. This kind of study has been developed in the last two decades, and further progress is to be expected. A recent monograph on these matters is that of Webb [3]. See also Cushing [3].

Let us also mention the applications that the Volterra equations have found in Economic theory. An adequate reference in this respect is Samuelson [1].

In 1929, Tonelli [2] introduced the concept of a Volterra operator, as an operator U acting between function spaces, such that $x(t) = y(t)$ for $t \leq T$ implies $(Ux)(T) = (Uy)(T)$. Such operators are also known as *causal* operators or *nonanticipative* operators. Since most phenomena investigated by means of mathematical models are causal phenomena, the importance of this class of operators is obvious. Followers of Tonelli, such as Graffi [1] and Cinquini [1], produced some of the first contributions towards the foundations of the theory of abstract Volterra operators. In 1938, Tychonoff [2] emphasized again the significance of the theory of abstract Volterra operators in connection with the numerous applications in mathematical physics. He advanced this theory and introduced a more modern approach which contributed to its development, and served as a model for future investigations. The theory of abstract Volterra operators and equations has made substantial progress during the last 10–15 years, as will be illustrated in this book (see Chapters 2 and 3, particularly). The further development of this theory is one of the major problems in the theory of integral equations, viewed as an extension of the classical theory. Without such development, it seems unsatisfactory to pretend that the mathematical tools available for the investigation of phenomena in which heredity (hysteresis) occurs are powerful enough.

Before we move on to the examination of the theory of integral equations in more recent times, we take this opportunity to mention the contributions of Carleman [1] from the 1920s, and von Neumann [1] from the 1930s. Carleman's contribution should be regarded as the beginning of the theory of unbounded linear integral operators. This theory is still under development, and fruits are likely to follow. A recent monograph on Carleman's operators is due to Korotkov [1]. The spectral theory of integral operators is another story which holds great promise for the future. Recent noteworthy contributions are due to Pietsch [1], [2] and to Elstner and Pietsch [1].

During the 1940s there was little progress in the theory of integral equations. One remarkable contribution was due to Dolph [1], who made a substantial addition to the nonlinear theory of Hammerstein equations. The interaction between the nonlinearity of the equation and the spectrum of the linear integral operator involved in the equation is illustrated in a simple manner. Another basic contribution worthy of mention was due to Akhiezer [1] and relates to the theory of Carleman operators.

The 1950s were more eventful with regard to the theory of integral equations. In 1953, Sato [1] dealt with Volterra nonlinear equations from the standpoint of qualitative theory. In 1956 the Russian edition of the book by Krasnoselskii [1] was published. In the late 1950s M. G. Krein and I. C. Gohberg started a series of research efforts directed towards the build-up of a theory for classes of convolution equations on a half-axis, or on the entire real axis. Resolvent kernels were constructed and the behavior of solutions was investigated in a systematic manner for equations that, in general, possess a continuous spectrum. An account of this theory is given in my book (C. Corduneanu [4], Ch. 4, Wiener–Hopf equations). More developments of this theory are included in the monograph by Gohberg and Feldman [1]. More recent results in this direction can be found in Gohberg and Kaashoek [1]. Also in the late 1950s, the first work by V. M. Popov was published relating to the use of integral equations occuring in the theory of feedback systems. These are Volterra equations of convolution type, containing one or more nonlinearities. Fundamental stability results were obtained within a short time by Popov and many of his followers. In my book (C. Corduneanu [4], Ch. 3), most of the results obtained before 1970 in this direction have been included or reviewed. The research work in this field has been continued by many authors (particularly in Soviet Union and Romania). As recently as 1976, Nohel and Shea [1] published results in this line (frequency domain criteria of stability or other kinds of asymptotic behavior of solutions).

From the 1960s onwards, interest in the theory of integral equations has reached a level of concentration unknown since the years following Fredholm's investigations of this topic. Many research schools in the United States, Soviet Union, Italy, India, Japan, Finland, Romania, Poland, Israel, and other countries are directing their efforts towards the investigation of various problems related to the theory of integral operators and integral equations. While this growing interest is motivated in part by the numerous applications that integral equations have found in the mathematical modeling of phenomena and processes occuring in various areas of contemporary research, it should also be stressed that the development of the methods in nonlinear analysis has made possible the successful investigation of this kind of problem. There are presently several journals dedicated to the theory of integral equations and operators: the *Journal of Integral Equations and Applications* (the new series being published by Rocky Mountains Mathematical Consortium), *Integral Equations and Operator Theory* (Birkhauser), as well as the semiperiodic publication *Investigations on Integro-differential Equations in Kirkizia* (Russian). Journals such as *Differential Equations* (transl.), the *Journal of Mathematical Analysis and Applications*, and the *Journal of Differential Equations* contain numerous contributions dedicated to the theory of integral equations. Mathematical Reviews, *Zentralblatt für Mathematik* and *Referativnyi Žurnal* (*Matematika*), insert annually over 1000 reviews dedicated to integral equations and operators (not counting those papers in which integral equations appear only casually). It is very difficult to sketch adequately the contemporary picture of the research field of integral equations and operators. This should really be the work of a whole team of investigators. I will, however, make an attempt, although convinced that serious shortcomings might occur.

In the early 1960s, a series of research papers due to J. J. Levin and J. A. Nohel pointed out the role of integral equations as tools in the study of the stability of nuclear reactors. Unlike Popov [1], Levin and Nohel based their investigation on the so-called 'energy method'. In other words, a kind of Liapunov functional was used to derive information about the solutions. The research work started by Levin and Nohel at the University of Wisconsin at Madison has been continued by attracting the attention of numerous other researchers. During the 1970s and the 1980s the Madison school concentrated its efforts on problems occuring in continuum mechanics, particularly in viscoelasticity. The recent monograph by Renardy, Hrusa and Nohel [1] provides a picture of the research activities conducted at Madison, though not a complete one.

About the same time, the School of Continuum Mechanics at Carnegie-Melon University in Pittsburgh was directing its efforts towards the foundations of this discipline, making systematic use of the theory of integral equations and the theory of equations with infinite delay. For contributions from this school see, in the list of references, papers under the names of MacCamy, Mizel, Hrusa, and their co-workers (Marcus, Leitman, and others). In relation to this school and contributions to the topic of integral equations or equations with delay we should also add the names of Coleman, Gurtin and Noll.

At Brown University in Providence, Rhode Island, J. K. Hale and many co-workers have made numerous and substantial contributions to the field under discussion. Also, C. M. Dafermos, mostly from the point of view of a researcher in continuum mechanics, has brought valuable contributions to the theory of integral equations (more precisely, integro-partial differential equations).

Another group of researchers interested in the theory of integral equations is the one at Virginia Polytechnic Institute and State University (K. Hannsgen, R. L. Wheeler, T. L. Herdman, M. Renardy and others).

At the Southern Illinois University in Carbondale, a group of researchers are actively participating in the development of the theory of integral and related equations, including investigations in the theory of control systems described by means of integral equations (T. A. Burton, D. Carlson, R. C. Grimmer, C. E. Langenhop).

A considerable number of isolated researchers have conducted valuable work in the field of integral and related equations, in the United States: F. Bloom, F. E. Browder, J. R. Cannon, D. Colton, J. M. Cushing, H. Engler, W. E. Fitzgibbon, H. E. Gollwitzer, C. W. Groetsch, M. L. Heard, A. J. Jerri, G. S. Jordan, R. K. Miller, M. Milman, K. S. Narendra, M. Z. Nashed, A. G. Ramm, W. J. Rugh, I. W. Sandberg, A. Schep, M. Schetzen, V. S. Sunder, and G. F. Webb.

In the Soviet Union, at least four schools of research in integral equations and integral operators have contributed remarkably to the progress of this field during the last three decades: the Krein-Gohberg school, the Krasnoselskii school in Voroneż (and then in Moscow), the school in Novosibirsk (Korotkov and his followers) and the school grouped around N. V. Abelev and Z. B. Caljuk (with ramifications in various academic centers of the USSR).

The Krein-Gohberg school had many followers in Odessa, Kishinev, and other centers in the Soviet Union. In the 1970s, I. C. Gohberg emigrated to Israel, where he continues his work on Wiener–Hopf techniques and their generalizations. In particular, the factorization problem has been worked on by Gohberg and his followers from Israel, the United States and the Netherlands. The Kishinev group also continued their research activities, more or less on the same lines.

The school created by M. A. Krasnoselskii in Voroneż has also spread to various centers. Among the most remarkable achievements of this group we mention the joint work by Krasnoselskii, Zabreyko, Pustylnik and Sobolevskii [1], as well as basic results obtained by P. P. Zabreyko on nonlinear integral operators. In Moscow, Krasnoselskii and Pokrovskii [1] are conducting research work on 'systems with hysteresis'.

The school in Novosibirsk is concentrating on the theory of Carleman (unbounded) integral operators. This direction is very promising for the near future, when we can expect some applications to the theory of integral equations. So far, only sporadic results have been obtained (see, for instance, Korotkov [2]).

Various interesting results concerning integral and related equations have been obtained by the group led by N. V. Azbelev and Z. B. Caljuk: stability problems, boundary value problems for functional-differential equations, integral representation of solutions, and many other aspects have been emphasized.

In many other academic centers in the USSR, the integral and related equations are cultivated both from a theoretical point of view, and from the point of view of their applications (we are not concerned here with the so-called singular equations which do appear in mechanical problems, and for which a vast literature is available in Russian).

In Italy, where there has always been interest in Volterra equations and their generalizations and applications, a group of scholars in Pisa, Rome and Trento are heavily engaged in developing the theory of abstract Volterra equations (the term abstract should be interpreted here as Banach-space valued), together with their applications to mechanics or population dynamics. Most of the researchers are students of G. DaPrato (M. Iannelli, A. Lunardi, E. Sinestrari, G. DiBlasio, and others). An important reference not belonging to the above category is due to Fichera [1], who has also made other significant contributions to the field.

As mentioned above, other countries participated during the last three decades to the advancement of the theory of integral equations: in Finland (S.-O. Londen, O. J. Staffans, G. Gripenberg) the theory of convolution equations has seen a real development. By means of classical and functional analytic methods, including semigroup theory and Laplace–Fourier transform theory, a large variety of qualitative problems have been successfully investigated. A book by Finnish mathematicians, probably dealing with this kind of investigation, is in preparation. In Japan, special attention has been paid to the theory of equations with infinite delays (J. Kato, T. Naito, Y. Hino, S. Murakami), and functional equations involving integral operators have been investigated by many authors (for instance, N. Hirano, Nobuyuki Kato).

While most contributions to the theory of integral and related equations, published during the last three decades, deal with various applications, it is worth while mentioning the fact that in their attempt to solve such applied problems the investigators made use of quite recent tools created by basic mathematical research (monotone operators, linear and nonlinear semigroups of transformation, as well as many other functional analytic methods). In this way, the theory of integral equations has been considerably developed and enriched.

I will now briefly describe the structure of the book, pointing out directions in which the topics under discussion could be further developed.

The first chapter is introductory, and is aimed at emphasizing the fact that integral equations/operators occur either directly in the description of certain phenomena, or indirectly – when processing other types of functional equations. Also, it contains a rather elementary introduction to the theory of Volterra equations, as well as the Fredholm theory of linear integral equations. The last section of the first chapter deals with Hammerstein equations, basically under the hypotheses adopted by Dolph in his thesis at Princeton (1944). This theory is nonlinear, and it can be dealt with using fairly classical tools. The reader might be surprised by the fact that very little is included in relation to the Hilbert (or Hilbert–Schmidt) theory of symmetric kernels. Indeed, this is one of the most salient parts of the classical theory of linear integral equations, and it should certainly be included in any book on this subject. In order to keep the size of the volume within reasonable dimensions, however, I decided not to include this classical chapter of the theory of integral equations. The interested reader can find numerous sources for this theory, under various basic assumptions on the kernel (besides its symmetry): Cochran [1], C. Corduneanu [6], Fenyo and Stolle [1], Goursat [1], Hamel [1], Hochstadt [1], Hoheisel [1], Kanwal [1], Lalesco [1], Lovitt [1], Mikhlin [1], Petrovskii [1], Schmeidler [1], Smithies [1], Tricomi [1], Vivanti [1].

The second chapter is an auxiliary for the following chapters and contains a series of definitions and properties of some function spaces to be used in this book, some properties of certain integral operators acting on the function spaces introduced earlier, as well as the statements (with indications for the proofs) of several basic results relating to fixed point theorems or monotone operators. This chapter is not intended to be a complete presentation of the topics which are considered, and its only role should be to enable readers to become acquainted with some of the methods and tools to be used in later chapters. On the other hand, special results from functional analysis that are needed only in connection with a single result in the book, have been stated (and reference provided in book form) in subsequent chapters.

The third chapter of the book is entirely dedicated to the (mostly but not only local) theory of Volterra equations, including functional-differential equations that can be reduced to Volterra ones. It turns out that, in the framework of abstract Volterra equations, one can encompass practically all types of equations with delayed argument. I have particularly illustrated the case of equations with infinite (unbounded) delays, but the theory of equations with finite delay is also a special case of abstract functional-differential equations with Volterra (causal) operators. This approach will probably change the way most particular problems are addressed nowadays, introducing a broad unifying idea and providing more generality. There is more to be done in this respect, and not only in connection with the general theory of existence, uniqueness, continuous dependence, etc. Applications of the abstract approach will certainly be extended to the theory of control processes described by this type of equation. The last section of the chapter is devoted to the presentation of an approach based on the singular perturbation method, also in the case of abstract Volterra equations. While this method has been used by several authors in connection with particular classes of Volterra equations, none of the contributions has dealt with the abstract case.

The fourth chapter is a collection of results related to both Volterra and Fredholm equations (particularly in the form $x = f + KNx$, with K a linear integral operator and N a nonlinear Niemytskii operator), which tries to emphasize the connection of these equations to various problems in nonlinear analysis, such as boundary value problems for ordinary differential equations. Admissibility techniques, construction of resolvent kernels with preassigned properties, Hammerstein equations in spaces of measurable functions, asymptotic behavior of solutions for integrodifferential equations on a half-axis, perodic and almost periodic solutions, multivalued integral equations (inclusions) are considered in this chapter. These topics have been recently discussed in the mathematical literature on this subject, and they certainly reflect some current preoccupations in this field of research. Of course, the list of topics could be considerably extended, owing to the significant amount of research work currently being carried out in this field. Let us point out that most of the material included in the fourth chapter is formed on the pattern of finite-dimensional theory, but some topics are suitable for generalization to the infinite-dimensional case.

In the fifth chapter some problems pertaining to the theory of integral equations in a Banach (Hilbert) space are discussed, while some methods like the semigroup theory and monotone operators are illustrated in connection with the theory of integral or related equations. In some cases, as for example in the second section, a general operator approach has been adopted for the Amann's generalization of Hammerstein theory, and then applications are considered to integral equations. The semigroup method is presented in connection with the problem of existence of the resolvent kernel (which is still far from a satisfactory solution in the infinite-dimensional case). The nonlinear semigroups appear naturally in discussing the dynamics described by a nonlinear time-invariant integral equation of Volterra type. Integrodifferential equations in Hilbert space are also discussed in this chapter, with applications to the existence theory for integro-partial differential equations (as those occurring in continuum mechanics). Much more remains to be done in the case of integral and related equations in Banach/Hilbert spaces, since the existing results belong either to classes of rather special equations, or relate to equations with bounded operators for which applications appear very seldom.

The last chapter of the book is devoted to some applications that the integral or related equations have found during the last three decades. We did not deal with such classical applications as those in population dynamics or continuum mechanics, which can be traced back to Volterra, and for which monographs have recently appeared. Instead, we have considered applications to transport theory (coagulation processes), as they appear in one of the pioneering papers on the subject (see Melzak [1]). Also, we have included a maximum principle result for optimal control of processes described by Volterra nonlinear equations (see Carlson [1]), from which the classical Pontrjagin's principle can be obtained as a special case. The stability of nuclear reactors is discussed in the last two sections of Chapter 6, using a model based on integrodifferential equations. It is worth while pointing out the fact that this problem has generated a good deal of research and contributed to the progress of integral and related equations during the last three decades. Other applications will be reviewed in the reference section to this chapter, with adequate sources indicated.

The list of references contains more than 500 entries. I have tried to include in the list all books and monographs published in the field. The first book on integral equations was by Maxime Bocher, published in 1909, by Cambridge University Press. Shortly after that, books by D. Hilbert, Maurice Frechet, and T. Lalesco came out, making the theory of integral equations a rather popular topic among mathematicians. The journal papers quoted in the list of references have been selected in accordance with the topics discussed in this book. Nevertheless, some of the classical contributions to the field, such as those due to I. Fredholm, E. Schmidt, L. Tonelli, A. Hammerstein, do appear in the list, because of their particular significance in the development of the theory of integral equations.

The list of references does not include titles related to the following important directions in the study of integral equations, which we have not covered in this book. First, the so-called theory of singular integral equations occurring in various problems of continuum mechanics. Some coverage of these topics is provided in the volume *Integral Equations – A Reference Text* by Zabreyko, Koshelev, Krasnoselskii, Mikhlin, Rakovshchik and Stetsenko [1]. Basic references on the subject are provided. Second, apart from some books that have been published relatively recently, and a few recent journal papers, there are no references concerned with numerical analysis of integral equations. This topic has acquired an impressive role in mathematical research during the last decade, and many more books have been published recently on the numerical treatment of integral equations than on the basic theory. See the references under the names of Anderssen *et al.*; Baker; Brunner and van der Houwen (this book contains a good historical sketch and a rich bibliography of integral equations); Delves and Mohamed; Delves and Walsh; Golberg (ed.); Ivanov (also including singular equations); Linz, Mohamed and Walsh; Paddon and Holstein; Reinhardt; te Riele; Kantorovich and Krylov (which is not entirely dedicated to integral equations but is one of the first references in this area). Third, random integral equations have not been covered in the book, despite an increased interest in this subject. See the books by Bharucha-Reid, and Tsokos and Padgett, as well as journal papers by Ahmed, Lewin and Tudor.

1

Introduction to the theory of integral equations

This introductory chapter is aimed at familiarizing the reader with the concept of *integral equations*. It will be shown that various classical problems in the theory of differential equations (ordinary or partial) lead to integral equations and, in many cases, can be dealt with in a more satisfactory manner using these than directly with differential equations. Also, various problems in applied science are conducive to integral equations in a natural way, these equations thus emerging as competent mathematical tools in modelling phenomena and processes encountered in those fields of investigation.

Certain basic results, mostly relating to the classical heritage, will be discussed in this chapter. By doing this, we hope to help the reader to prepare for more challenging problems that will be considered in subsequent chapters.

1.1 Integral equations generated by problems in the theory of ordinary differential equations

Several problems in the theory of ordinary differential equations are related to systems of the form

$$\dot{x}(t) = A(t)x(t) + f(t, x(t)), \tag{1.1.1}$$

in which the vector-valued unknown function $x(t)$ takes values in a finite-dimensional linear space, say R^n, $A(t)$ is a square matrix of order n with real-valued entries, while $f(t, x)$ stands for a vector-valued function with values in R^n, defined on some interval $[t_0, T)$, $T \leq \infty$, and $x \in R^n$ (or a subset of R^n).

If we are interested to prove the existence of a solution to (1.1.1), satisfying the initial condition

$$x(t_0) = x^0, \quad x^0 \in R^n, \tag{1.1.2}$$

then it is convenient to transform the problem (1.1.1), (1.1.2) into an integral equation with the same unknown function $x(t)$.

13

The basic tool in obtaining such an integral equation, equivalent to (1.1.1), (1.1.2), is the variation of constants formula for linear differential systems of the form

$$\dot{x}(t) = A(t)x(t) + f(t), \tag{1.1.3}$$

which states that the solution of (1.1.3), under initial condition (1.1.2), can be represented by means of the formula

$$x(t) = X(t)X^{-1}(t_0)x^0 + \int_{t_0}^{t} X(t)X^{-1}(s)f(s)\,ds, \tag{1.1.4}$$

where $X(t)$ is determined from $\dot{X}(t) = A(t)X(t)$ on $[t_0, T)$, $\det X(t_0) \neq 0$, up to a constant (matrix) factor. Instead, the product $X(t)X^{-1}(s)$, $t \geq s \geq t_0$, is uniquely determined (sometimes it is called the transition operator along the solutions of the homogeneous equation associated to (1.1.3)).

By means of formula (1.1.4), the following integral equation is obtained from (1.1.1), (1.1.2):

$$x(t) = X(t)X^{-1}(t_0)x^0 + \int_{t_0}^{t} X(t)X^{-1}(s)f(s, x(s))\,ds, \tag{1.1.5}$$

which constitutes a particular case of a more general integral equation, namely

$$x(t) = f(t) + \int_{t_0}^{t} k(t, s, x(s))\,ds. \tag{1.1.6}$$

In equation (1.1.6), x, f and k are vector-valued functions, with values in R^n.

Of course, certain conditions have to be assumed in order to show the equivalence of (1.1.6) with (1.1.1), (1.1.2). For instance, the continuity of the matrix-valued function $A(t)$, and that of the vector-valued function $f(t, x)$ will suffice in order to be able to operate as shown above. But Carathéodory type conditions are also acceptable, the differentiability being then assumed only almost everywhere. The equivalence is usually established within the class of continuous (or even absolutely continuous) functions.

From (1.1.5), assuming only the continuity of the solution $x(t)$, one can easily obtain both (1.1.1) and (1.1.2). Hence, a continuous solution of the integral equation (1.1.5) is necessarily differentiable and satisfies both (1.1.1) and (1.1.2). The fact that for the integral equation we have to look only for continuous solutions, and are not therefore concerned with regularity properties such as the differentiability, is a major advantage in terms of the technicalities involved.

Let us mention the fact that an integral equation of the form (1.1.6) is known as a *Volterra integral equation* of the second kind. At least one of the limits of integration must be variable (not necessarily identical to the independent variable as in the case above, but depending in some way on t).

In the special case $A(t) \equiv 0$ (the zero matrix of order n), one has $X(t)X^{-1}(s) \equiv I$

(the unit matrix of order n), and (1.1.5) becomes

$$x(t) = x^0 + \int_{t_0}^{t} f(s, x(s)) \, ds, \tag{1.1.7}$$

which is known to be equivalent to the equation

$$\dot{x}(t) = f(t, x(t)), \tag{1.1.8}$$

with the initial condition (1.1.2). Most methods of proof for the existence of solutions deal directly with the integral equation (1.1.7), instead of (1.1.8), (1.1.2).

Let us consider now the differential equation (system) (1.1.8), under some requirements for the solution, more sophisticated than (1.1.2). Such requirements are usually known as *boundary value conditions*.

If we assume, for instance, that $f(t, x)$ in equation (1.1.8) is a continuous map from $[a, b] \times D$, with $D \subset R^n$, into the space R^n, then we look for solutions of (1.1.8) which satisfy

$$\int_{a}^{b} [dH(t)] x(t) = \int_{a}^{b} F(t, x(t)) \, dt, \tag{1.1.9}$$

where $H(t)$ is a square matrix of order n whose entries are real-valued functions with bounded variation on the interval $[a, b]$, and $F(t, x)$ is like the function $f(t, x)$ in equation (1.1.8).

In order to obtain from (1.1.9) the usual initial condition $x(a) = x^0$ as a special case, we have to choose $H(t)$ to be the diagonal matrix whose elements on the main diagonal are identical to $h(t)$: $h(a) = 0$, $h(t) = 1$, $a < t \leq b$, and $F(t, x) = x^0 (b - a)^{-1}$.

We shall prove now that the problem (1.1.8), (1.1.9) is equivalent to an integral equation for the unknown function $x(t)$. Indeed, from (1.1.8) we obtain

$$x(t) = x(a) + \int_{a}^{t} f(s, x(s)) \, ds, \quad t \in [a, b], \tag{1.1.10}$$

which is actually equivalent to (1.1.8). If we can 'determine' $x(a)$ from the condition (1.1.9), then we shall have an integral equation instead of (1.1.10). It is easy to see then that (1.1.8), (1.1.9) are equivalent to the integral equation resulting from (1.1.10).

Notice that the integration by parts formula in the Stieltjes integral yields the relation

$$\int_{a}^{b} [dH(t)] x(t) = H(b) x(b) - H(a) x(a) - \int_{a}^{b} H(t) \dot{x}(t) \, dt, \tag{1.1.11}$$

if we assume $x(t)$ to be continuously differentiable (or at least absolutely continuous). If $x(t)$ is a solution of (1.1.8), (1.1.9), then (1.1.11) leads to

$$H(b)x(b) - H(a)x(a) - \int_a^b H(t)f(t,x(t))\,dt = \int_a^b F(t,x(t))\,dt. \quad (1.1.12)$$

But

$$x(b) = x(a) + \int_a^b f(t,x(t))\,dt,$$

which, substituted in (1.1.12), leads to the relation

$$[H(b) - H(a)]x(a) = [H(t) - H(b)]\int_a^b f(t,x(t))\,dt + \int_a^b F(t,x(t))\,dt. \quad (1.1.13)$$

Let us now assume that $H(t)$ satisfies the condition

$$\det[H(b) - H(a)] \neq 0. \quad (1.1.14)$$

Then (1.1.13) yields

$$x(a) = [H(b) - H(a)]^{-1}\left\{[H(t) - H(b)]\int_a^b f(s,x(s))\,ds + \int_a^b F(s,x(s))\,ds\right\},$$
$$(1.1.15)$$

which, substituted in (1.1.10), yields

$$x(t) = \int_a^b G(t,s,x(s))\,ds, \quad (1.1.16)$$

where

$$G(t,s,x) = \begin{cases} [H(b) - H(a)]^{-1}\{[H(t) - H(a)]f(s,x) + F(s,x)\}, & a \le s \le t, \\ [H(b) - H(a)]^{-1}\{[H(t) - H(b)]f(s,x) + F(s,x)\}, & t \le s \le b. \end{cases}$$
$$(1.1.17)$$

The integral equation (1.1.16), with the unknown function $x(t)$, turns out to be equivalent to the boundary value problem (1.1.8), (1.1.9). Indeed, if we assume the existence of a continuously differentiable solution of (1.1.8), (1.1.9), we obtain as seen above the integral equation (1.1.16). On the other hand, if $x(t)$ is a continuous solution of (1.1.16), then noticing the equivalence of (1.1.8), (1.1.15) with (1.1.16), one obtains the converse implication.

An equation of the form (1.1.16) is called an *integral equation of Fredholm type*. Solving such equations and proving the existence of their solutions is usually a much more complicated problem than in the case of Volterra type equations.

For the reader familiar with the Green's function method in the theory of boundary value problems, obtaining the integral equation (1.1.16) from (1.1.8), (1.1.9) is, in fact, a repetition of the procedure leading to the construction of such a function.

The theory of partial differential equations is another important source of integral equations. For instance, the first-order partial differential equation

$$u_t + \sum_{1}^{n} f_i(t, x) u_{x_i} = g(t, x, u), \tag{1.1.18}$$

with an initial condition of the form

$$u(0, x) = \phi(x), \quad x \in D \subset R^n, \tag{1.1.19}$$

can be reduced to an integral equation of Volterra type, if appropriate conditions are satisfied.

First, let us notice that the right-hand side of (1.1.18) represents the derivative of the function $u(t, x)$ along the trajectories of the differential system (1.1.8). In other words, if in (1.1.18) we substitute x by $x(t)$ – a solution of the system (1.1.8) – then (1.1.18) becomes

$$\frac{d}{dt} u(t, x(t)) = g(t, x(t), u(t, x(t))), \tag{1.1.20}$$

or, denoting $u(t, x(t))$ by $U(t)$,

$$\frac{d}{dt} U(t) = g(t, x(t), U(t)), \tag{1.1.21}$$

This constitutes an ordinary differential equation in U, for each trajectory $x(t)$ of the differential system (1.1.8). Therefore, we can attach to (1.1.21) the integral equation

$$U(t) = \phi(x(0)) + \int_0^t g(s, x(s), U(s)) \, ds, \tag{1.1.22}$$

which is of Volterra type. It is useful to notice that, based on the notation adopted, $U(0) = \phi(x(0))$. More precisely, each trajectory of (1.1.8) generates an equation of the form (1.1.22).

But it is easy to transform the integral equation (1.1.22) into an integral equation depending on the parameter x, provided uniqueness holds true for the system (1.1.8). Indeed, if $x = F(t; 0, x^0)$ denotes the unique solution of (1.1.8), satisfying the initial condition $x(0) = x^0 \in R^n$, then $x^0 = F(0; t, x)$. The equation (1.1.22) can be rewritten in the form

$$u(t, x) = \phi(F(0; t, x)) + \int_0^t g(s, F(s; t, x), u(s, F(s; t, x))) \, ds, \tag{1.1.23}$$

which can be easily manipulated in order to determine the existence of solutions. Of course, $x = F(s; t, x)$ represents the solution of (1.1.8), such that $F(t; t, x) \equiv x$ and, since we assume existence and uniqueness for (1.1.8), the properties of $F(s; t, x)$ are known (i.e., continuous dependence, even differentiability, etc.). In this manner, existence for (1.1.22) or (1.1.23) implies existence for (1.1.18). Regularity problems, such as the existence of the derivatives involved in (1.1.18), must be discussed if we want a classical solution of that equation. Otherwise,

(1.1.22) or (1.1.23) will provide only generalized (mild) solutions for the partial differential equation (1.1.18).

Let us point out that the method sketched above is known as the *method of characteristics* (i.e., the trajectories of the system (1.1.8)). Its success relies on the possibility of having as much information as possible about the function $F(s; t, x)$, which represents the family of trajectories of (1.1.8). These trajectories are known as the characteristic curves of the equation (1.1.18).

The second-order partial differential equations also lead to integral equations if we deal with the existence problems. It is common knowledge that Dirichlet problems or Neumann type problems for the Laplace equation

$$\Delta u = u_{xx} + u_{yy} + u_{zz} = 0 \quad \text{in } D \subset R^3, \tag{1.1.24}$$

can be reduced to integral equations of the form

$$u(P) = f(P) + \int_S k(P, Q) u(Q) \, dQ, \tag{1.1.25}$$

where $S = \partial D$ is assumed to be smooth enough, and $f(P)$ and $k(P, Q)$ are known functions on S and $S \times S$ respectively. For details on these matters, see the book by Sobolev [1], or A. Friedman [1; Section 5.4].

It is relevant to mention the fact that the integral equations obtained in relation to these problems are of Fredholm type, and that the so-called Fredholm alternative plays the central role in their discussion. These equations will be investigated in Section 1.3.

In the case of parabolic equations which describe diffusion or heat transfer phenomena, the integral equations also occur in a natural way. For instance, the solution of the boundary value problem

$$u_t = \Delta u + f(t, x), \quad t > 0, x \in R^3, \tag{1.1.26}$$

$$u(0, x) = u_0(x), \quad x \in R^3, \tag{1.1.27}$$

can be represented under appropriate conditions by the formula

$$u(t, x) = \int_{R^3} G(t, x - \xi) u_0(\xi) \, d\xi + \int_0^t \left\{ \int_{R^3} G(t - s, x - \xi) f(s, \xi) \, d\xi \right\} ds, \tag{1.1.28}$$

where $G(t, x)$ is the fundamental solution (or Green's function) given by

$$G(t, x) = (2\sqrt{(\pi t)})^{-3} \exp\{-|x|^2/4t\}, \quad t > 0, x \in R^3. \tag{1.1.29}$$

If we modify equation (1.1.26), letting f also be dependent on u, then from (1.1.28) we obtain for the solution of the nonlinear equation

$$u_t = \Delta u + f(t, x, u), \quad t > 0, x \in R^3, \tag{1.1.30}$$

under initial condition (1.1.27), the following nonlinear integral equation:

$$u(t, x) = \int_{R^3} G(t, x - y) u_0(y) \, dy + \int_0^t \left\{ \int_{R^3} G(t - s, x - y) f(s, y, u(s, y)) \, dy \right\} ds.$$

(1.1.31)

Equation (1.1.31) appears to be of Volterra type in t, and of Fredholm type with respect to x. We can view it as a Volterra integral equation if we agree to consider $u(t, \cdot)$ as a variable whose values belong to a function space of functions defined on the whole x-space R^3. To be more specific, for every $t > 0$, $u(t, x)$ is supposed to belong to the function space of bounded uniformly continuous functions on R^3, to which usually $u_0(x)$ is assumed to belong. Again, the advantage of dealing with an integral equation like (1.1.31) is that the solution can be sought in a function space whose elements do not necessarily possess strong regularity properties, as required when we deal directly with partial differential equations. On the other hand, if a merely continuous solution can be proved to exist for an integral equation of the form (1.1.31), this will automatically imply the properties of regularity required by equation (1.1.30).

Another interesting example is provided by the boundary value problem

$$u_t = u_{xx}, \quad u(x, 0) = 0, \quad u_x(0, t) + f(u(0, t)) = 0, \tag{1.1.32}$$

in the first quadrant $x \geq 0$, $t \geq 0$.

We shall prove that the solution of the problem (1.1.32) can be obtained by solving an integral equation. It is assumed that f stands for a nonlinear function, a feature which corresponds to such phenomena as the Stefan–Boltzman radiation condition. The equation we shall obtain for $u(0, t)$, $t > 0$, will be a Volterra type equation with singular integral.

Since the solution of $u_t = u_{xx}$, with the initial condition $u(x, 0) = 0$, and the boundary value condition $u_x(0, t) + g(t) = 0$ for $t > 0$, is given by

$$u(x, t) = \int_0^t g(s) \{\pi(t - s)\}^{-1/2} \exp\{-x^2/4(t - s)\} \, ds, \tag{1.1.33}$$

for positive t, we realize that the problem (1.1.32) can be reduced to the integral equation

$$u(x, t) = \int_0^t f(u(0, s)) \{\pi(t - s)\}^{-1/2} \exp\{-x^2/4(t - s)\} \, ds. \tag{1.1.34}$$

The formula (1.1.34) shows that it is sufficient to know $u(0, t)$, in order to know $u(x, t)$ in the whole of the first quadrant. But from (1.1.34) one obtains

$$u(0, t) = \int_0^t f(u(0, s)) \{\pi(t - s)\}^{-1/2} \, ds, \quad t > 0. \tag{1.1.35}$$

Therefore, if we can show the existence of a solution to equation (1.1.35), formula (1.1.34) will provide the solution of the boundary value problem (1.1.32) in the whole quadrant. Of course, various assumptions have to be made in order to validate this procedure. See Saaty [1] for more comments on this problem and similar ones.

Let us consider now the quasilinear hyperbolic equation in two variables

$$u_{xt} + a_0(x,t)u_x + b_0(x,t)u_t = c_0(x,t,u), \qquad (1.1.36)$$

in the semi-strip (Δ)

$$0 \leq x \leq l, \quad 0 \leq t < \infty,$$

with the characteristic data

$$u(x,0) = u_1(x), \quad u(0,t) = u_0(t). \qquad (1.1.37)$$

The coefficients $a_0(x,t)$, $b_0(x,t)$ are assumed continuous in Δ, together with the derivative $\partial b_0/\partial t$, while $c_0(x,t,u)$ is a continuous map from $\Delta \times R$ into R.

It is easily seen that by means of the substitution

$$u = v \exp \left\{ -\int_0^x b_0(y,t)\,dy \right\}, \qquad (1.1.38)$$

equation (1.1.36) takes the form

$$v_{xt} + a(x,t)v_x = c(x,t,v), \qquad (1.1.39)$$

in which $a(x,t)$ and $c(x,t,u)$ are like $a_0(x,t)$ and $c_0(x,t,u)$. The data on the characteristics preserve their form, more precisely

$$v(x,0) = u_1(x) \exp \left\{ \int_0^x [\partial b_0/\partial t]\,dy \right\} = v_1(x), \quad v(0,t) = u_0(t). \quad (1.1.40)$$

To reduce (1.1.39), (1.1.36) to an integral equation, we look at (1.1.36) as a first-order equation in v_x. This allows us to write

$$v_x(x,t) = \exp \left\{ -\int_0^t a(x,s)\,ds \right\} \left[v_1'(x) + \int_0^t c(x,\sigma,u(x,\sigma)) \exp \left\{ \int_0^\sigma a(x,s)\,ds \right\} d\sigma \right]$$
$$(1.1.41)$$

taking into account the first characteristic condition. Integrating both sides of (1.1.41) with respect to x, from 0 to x, we obtain

$$v(x,t) = u_0(t) + \int_0^x v_1'(y) \exp \left\{ -\int_0^t a(y,s)\,ds \right\} dy$$
$$+ \int_0^x \int_0^t c(y,\sigma,u(y,\sigma)) \exp \left\{ -\int_\sigma^t a(y,s)\,ds \right\} d\sigma\,dy, \qquad (1.1.42)$$

which belongs to the type

$$v(x,t) = f(x,t) + \int_0^x \int_0^t g(x,t;y,s;u(y,s))\,ds\,dy. \qquad (1.1.43)$$

Equation (1.1.43) is a *Volterra equation* in two independent variables x, t.

To conclude this section we will consider the abstract semilinear initial value problem

$$\dot{x}(t) + Ax(t) = f(t, x(t)), \quad x(t_0) = x^0 \in X, t \in [t_0, T], \qquad (1.1.44)$$

where x and f take values in the Banach space X. The following conditions will be assumed: (1) The operator $-A$ is the infinitesimal generator of a C_0-semigroup $\{T(t); t \geq 0\}$ of bounded linear operators on X; (2) The map $f: [t_0, T] \times X \to X$ is continuous in t, and Lipschitz continuous in x.

Then (see A. Pazy [1]), if the problem (1.1.44) has a classical (or a strong) solution $x(t)$, this solution will also satisfy the integral equation

$$x(t) = T(t - t_0)x^0 + \int_0^t T(t - s)f(s, x(s)) \, ds. \qquad (1.1.45)$$

The converse property is not necessarily true. A continuous solution of (1.1.45) may not be a solution to equation (1.1.44). In the last case, it is usually considered a *generalized* solution for that problem (also called a *mild* solution).

Owing to the properties of C_0-semigroups, in particular the estimate $|T(t)| \leq \exp(\omega t)$, $t \geq 0$), it is an elementary task to establish the existence and uniqueness of a continuous solution to (1.1.45). The method of successive approximations leads immediately to this result (see again Pazy [1]).

Finally, another significant source of integral equations, especially of Volterra type, is the so-called *inverse problems* in the theory of partial differential equations (see for instance the books by Romanov [1], Lavrentiev, Romanov and Shishatskii [1], Cannon [1]).

Most of the equations obtained in this way can be represented in the form

$$x(t) = f(t) + \int_0^t (Gx)(s) \, ds, \quad 0 \leq t \leq T, \qquad (1.1.46)$$

where G stands for a Volterra operator (a causal or nonanticipative operator, i.e., $x(s) = y(s)$ for $s \leq t$, implies $(Gx)(t) = (Gy)(t)$).

Equations like (1.1.46) can be dealt with by means of quite elementary methods. In Section 3.4 we will discuss problems of this nature, showing local or global existence of solutions under fairly general assumptions.

1.2 Integral equations occurring in the mathematical description of applied problems

A problem which appears in many mechanics textbooks deals with the motion of a chain sliding off a table. We assume the total length of the chain is a, $a > 0$, and its structure is flexible so that when sliding off the table if can slide along a given curve. The friction forces are neglected. The linear density of the chain need not be a constant, and actually we want to determine this density in such a way

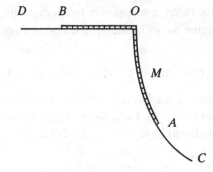

Figure 1

that the motion of the chain is preassigned. This is a kind of inverse problem in connection with the sliding chain.

In the plane in which the motion of the chain is taking place, we choose a coordinate system as indicated in Figure 1 and denote by s the length of the arc of the curve on which the motion is performed, with origin at O. If z denotes the coordinate of the generic point M on the chain, then the equation of the curve OC can be written as

$$z = \phi(s), \tag{1.2.1}$$

where $\phi(s)$ is supposed to be a smooth function satisfying $\phi(0) = 0$.

Let σ be the length of the arc of the curve OC between the points O and A, with A the end of the sliding chain. It is obvious that the position of the chain is known as soon as σ is known. Denote by λ the length of the arc AM of the curve OC. Then $s = \sigma - \lambda$, and consequently (1.2.1) leads to

$$z = \phi(\sigma - \lambda). \tag{1.2.2}$$

If $f(\lambda)$ is the linear density of the chain at the point M, then $f(\lambda)\,d\lambda$ is the mass of the portion of the chain about M, of length $d\lambda$. The work corresponding to the sliding of the portion of the chain from z to $z + dz$ will then be

$$gf(\lambda)\,d\lambda\,dz = g\phi'(\sigma - \lambda)f(\lambda)\,d\lambda\,dz, \tag{1.2.3}$$

if we take (1.2.2) into account, and denote the acceleration of gravity by g as usual. Therefore, the work associated with the whole portion of the chain below the table is given by

$$g\,d\sigma \int_0^\sigma \phi'(\sigma - \lambda)f(\lambda)\,d\lambda. \tag{1.2.4}$$

Let us estimate now the variation of the kinetic energy of the chain in the interval of time from t to $t + dt$ (over the period necessary to move z to $z + dz$). Since

$$T = \frac{1}{2}Mv^2, \tag{1.2.5}$$

with $M = \int_0^a f(\lambda)\,d\lambda$ the total mass of the chain, and since $v = ds/dt = d\sigma/dt$, we obtain

$$dT = M\frac{d\sigma}{dt}\frac{d^2\sigma}{dt^2}\,dt = M\frac{d^2\sigma}{dt^2}\,d\sigma. \tag{1.2.6}$$

To obtain the equation of motion, we combine (1.2.4) and (1.2.6), which gives

$$M\frac{d^2\sigma}{dt^2} = g\int_0^\sigma \phi'(\sigma - \lambda)f(\lambda)\,d\lambda. \tag{1.2.7}$$

Equation (1.2.7) is an integrodifferential equation in $\sigma = \sigma(t)$, assuming f and ϕ to be known. With assigned initial data σ_0 and v_0, we can obtain a unique solution of (1.2.7), which completely solves the problem of determining the motion of a sliding chain.

Instead of dealing with the direct mechanical problem, we shall consider, as stated above, an inverse problem. Namely, if $f(\lambda)$ is unknown and has to be determined in such a way that the motion of the chain is the one assigned in advance, i.e., when $\sigma = \sigma(t)$ is known, or the law of the motion is known to be $d^2\sigma = F(\sigma)\,dt^2$, then (1.2.7) leads to the equation

$$\int_0^\sigma \phi'(\sigma - \lambda)f(\lambda)\,d\lambda = g^{-1}MF(\sigma), \tag{1.2.8}$$

which constitutes an integral equation of Volterra type, and the first kind.

Under assumptions that we are not stating here, equation (1.2.8) has a unique solution $f = f(\lambda)$. In other words, the inverse problem of the sliding has a unique solution.

The situation becomes completely different if the whole chain has slid off the table and occupies a certain position on the curve OC. In this case equation (1.2.8) must be replaced by the similar equation

$$\int_0^a \phi'(\sigma - \lambda)f(\lambda)\,d\lambda = g^{-1}MF(\sigma), \quad 0 \le \sigma \le a, \tag{1.2.9}$$

which is an integral equation of the Fredholm type and the first kind. In general, such equations do not possess solutions (unless the right-hand side belongs to the range of the integral operator appearing in the left-hand side of the equation).

A case which illustrates this situation corresponds to $\phi(s) = s^2/8R$ (i.e., the curve OC is a cycloid), when (1.2.9) takes the form

$$\int_0^a (\sigma - \lambda)f(\lambda)\,d\lambda = 4g^{-1}RMF(\sigma). \tag{1.2.10}$$

It is easy to notice that equation (1.2.10) has no solutions at all if $F(\sigma)$ is not linear in σ. On the other hand, if F is linear, then one can easily construct more than one solution.

Next, we shall consider a problem which is related to temperature control by means of a thermostat. This problem leads to an integral equation, which also involves some discontinuous functions. To make these functions 'smoother', we shall consider an *integral inclusion* which can be naturally associated to the integral equation obtained initially. This example is covered in the paper by Glashoff and Sprekels [1].

The heat transfer process which is controlled by the thermostat is supposed to take place in a rod that is perfectly insulated from the surroundings, except at one of its ends where the heater is placed. The reading of the temperature is taken at the other end of the rod. If we agree to take the length of the rod equal to 1, then $T(x, t)$ will denote the temperature in the cross-section of the rod at x, and at time t ($t > 0, 0 \leq x \leq 1$). Since no sources of heat are located in the rod, the heat transfer equation

$$T_t = T_{xx}, \quad 0 \leq x \leq 1, \quad t \geq 0, \tag{1.2.11}$$

provides one of the conditions to which $T(x, t)$ is subject during the process. At the end $x = 0$, because of the insulation, we must have

$$T_x(0, t) = 0, \quad t \geq 0. \tag{1.2.12}$$

At the end $x = 1$, where the heater is located, and where the temperature of the surrounding medium is $u(t)$ at time t, by Newton's law of cooling, we must have

$$\alpha T_x(1, t) + T(1, t) = u(t), \quad t \geq 0, \tag{1.2.13}$$

where α is a positive constant. Of course, $u(t)$ denotes the temperature resulting from the switching on of the heater, which in turn is determined by the temperature at $x = 0$ (where the reading device is located). When $u(t)$ is given (which is not the case now), equation (1.2.11), with the boundary value conditions (1.2.12), (1.2.13), and under the initial condition

$$T(x, 0) = 0, \quad 0 \leq x \leq 1, \tag{1.2.14}$$

will have a unique solution if adequate conditions are satisfied.

In our case, the objective is to maintain the temperature at $x = 0$ between certain limits, say T_0 and T_1, $T_0 < T_1$, by means of the action of the thermostat. In other words, when the reading device reads a temperature $T(0, t) < T_0$, the heater is activated, and it will continue working until $T(0, t) = T_1$. At that time the heater is switched off. Let $y(t) = T(0, t)$ for positive t. Then the switching process can be described by means of the following function:

$$f(y, \dot{y}) = \tfrac{1}{4}[1 - \text{sign}(y - T_0)](1 - \text{sign}\,\dot{y}) + \tfrac{1}{4}[1 - \text{sign}(y - T_1)](1 + \text{sign}\,\dot{y}), \tag{1.2.15}$$

which means that $f(y, \dot{y})$ takes the value 1 when the heater has to be on, and the value 0 otherwise. Of course, $f(y, \dot{y})$ is a discontinuous function.

The temperature $u(t)$ at the end $x = 1$ of the rod is the result of burning the fuel in the heater. If $v(t)$, $0 \leq v(t) \leq 1$, denotes the fuel supply at time t, then for some $\beta > 0$ is assumed that

$$\beta \dot{u}(t) + u(t) = v(t), \quad t > 0, \tag{1.2.16}$$

with $u(0) = 0$, in order to agree with (1.2.14).

Since from the equations (1.2.11)–(1.2.14) we derive the formula

$$T(x, t) = \int_0^t K(x, t - s)u(s)\,ds, \tag{1.2.17}$$

with the kernel $K(x, t)$ uniquely determined by (1.2.11)–(1.2.13), we can get, taking into account (1.2.16),

$$T(x, t) = \int_0^t \bar{K}(x, t - s)v(s)\,ds, \tag{1.2.18}$$

where the kernel $\bar{K}(x, t)$ is easily obtained by the composition of $K(x, t)$ and $\beta^{-1}\exp\{-\beta^{-1}t\}$. The latter appears when expressing $u(t)$ from (1.2.16) in terms of $v(t)$. From (1.2.18) we derive at $x = 0$ the relationship

$$y(t) = \int_0^t G(t - s)v(s)\,ds, \tag{1.2.19}$$

with $G(t) = \bar{K}(0, t)$. Since the feedback equation of the system is

$$v(t) = f(y(t), \dot{y}(t)), \tag{1.2.20}$$

we obtain from (1.2.19) and (1.2.20) the following integrodifferential equation:

$$y(t) = \int_0^t G(t - s)f(y(s), \dot{y}(s))\,ds. \tag{1.2.21}$$

Equation (1.2.21) in $y(t) = T(0, t)$ is obviously of convolution type. We can replace it by an inclusion involving multivalued functions on the right hand side, to avoid dealing with the discontinuities of $f(y, \dot{y})$. To this end, we consider the so-called *convexification* of the function $f(y, \dot{y})$, which is given by the formula

$$F(y, \dot{y}) = \begin{cases} [0, 1] & \text{when } y = T_0 \text{ and } \dot{y} \leq 0, \text{ or } T_0 < y < T_1 \text{ and } \dot{y} = 0, \\ & \text{or } y = T_1 \text{ and } \dot{y} \geq 0, \\ \{f(y, \dot{y})\} & \text{otherwise.} \end{cases} \tag{1.2.22}$$

Of course, F takes R^2 into the set of parts of R, say $\mathscr{P}(R)$. These parts happen to be convex sets of R.

Instead of the equation (1.2.21), we shall now consider the integral inclusion

$$y(t) \in \int_0^t G(t - s)F(y(s), \dot{y}(s))\,ds, \tag{1.2.23}$$

which constitutes a generalization of the concept of an integral equation. We will investigate some integral inclusions in Section 4.7.

Remark.   The integral representation of the solution $T(x, t)$ of the heat equation (1.2.17), can be derived by using the Laplace transform method. Indeed, equation (1.2.11), with the initial condition (1.2.14), leads to the equation $\tilde{T}_{xx}(x, s) = s\tilde{T}(x, s)$, where $\tilde{T}(x, s)$ denotes the Laplace transform of $T(x, t)$, with respect to the second argument. Therefore, we obtain $\tilde{T}(x, s) = A(s) \cosh\sqrt{s}x + B(s) \sinh\sqrt{s}x$. The arbitrary functions $A(s)$ and $B(s)$ in the formula for the function $\tilde{T}(x, s)$ can be determined from the equations obtained by taking the Laplace transform of both sides in equations (1.2.12) and (1.2.13). In particular, one obtains $B(s) = 0$. Since the final form of the transform $\tilde{T}(x, s)$ is $\tilde{K}(x, s)\tilde{u}(s)$, the formula (1.2.17) follows immediately. We are not interested here in the expressions for $\tilde{K}(x, s)$ or $K(x, t)$. Readers interested in these details can carry out the necessary calculations themselves.

In system theory (engineering systems), an open loop input–output linear system is described in many circumstances by the integral relationship

$$x(t) = x_0(t) + \int_0^t K(t, s)u(s)\,ds, \quad t \geq 0, \tag{1.2.24}$$

where u stands for the input, and x for the output. We assume u to be an m-vector-valued function, while x is an n-vector-valued function. $K(t, s)$ is a matrix-valued function of type n by m, describing the linear plant, and $x_0(t)$ represents the particular state (output) corresponding to the zero input.

If the system (1.2.24) is controlled by a feedback law of the form

$$u(t) = f(t, x(t)), \tag{1.2.25}$$

then (1.2.24) and (1.2.25) imply

$$x(t) = x_0(t) + \int_0^t K(t, s)f(s, x(s))\,ds. \tag{1.2.26}$$

Equation (1.2.26) is a nonlinear integral equation in the variable $x(t)$, and is commonly known as a Volterra–Hammerstein equation. It is somewhat more general than equation (1.1.5).

As an illustration of the considerations leading to (1.2.26), let us deal with the control system described by the differential equation $\dot{x} = Ax + bu$, $x(0) = x^0$, under the feedback law (1.2.25). Equation (1.2.26) now takes the form

$$x(t) = e^{At}x^0 + \int_0^t e^{A(t-s)}f(s, x(s))\,ds, \tag{1.2.27}$$

which is a special case of equation (1.1.5).

One more example of a problem in mechanics leading to an integrodifferential equation with partial derivatives will be presented. It is one of the simplest problems in viscoelasticity, which is a major source of integral or integrodifferential equations.

Let B be an isotropic homogeneous body with a linear viscoelastic behavior of creep type. We assume one of the dimensions of B is predominant, and consequently we can place ourselves in the framework of the one-dimensional linear theory of elasticity. Let $u = u(x, t)$ denote the (scalar) displacement field, with respect to an undeformed reference configuration, while $\sigma = \sigma(x, t)$ will designate the stress force (just one scalar component for the stress tensor). We denote by $e = e(x, t)$ the strain force (also a unique scalar component of the strain tensor), and $f(x, t)$ will represent the volume density of the external forces. If ρ is the (constant) mass density, then Newton's law of dynamics leads to the following equation:

$$\rho u_{tt}(x, t)\Delta x = \sigma(x + \Delta x, t) - \sigma(x, t) + f(x, t)\Delta x, \qquad (1.2.28)$$

if we consider only a thin 'slice' of the body B, between x and $x + \Delta x$. From (1.2.28) we derive, after dividing by Δx and letting $\Delta x \to 0$,

$$\rho u_{tt}(x, t) = \sigma_x(x, t) + f(x, t). \qquad (1.2.29)$$

From Hooke's law in elasticity we can write

$$e(x, t) = u_x(x, t), \qquad (1.2.30)$$

this relationship being sometimes taken as the definition of the strain force.

Finally, the elastic heredity of creep type is described by the equation relating the strain and the *stress*, namely

$$e(x, t) = J(0)\left[\sigma(x, t) + \int_{-\infty}^{t} g(t - s)\sigma(x, s)\,\mathrm{d}s\right], \qquad (1.2.31)$$

where $J(t)$ is the creep compliance, and $g(t) = [J(0)]^{-1}\,\mathrm{d}J(t)/\mathrm{d}t$. In some cases instead of (1.2.31) we consider a more general relationship of the form

$$\sigma(x, t) = \int_{-\infty}^{t} e(x, t - s)\,\mathrm{d}G(s), \qquad (1.2.32)$$

where $G(t)$ is a function of bounded variation on the entire real axis. We shall see that (1.2.31) can be written in the form (1.2.32), while the converse is not necessarily true.

Considering now (1.2.31), and taking into account (according to Volterra) an initial condition of the form $\sigma(x, s) = \phi(x, s)$, $-\infty < s \leq 0$, which means that the whole past history of the phenomenon must be known, one obtains

$$\sigma(x, t) + \int_{0}^{t} g(t - s)\sigma(x, s)\,\mathrm{d}s = [J(0)]^{-1}e(x, t) - \int_{-\infty}^{0} g(t - s)\phi(x, s)\,\mathrm{d}s. \qquad (1.2.33)$$

The linear integral operator on the left hand side of (1.2.33) is of Volterra convolution type (on a finite interval!), which means that (1.2.33) can always be inverted. (We shall discuss this matter in the next section.) Therefore, we can write

$$\sigma(x,t) = [J(0)]^{-1} \left[e(x,t) + \int_0^t \bar{g}(t-s)e(x,s)\,ds \right] + \bar{\phi}(x,t), \quad (1.2.34)$$

with $\bar{\phi}(x,t)$ easily obtainable from (1.2.33). From (1.2.34) we can obtain $\sigma_x(x,t)$ in terms of $e_x(x,t)$ (just differentiating under the integral), and since (1.2.30) implies $e_x(x,t) = u_{xx}(x,t)$, we obtain from (1.2.29):

$$u_{tt}(x,t) - \left[c^2 u_{xx}(x,t) + \int_0^t \bar{g}(t-s)u_{xx}(x,s)\,ds \right] = \bar{f}(x,t), \quad (1.2.35)$$

where $c^2 = [\rho J(0)]^{-1}$, and

$$\bar{f}(x,t) = \rho^{-1}f(x,t) + \bar{\phi}_x(x,t). \quad (1.2.36)$$

Equation (1.2.35) constitutes the law of motion of the viscoelastic medium B and, as we can see, it is an integrodifferential equation with partial derivatives of the second order.

Of course, if we use (1.2.32) instead of (1.2.31), we obtain in a similar way the integrodifferential equation

$$u_{tt}(x,t) - \int_{-\infty}^t \tilde{g}(t-s)u_{xx}(x,s)\,ds = \tilde{f}(x,t), \quad (1.2.37)$$

with obvious notations. This last equation appears to be more complicated than equation (1.2.35), and very few results are available in the literature.

If the elastic heredity is of nonlinear type, for instance such that the relationship between stress and strain is represented by the equation

$$\sigma(x,t) = \Phi(e(x,t)) + \int_0^\infty m(s)\Psi(e(x,t-s))\,ds, \quad (1.2.38)$$

then by eliminating e and σ from (1.2.29), (1.2.30) and (1.2.38) we obtain a nonlinear integrodifferential equation:

$$\rho u_{tt} = \frac{\partial}{\partial x}\Phi(u_x(x,t)) + \int_0^\infty m(s)\Psi'(u_x(x,t-s))u_{xx}(x,t-s)\,ds + f(x,t). \quad (1.2.39)$$

This equation is even more complicated than equation (1.2.37).

We will conclude the series of examples which illustrate how integral or integrodifferential equations appear in applied areas of investigation by considering a problem in population dynamics. The population need not necessarily be human, but various biological populations, including cell populations, can be discussed using the same mathematical apparatus. We follow here Thieme [1], [2].

The population under investigation is described by means of several charac-teristics listed below.

First, we will assume the population has its habitat in a certain (large) domain of the space R^3, such that in our description we can consider the whole space R^3 as the underlying space. The population density will then depend on x, $x \in R^3$, and t, $t \geq 0$. Since we are going to count only the mature individuals in the population, we denote by $u(t, x)$ the number of such individuals per unit volume (or per unit area if we replace R^3 by R^2) at time t, and place x.

Second, we assume that each individual becomes reproductive after a certain period of time $\sigma > 0$, following the birth. The migration of the juvenile individuals is neglected, in comparison with the migration of the reproductive ones.

Third, the migration is supposed to be radially symmetric, and therefore we can describe it by means of a function $\tilde{k}(|x|, t)$, which will act as a migration law, and will appear as an integral kernel in the final count of reproductive indi-viduals. More precisely, $\tilde{k}(|x|, t)$ gives the rate at which an individual starting at the origin $x = 0$, at time $t = 0$, will arrive at place x at time $t > 0$, without losing its character of being reproductive. Moreover, the migration process is supposed to be homogeneous. In the case of a numerous bacterial population for instance, one could assimilate the migration process with a process of diffusion, and choose $\tilde{k}$ to be the fundamental solution of the heat-diffusion operator $\partial_t - a^2 \Delta_x$.

Fourth, we assume that the fertility of a reproductive individual depends on the number of conspecifics living at the same place, at the same time. This dependence is expressed in terms of a so-called reproductive curve (function) $g(u)$, which is subject to several restrictions: (a) $g(0) = 0$, which means that if there are no reproductive individuals there are no offsprings; (b) $g(u) > 0$ for $u > 0$, which means that reproduction does not completely cease; (c) $g(u)/u \rightarrow 0$ as $u \rightarrow \infty$, which expresses the effect of overcrowding.

In order to derive the integral equation which is satisfied by the function $u(x, t)$, we notice that the total number of reproductive individuals, at place x and time t, is the sum of two numbers: the number of reproductive individuals at (x, t) that have been reproductive from the beginning, say $\tilde{u}_0(x, t)$; and the number of individuals becoming reproductive anywhere, at any moment before t, and migrating at (x, t). If $v(x, t)$ is the number of individuals becoming reproductive at (x, t), then according to our assumptions this number is given by

$$\int_0^t \int_{R^3} \tilde{k}(|y|, s) v(x + y, t - s) \, ds \, dy. \qquad (1.2.40)$$

Taking this expression into account, the relationship expressing the equality of the two different counts for the reproductive individuals is

$$u(x, t) = u_0(x, t) + \int_0^t \int_{R^3} \tilde{k}(|y|, s) v(x + y, t - s) \, ds \, dy. \qquad (1.2.41)$$

As seen above in our discussion, $v(x, t)$ is related to $u(x, t)$ by

$$v(x, t) = g(u(x, t - \sigma)), \quad t > \sigma. \tag{1.2.42}$$

One must have

$$v(x, t) = v_0(x, t), \quad t \geq \sigma, \tag{1.2.43}$$

where $v_0(x, t)$ is known (say, determined experimentally). If we now consider (1.2.41)–(1.2.43), we obtain the final equation for $u(x, t)$:

$$u(x, t) = u_0(t) + \int_0^t \int_{R^3} k(|y|, s) g(u(x + y, t - s)) \, ds \, dy, \tag{1.2.44}$$

where

$$k(r, t) = \begin{cases} 0 & 0 \leq t \leq \sigma, r \geq 0, \\ \tilde{k}(r, t - \sigma) & t > \sigma, r \geq 0, \end{cases} \tag{1.2.45}$$

and

$$u_0(x, t) = \tilde{u}_0(x, t) + \int_0^t \int_{R^3} \tilde{k}(|y|, s) v_0(x + y, t - s) \, ds \, dy. \tag{1.2.46}$$

The integral equation (1.2.44) involves the known functions $u_0(x, t)$, $k(|x|, t)$, and $g(u)$, as well as the unknown function $u(x, t)$. It is like equation (1.1.31) which was obtained in relation to heat-transfer (diffusion) processes.

As possible examples of function $g(u)$ one can take $g(u) = \gamma u \exp(-\beta u)$, or $g(u) = \alpha u/(\beta + u)$, where all the constants involved are positive.

Other mathematical models for population dynamics have been discussed by many authors since 1970. The bibliographical notes at the end of this chapter will provide more information about this topic.

1.3 Basic problems for integral equations of Volterra type (elementary approach)

Among the various integral equations we have encountered in the preceding sections of this chapter, let us consider first the nonlinear Volterra equation (1.1.6),

$$x(t) = f(t) + \int_0^t k(t, s, x(s)) \, ds, \quad t \in [t_0, t_0 + a]. \tag{1.3.1}$$

More precisely, we assume f and k to be given n-vector-valued functions, while x is the unknown n-vector function. It is our aim to prove the existence of solutions to (1.3.1), at least in a subinterval of $[t_0, t_0 + a]$, say $[t_0, t_0 + \delta]$, with $\delta \leq a$. To achieve this goal, we shall use the classical method of successive approximations. But, first, let us state adequate conditions on (1.3.1), that will enable us to carry out the proof of the existence and uniqueness of the solution. These conditions are:

(1) f is a continuous map from $[t_0, t_0 + a]$ into R^n;
(2) k is a continuous map from $\Delta \times B_r$ into R^n, where

$$\Delta = \{(t, s); t_0 \leq s \leq t_0 + a\} \qquad (1.3.2)$$

and

$$B_r = \{x; x \in R^n, |x| \leq r\}; \qquad (1.3.3)$$

(3) k satisfies in $\Delta \times B_r$ the Lipschitz condition

$$|k(t, s, x) - k(t, s, y)| \leq L|x - y|, \qquad (1.3.4)$$

for some $L > 0$;

(4) if $b = \sup|f(t)|$, $t \in [t_0, t_0 + a]$, then

$$b < r. \qquad (1.3.5)$$

The method of successive approximations consists in constructing the sequence of continuous functions

$$x_0(t) = f(t), \quad x_m(t) = f(t) + \int_{t_0}^{t} k(t, s, x_{m-1}(s)) \, ds, \quad m \geq 1, \qquad (1.3.6)$$

on $[t_0, t_0 + a]$ when possible, or at least on some subinterval $[t_0, t_0 + \delta]$, $\delta < a$.

On the basis of our assumptions one easily obtains from (1.3.6) that $x_0(t)$ and $x_1(t)$ are defined and continuous on $[t_0, t_0 + a]$. In order to ascertain the fact that $x_2(t)$ is defined on some interval, we have to check whether the values of $x_1(t)$ do belong to the domain of definition of k, in other words, whether

$$|x_1(t)| \leq r. \qquad (1.3.7)$$

If we set

$$M = \sup|k(t, s, x)| \quad \text{in } \Delta \times B_r. \qquad (1.3.8)$$

then (1.3.6) yields

$$|x_1(t)| \leq b + M(t - t_0), \quad t \in [t_0, t_0 + a]. \qquad (1.3.9)$$

But condition (1.3.7) has to be verified, which means that $b + M(t - t_0) \leq r$, or

$$t - t_0 \leq \frac{r - b}{M}. \qquad (1.3.10)$$

With

$$\delta = \min\left\{a, \frac{r - b}{M}\right\}, \qquad (1.3.11)$$

we see that $x_2(t)$ is certainly defined on $[t_0, t_0 + \delta]$. By induction, we find out that all terms in the sequence $\{x_m(t)\}$ defined by (1.3.6) are continuous on $[t_0, t_0 + \delta]$.

We shall now prove that the sequence (1.3.6) is uniformly convergent on the interval $[t_0, t_0 + \delta]$ to some continuous function $x(t)$:

$$\lim_{m \to \infty} x_m(t) = x(t), \quad \text{uniformly on } [t_0, t_0 + \delta]. \tag{1.3.12}$$

Since

$$x_m(t) = x_0(t) + \sum_{k=1}^{m} [x_k(t) - x_{k-1}(t)],$$

we shall prove the uniform convergence on $[t_0, t_0 + \delta]$ of the series

$$\sum_{k=1}^{m} [x_k(t) - x_{k-1}(t)]. \tag{1.3.13}$$

But we obviously have

$$|x_1(t) - x_0(t)| \leq M(t - t_0), \tag{1.3.14}$$

and for $k > 1$ we obtain

$$|x_k(t) - x_{k-1}(t)| \leq L \int_{t_0}^{t} |x_{k-1}(s) - x_{k-2}(s)| \, ds, \tag{1.3.15}$$

if we take into account (1.3.4) and (1.3.6). From (1.3.14) and (1.3.15) one easily derives

$$|x_m(t) - x_{m-1}(t)| \leq \frac{M}{L} \frac{(L(t - t_0))^m}{m!}, \tag{1.3.16}$$

which guarantees the uniform convergence of the sequence $\{x_m(t)\}$ on $[t_0, t_0 + \delta]$.

Since

$$|k(t, s, x(s)) - k(t, s, x_m(s))| \leq L|x(s) - x_m(s)|, \tag{1.3.17}$$

we obtain

$$\lim_{m \to \infty} k(t, s, x_m(s)) = k(t, s, x(s)), \tag{1.3.18}$$

uniformly on the set $\{(t, s); t_0 \leq s \leq t \leq \delta\}$.

Now letting $m \to \infty$ in the second equation in (1.3.6), and taking into account (1.3.18), one obtains for $x(t)$ defined by (1.3.12) the Volterra equation (1.3.1).

The uniqueness of the solution constructed above, under conditions (1)–(4), can be also obtained by standard arguments. Let $y(t)$ be a continuous solution of (1.3.1) on $[t_0, t_0 + \delta]$. Then for $m \geq 1$ (1.3.6) leads to the inequality

$$|y(t) - x_m(t)| \leq L \int_{t_0}^{t} |y(s) - x_{m-1}(s)| \, ds. \tag{1.3.19}$$

Since $|y(t) - x_0(t)| \leq M(t - t_0)$, as seen from (1.3.1) when $x(t)$ is substituted by $y(t)$, we obtain from above (by induction on m):

$$|y(t) - x_m(t)| \leq \frac{M}{L} \frac{(L(t - t_0))^{m+1}}{(m + 1)!}. \tag{1.3.20}$$

From (1.3.20) we see that $y(t)$ is also the uniform limit of the sequence $\{x_m(t)\}$ and since the limit is unique we obtain $y(t) \equiv x(t)$ on $[t_0, t_0 + \delta]$.

The above discussion leads to the following result about the existence and uniqueness of the solution of the integral equation (1.3.1):

Theorem 1.3.1. *If we assume that conditions (1)–(4) hold, then there exists a unique continuous solution of equation (1.3.1), defined on the interval $[t_0, t_0 + \delta]$, with δ given by (1.3.11). Moreover, this solution is the uniform limit of the sequence of successive approximations (1.3.6).*

Remark 1. When $f(t) = x^0 \in R^n$, and $k(t, s, x) \equiv g(s, x)$, then Theorem 1.3.1 yields the classical result of Picard concerning the existence and uniqueness of the solution to the Cauchy problem

$$\dot{x}(t) = g(t, x(t)), \quad x(t_0) = x^0. \tag{1.3.21}$$

Indeed, the continuous solution of the integral equation

$$x(t) = x^0 + \int_{t_0}^{t} g(s, x(s)) \, ds$$

is continuously differentiable.

Remark 2. As seen from (1.3.11), we have to restrict (in general) the interval on which the solution of the equation (1.3.1) is defined. In other words, Theorem 1.3.1 provides a result about *local existence*. If we assume that k is defined on $\Delta \times R^n$, instead of $\Delta \times B_r$, it becomes clear that there is no need to restrict the initial interval $[t_0, t_0 + a]$ on which all the successive approximations are defined. The convergence of this sequence is uniform on $[t_0, t_0 + a]$, and the solution $x(t)$ is defined on the same interval. In this case we obtain a *global existence* result.

Remark 3. If condition (4) does not hold, but $|f(t_0)| < r$, then from the continuity of $f(t)$ one derives $|f(t)| \leq b < r$ on some interval $[t_0, t_0 + a']$, with $a' \leq a$. Therefore, a local existence result is still valid.

Remark 4. Existence results for equation (1.3.1) when the Lipschitz condition is not satisfied, as well as the existence of a solution in spaces of measurable functions will be obtained in Chapter 3.

Let us now consider a special case of equation (1.3.1), namely, the linear case: $k(t, s, x) = k(t, s)x$, where $k: \Delta \to \mathcal{L}(R^n, R^n)$ is continuous. Since $k(t, s, x)$ is now defined on $\Delta \times R^n$, we are obviously under the conditions described in Remark 2. Therefore, for the linear Volterra equation

$$x(t) = f(t) + \int_{t_0}^t k(t, s)x(s)\,ds, \quad t \in [t_0, t_0 + a], \tag{1.3.22}$$

we can assert the global existence (and uniqueness) of the solution $x(t)$, defined as the limit of sucessive approximation constructed by means of the formulas

$$x_0(t) = f(t), \quad x_m(t) = f(t) + \int_{t_0}^t k(t, s)x_{m-1}(s)\,ds, \quad m \geq 1. \tag{1.3.23}$$

If $f(t)$ is continuous on the interval $[t_0, T)$, $T \leq \infty$, and $k(t, s)$ is continuous on

$$\Delta' = \{(t, s); t_0 \leq s \leq t < T\}, \tag{1.3.24}$$

then the successive approximations (1.3.23) do converge toward the solution $x(t)$, which is continuous on $[t_0, T)$, uniformly on any compact subinterval $[t_0, t_1] \in [t_0, T)$. Indeed, $|k(t, s)|$ is bounded on any subset of Δ' of the form $t_0 \leq s \leq t \leq t_1 < T$. Therefore, $k(t, s)x$ verifies a Lipschitz condition which is needed in obtaining the estimate (1.3.16) on $[t_0, t_1]$.

We shall now obtain a formula expressing the (unique) solution of the linear equation (1.3.22). The so-called *associate* or *resolvent* kernel corresponding to the kernel $k(t, s)$ will be constructed, and its important role will be illustrated.

Let us notice that the successive approximations (1.3.23), allow us to write

$$x_m(t) = f(t) + \int_{t_0}^t \left\{ \sum_{j=1}^m k_j(t, s) \right\} f(s)\,ds, \tag{1.3.25}$$

where

$$k_1(t, s) = k(t, s), \quad k_m(t, s) = \int_s^t k_{m-1}(t, u)k(u, s)\,du, \quad m \geq 2. \tag{1.3.26}$$

From (1.3.26) one can derive the following estimates for the 'iterated' kernels $k_m(t, s)$:

$$|k_m(t, s)| \leq K_1^m \frac{(t - s)^{m-1}}{(m - 1)!}, \tag{1.3.27}$$

with $K_1 = \sup |k(t, s)|$, $t_0 \leq s \leq t \leq t_1 < T$. The estimate (1.3.27) implies the uniform convergence of the series

$$\sum_{m=1}^\infty k_m(t, s) = \tilde{k}(t, s), \tag{1.3.28}$$

in each compact subset of Δ'.

If we take into account (1.3.25), then the following formula is validated:

$$x(t) = f(t) + \int_{t_0}^{t} \tilde{k}(t, s) f(s) \, ds. \tag{1.3.29}$$

This formula provides the solution of the linear integral equation (1.3.22), for any continuous $f(t)$. It can easily be seen that the formula makes sense for some functions $f(t)$ which are not necessarily continuous. Once the *resolvent kernel* $\tilde{k}(t, s)$ has been constructed according to formula (1.3.27), it can be used to express in a simple fashion the solution of the equation (1.3.22), regardless of the particular choice of the function $f(t)$.

From (1.3.28) and (1.3.26) one easily obtains the *integral equation of the resolvent kernel*:

$$\tilde{k}(t, s) = k(t, s) + \int_{s}^{t} \tilde{k}(t, u) k(u, s) \, du.$$

Another similar equation is obtained if we notice first that

$$k_m(t, s) = \int_{s}^{t} k(t, u) k_{m-1}(u, s) \, du, \quad m > 1,$$

and then multiply both sides of (1.3.28), from the left, by $k(u, t)$, and integrate both sides with respect to t, from s to u:

$$\tilde{k}(t, s) = k(t, s) + \int_{s}^{t} k(t, u) \tilde{k}(u, s) \, du.$$

An immediate application of the concept of a resolvent kernel, and of the representation (1.3.29), concerns *linear integral inequalities* (of course, of Volterra type).

If we assume that all intervening functions are real-valued, and we consider the inequality

$$y(t) \leq f(t) + \int_{t_0}^{t} k(t, s) y(s) \, ds, \quad t \in [t_0, T) \tag{1.3.30}$$

where $f(t)$ is continuous on $[t_0, T)$, k is continuous on Δ', and

$$k(t, s) \geq 0 \quad \text{in } \Delta', \tag{1.3.31}$$

then we can easily prove that

$$y(t) \leq f(t) + \int_{t_0}^{t} \tilde{k}(t, s) f(s) \, ds, \tag{1.3.32}$$

which shows that the (continuous) solution of the inequality (1.3.30) is dominated by the solution of the corresponding equation (1.3.22).

Indeed, if we change t to s and s to u in (1.3.30), then multiply by $k(t, s)$ – which is nonnegative – and integrate the inequality with respect to s from t_0 to t, we obtain after adding $f(t)$ to both sides

$$f(t) + \int_{t_0}^{t} k(t,s)y(s)\,ds \leq f(t) + \int_{t_0}^{t} k(t,s)f(s)\,ds + \int_{t_0}^{t} k_2(t,s)y(s)\,ds. \quad (1.3.33)$$

But (1.3.30) shows that (1.3.33) implies

$$y(t) \leq f(t) + \int_{t_0}^{t} k(t,s)f(s)\,ds + \int_{t_0}^{t} k_2(t,s)y(s)\,ds, \quad (1.3.34)$$

on the whole interval $[t_0, T]$. If we now multiply both sides of (1.3.34) by $k(t,s)$, after changing t to s and s to u in (1.3.34), etc., we obtain by induction

$$y(t) \leq f(t) + \int_{t_0}^{t} \left\{ \sum_{j=1}^{m} k_j(t,s) \right\} f(s)\,ds + \int_{t_0}^{t} k_{m+1}(t,s)y(s)\,ds. \quad (1.3.35)$$

Taking into account the definition of the resolvent kernel and (1.3.27), (1.3.35) leads to (1.3.32), as $m \to \infty$.

Special cases of the inequality (1.3.30) are often encountered in the theory of ordinary differential equations. The well-known Gronwall–Bellman inequality

$$y(t) \leq C + \int_{t_0}^{t} k(s)y(s)\,ds, \quad t \geq t_0, \quad (1.3.36)$$

with $k(t)$ continuous and nonnegative leads to

$$y(t) \leq C \exp\left(\int_{t_0}^{t} k(s)\,ds \right) \quad (1.3.37)$$

Indeed, the solution of the associated equation

$$x(t) = C + \int_{t_0}^{t} k(s)x(s)\,ds,$$

is the function in the right-hand side of (1.3.37).

It is possible to extend the theory of integral inequalities to the nonlinear case. Such generalizations will be considered in Chapter 3.

Besides the existence and the uniqueness problems for integral equations, there are many other basic problems which are significant with respect to the theory itself, or to applications of it. For instance, in the case of the integral equations we derived in the preceding sections, it is important to know what kind of behavior (at infinity, or at the end of the existence interval) they possess. And how is that kind of behavior affected by changes in the data?

In order to approach such problems, we will here confine our investigation to the case of Volterra integral equation (1.3.1), under the following basic assumptions:

(a) $f: [t_0, \infty) \to R^n$ is continuous;

(b) $k: \Delta' \times R^n \to R^n$ is continuous, and there exists a continuous nonnegative function $k_0(t, s)$, defined on Δ', such that

$$|k(t, s, x) - k(t, s, y)| \le k_0(t, s)|x - y| \tag{1.3.38}$$

in $\Delta' \times R^n$, and

$$k(t, s, 0) = 0 \quad \text{in } \Delta'. \tag{1.3.39}$$

Let us point out that (1.3.39) does not really restrict the generality. Indeed, equation (1.3.1) can be rewritten in the form

$$x(t) = f(t) + \int_{t_0}^{t} k(t, s, 0)\, ds + \int_{t_0}^{t} [k(t, s, x(s)) - k(t, s, 0)]\, ds,$$

which displays an integrand satisfying condition (1.3.39). Of course, the above form is equivalent to (1.3.1).

We would like to determine an adequate scalar function $g: [t_0, \infty) \to R_+$, continuous and such that the solution of (1.3.1) satisfies the estimate

$$|x(t)| \le Kg(t), \quad t \in [t_0, \infty), \tag{1.3.40}$$

where K is a positive constant.

An estimate like (1.3.40) provides some information on the growth, or behavior, of the solution $x(t)$ on the entire half-axis $t \ge t_0$. This kind of information is usually useful in applications.

The method we shall use in conducting this investigation will be the same method of successive approximations that we have already used, but in its generalized (abstract) form of the contraction mapping (or Banach fixed point) theorem:

Let (S, d) be a complete metric space, and $T: S \to S$ a mapping that satisfies

$$d(Tx, Ty) \le \alpha d(x, y), \quad x, y \in S, \tag{1.3.41}$$

where $0 < \alpha < 1$. Then, there exists a unique $\bar{x} \in S$, such that

$$\bar{x} = T\bar{x}. \tag{1.3.42}$$

Moreover, $\bar{x} = \lim x_n$ as $n \to \infty$, where $x_0 \in S$ is arbitrarily chosen, while $x_m = Tx_{m-1}$, $m \ge 1$.

The proof of this result can be found in many books. See, for instance, Krasnoselskii [1].

We shall choose now as underlying space for our existence problem related to the equation (1.3.1) the function space of maps from $[t_0, \infty)$ into R^n, such that an inequality of the form (1.3.40) holds. Of course, the constant K in (1.3.40) depends on the map x, and the real function g is yet to be determined in terms of our data. Since the set of those x satisfying (1.3.40) is obviously a linear space, it is enough to introduce a convenient norm. Let

$$|x|_g = \sup\{|x(t)|/g(t); t \in [t_0, \infty)\}. \tag{1.3.43}$$

The right-hand side in (1.3.43) is always finite, and it is an elementary exercise to show that $x \to |x|_g$ is a norm on the linear space of continuous maps satisfying an estimate of the form (1.3.40). The completeness of this space, which we shall denote by $C_g = C_g([t_0, \infty), R^n)$, follows from quite standard arguments. See C. Corduneanu [4] for details.

The operator T is now defined by

$$(Tx)(t) = f(t) + \int_{t_0}^{t} k(t, s, x(s))\, ds, \quad t \in [t_0, \infty). \tag{1.3.44}$$

While (1.3.44) shows that $(Tx)(t)$ is a continuous map from $[t_0, \infty)$ into R^n, for any $x \in C_g$, we need to prove the inclusion

$$TC_g \subset C_g. \tag{1.3.45}$$

For (1.3.45) to be true, it is obviously necessary to have $(T\theta)(t) = f(t) \in C_g$. In other words, we have to assume

$$f \in C_g([t_0, \infty), R^n). \tag{1.3.46}$$

On the other hand, from condition (b), we have

$$\left| \int_{t_0}^{t} k(t, s, x(s))\, ds \right| \le K \int_{t_0}^{t} k_0(t, s)g(s)\, ds$$

for some $K > 0$, depending upon $x \in C_g$. Therefore, if we assume

$$\int_{t_0}^{t} k_0(t, s)g(s)\, ds \le \alpha g(t), \quad t \in [t_0, \infty), \tag{1.3.47}$$

the second term in the right-hand side of (1.3.44) will be in C_g, for any x in C_g.

The integral inequality (1.3.47) is the only restriction we have to impose on $g(t)$, if we want (1.3.45) to be true.

It remains only to ensure the contraction of the mapping T. Since we have

$$|(Tx)(t) - (Ty)(t)| \le \int_{t_0}^{t} k_0(t, s)|x(s) - y(s)|\, ds,$$

from (1.3.43), (1.3.47) we obtain

$$|Tx - Ty|_g \le \sup_t \left\{ \frac{1}{g(t)} \int_{t_0}^{t} k_0(t, s)g(s)\, ds \right\} |x - y|_g \le \alpha|x - y|_g. \tag{1.3.48}$$

Therefore, if we assume $\alpha < 1$, the operator T will be a contraction on the space

C_g, and the only condition we have to retain for the function $g(t)$ is inequality (1.3.47), with $\alpha < 1$.

Consequently, in order to ensure the existence and uniqueness of a solution of equation (1.3.1) in the space C_g, we have to construct a positive solution $g(t)$ of inequality (1.3.47), for $\alpha < 1$.

Let us now solve the above problem for the function $g(t)$. We will show how a solution of (1.3.47) can be obtained. Define

$$\tilde{k}(t) = \sup k_0(u, s), \quad t_0 \le s \le u \le t,$$

and consider the 'smooth' function

$$k(t) = \int_t^{t+1} \tilde{k}(s)\, ds, \tag{1.3.49}$$

which obviously satisfies $k(t) \ge \tilde{k}(t) \ge k_0(t, s)$.

Inequality (1.3.47) will certainly be verified if the stronger inequality

$$k(t) \int_{t_0}^{t} g(s)\, ds \le \alpha g(t), \quad t \in [t_0, \infty), \tag{1.3.50}$$

is satisfied. But we can easily see that inequality (1.3.50) has the solution

$$g(t) = k(t) \exp\left\{\alpha^{-1} \int_{t_0}^{t} k(s)\, ds\right\}, \tag{1.3.51}$$

which obviously verifies $g(t) > 0$ if $k_0(t, s)$ does not vanish identically in any set $t_0 \le s \le t \le t_1, t_1 > t_0$. Of course, if we take (1.3.38) into account, we realize that the last requirement on $k_0(t, s)$ is not a restriction at all.

As a result of the discussion conducted above with regard to the integral equation (1.3.1), the following result can be stated.

Theorem 1.3.2. *Consider equation (1.3.1), and assume that the following conditions are satisfied*:

(1) $f \in C_g([t_0, \infty), R^n)$;
(2) $k: \Delta' \times R^n \to R^n$ *is a continuous map verifying a generalized Lipschitz condition of the form (1.3.38), where $k_0(t, s)$ is a continuous nonnegative function on Δ'*;
(3) g *is given by the formula (1.3.51)*.

Then there exists a unique solution $x \in C_g$ of equation (1.3.1). This solution can be obtained by the method of successive approximations, which converges in the space C_g (in particular, the method converges uniformly on any compact interval of the half-axis $[t_0, \infty)$).

Theorem 1.3.2 contains as special cases many results concerning boundedness of solutions or other kinds of asymptotic behavior of solutions to the equation (1.3.1). We shall illustrate here the applicability of this theorem to the stability theory of ordinary differential equations of the form (1.1.1). In order to derive the theorem of asymptotic stability for (1.1.1), we shall use the integral equation

$$x(t) = X(t)X^{-1}(t_0)x^0 + \int_{t_0}^{t} X(t)X^{-1}(s)f(s, x(s)) \, ds, \qquad (1.1.5)$$

and indicate the convenient functions $f(t)$, $k_0(t, s)$, and $g(t)$ which appear in the statement of Theorem 1.3.2.

We assume the uniform asymptotic stability of the zero solution of the differential system of the first approximation $\dot{x} = A(t)x$ (see, for instance, C. Corduneanu [6]), which leads to the following type of estimate:

$$|X(t)X^{-1}(s)| \leq K \exp\{-\beta(t-s)\}, \quad t \geq s \geq t_0 \qquad (1.3.52)$$

for convenient positive constants K and β. More precisely, this type of estimate characterizes the uniform asymptotic stability.

If $f(t, x)$ is continuous from $[t_0, \infty) \times R^n$ into R^n, $f(t, 0) = 0$, and the Lipschitz condition $|f(t, x) - f(t, y)| \leq L|x - y|$ holds, then we can take

$$k_0(t, s) = K \exp\{-\beta(t-s)\}, \quad t \geq s \geq t_0. \qquad (1.3.53)$$

In this case the integral inequality (1.3.47) becomes

$$KL \int_{t_0}^{t} g(s) \exp(\beta s) \, ds \leq \alpha g(t) \exp(\beta t), \quad t \geq t_0, \qquad (1.3.54)$$

and is obviously satisfied by $g(t) = \exp(-\gamma t)$, $0 < \gamma < \beta$, provided

$$L < \alpha(\beta - \gamma)K^{-1}. \qquad (1.3.55)$$

In remains only to check condition (1) from the statement of Theorem 1.3.2. This condition reduces to

$$|X(t)X^{-1}(t_0)x^0| \leq C \exp(-\gamma t), \quad t \geq t_0. \qquad (1.3.56)$$

To reconcile (1.3.56) and (1.3.52), we can proceed as follows:

$$|X(t)X^{-1}(t_0)x^0| \leq |X(t)X^{-1}(t_0)| \, |x^0| \leq K|x^0| \exp\{-\beta(t-t_0)\}$$
$$= K|x^0| \exp(\beta t_0) \exp(-\beta t) \leq K|x^0| \exp(\beta t_0) \exp(-\gamma t),$$

and take $C = K|x^0| \exp(\beta t_0)$.

Therefore, the solution $x(t)$ of (1.1.5), the integral form of (1.1.1), with initial condition $x(t_0) = x^0$, satisfies an exponential estimate of the form

$$|x(t)| \leq A \exp(-\gamma t),$$

which implies the asymptotic stability of the zero solution of the system (1.1.1).

Condition (1.3.55) shows that the asymptotic stability of the zero solution of the linear system of the first approximation is preserved only for small values of the Lipschitz constant of the perturbing term $f(t, x)$. Actually, since $f(t, 0) = 0$, we derive from the Lipschitz condition for $f(t, x)$ the estimate $|f(t, x)| \leq L|x|$, which shows that the perturbing term itself has to be small. This fact is in perfect agreement with the theory of stability of ordinary differential equations.

1.4   Fredholm theory of linear integral equations with square summable kernel

The Fredholm type equation which constitutes the analogue of equation (1.3.1) can be written as

$$x(t) = f(t) + \int_a^b k(t, s, x(s)) \, ds, \quad t \in [a, b], \tag{1.4.1}$$

where $f: [a, b] \to R^n$, $k: Q \times B_r \to R^n$, with $Q = [a, b] \times [a, b]$, and B_r is the ball of radius $r > 0$, centered at the origin of R^n.

One usually notices that for those kernels $k(t, s, x)$, for which $k(t, s, x) \equiv 0$ when $t \geq s$, equation (1.4.1) reduces to the Volterra integral equation (1.3.1):

$$x(t) = f(t) + \int_a^t k(t, s, x(s)) \, ds.$$

In many textbooks or monographs the Volterra type equation is only briefly discussed and categorized as a special case of Fredholm equations.

While this is formally correct, we should also point out that the Volterra equation has properties we do not encounter with Fredholm equations. For instance, as seen in Theorems 1.3.1 and 1.3.2, Volterra equations provide a valid example of functional equations for which, under quite mild assumptions, both local and global existence results can be obtained. This feature makes Volterra type equations a suitable tool in investigating evolution problems. If we notice that, for Fredholm type equations such as (1.4.1), local existence (i.e., existence in a small interval $[a, a + \delta]$) does not make sense, because the right-hand side involves the values of the solution x on the whole interval $[a, b]$, we realize that dealing with Fredholm equations is going to be more challenging than dealing with Volterra equations.

In general, we cannot expect a Fredholm equation to be solvable. One of the simplest examples, given by the equation

$$x(t) = t + \int_0^1 k(t, s) x(s) \, ds, \tag{1.4.2}$$

with $k(t, s)$ defined by

$$k(t,s) = \begin{cases} \pi^2 s(1-t) & 0 \le s \le t \\ \pi^2 t(1-s) & t \le s \le 1 \end{cases}$$

shows that the absence of solutions is quite a common feature for Fredholm equations.

Equation (1.4.2) leads immediately by differentiation to the second-order differential equation $x'' + \pi^2 x = 0$, as well as to the boundary value conditions $x(0) = 0$, $x(1) = 1$. It is easy to see that no solution of the obtained differential equation verifies these boundary value conditions. It is worth while noticing that the kernel $k(t, s)$ is not only continuous, but also piecewise continuously differentiable.

In relation to equation (1.4.1), we shall make the following assumptions:

(i) $f \in L^2([a, b], R^n)$;
(ii) $k: Q \times R^n \to R^n$ is measurable, $k(t, s, 0) \in L^2(Q, R^n)$, and

$$|k(t, s, x) - k(t, s, y)| \le k_0(t, s)|x - y|, \tag{1.4.3}$$

for some $k_0(t, s) \in L^2(Q, R_+)$.

Under the assumptions (i) and (ii), it is immediate that

$$(Tx)(t) = f(t) + \int_a^b k(t, s, x(s))\, ds \tag{1.4.4}$$

defines an operator which takes the function space $L^2([a, b], R^n)$ into itself. It is useful to notice that

$$|k(t, s, x(s))| \le |k(t, s, 0)| + k_0(t, s)|x(s)|, \tag{1.4.5}$$

almost everywhere (a.e.) on Q, which immediately leads to the conclusion that T is acting on the space $L^2([a, b], R^n)$.

Therefore, if we choose the space $L^2([a, b], R^n)$ as underlying space, it will suffice to find conditions under which the operator T, given by (1.4.4), is a contraction map on this space.

Since

$$|(Tx)(t) - (Ty)(t)| \le \int_a^b k_0(t, s)|x(s) - y(s)|\, ds, \tag{1.4.6}$$

for almost all $t \in [a, b]$, from (1.4.6) we derive immediately

$$|Tx - Ty|_{L^2} \le \alpha |x - y|_{L^2} \tag{1.4.7}$$

where

$$\alpha^2 = \int_a^b \int_a^b k_0^2(t, s)\, dt\, ds. \tag{1.4.8}$$

Consequently, taking into account inequality (1.4.7) and the contraction mapping principle, we can state the following result.

Theorem 1.4.1.   *Under the assumptions* (i) *and* (ii), *formulated above, the nonlinear Fredholm equation* (1.4.1) *has a unique solution in the space* $L^2([a,b], R^n)$, *provided*

$$\int_a^b \int_a^b k_0^2(t,s)\, dt\, ds < 1. \tag{1.4.9}$$

Corollary.   *The linear equation*

$$x(t) = f(t) + \int_a^b k(t,s)x(s)\, ds, \tag{1.4.10}$$

in which $f \in L^2([a,b], R^n)$, *and* $k: Q \to \mathscr{L}(R^n, R^n)$ *is measurable and satisfies*

$$\int_a^b \int_a^b |k(t,s)|^2\, dt\, ds < 1, \tag{1.4.11}$$

has a unique solution in the space $L^2([a,b], R^n)$.

Remark.   Conditions like (1.4.9) or (1.4.11) are certainly very restrictive. They do require that the kernel, or the function occurring in the generalized Lipschitz condition, be small in the L^2-norm. Generally speaking, this is a rather rare occurrence in the theory of Fredholm equations. At least in the linear case, much better results are available for Fredholm equations of the form (1.4.10). To be able to state such results, it is sufficient to introduce (following Fredholm) a parameter in equation (1.4.10), namely

$$x(t) = f(t) + \lambda \int_a^b k(t,s)x(s)\, ds, \tag{1.4.12}$$

and assume, instead of (1.4.11), the less restrictive condition

$$\int_a^b \int_a^b |k(t,s)|^2\, dt\, ds < +\infty. \tag{1.4.13}$$

Of course, the Corollary to Theorem 1.4.1 yields that equation (1.4.12) has a unique solution in $L^2([a,b], R^n)$, for every f belonging to this space, as soon as $|\lambda|$ is small enough. More precisely, it suffices to assume

$$|\lambda| < \left\{ \int_a^b \int_a^b |k(t,s)|^2\, dt\, ds \right\}^{-1/2}, \tag{1.4.14}$$

in order to secure existence and uniqueness in L^2, for every $f \in L^2$, for equation (1.4.12). As we know, condition (1.4.14) also suffices for the convergence in L^2 of successive approximations constructed in the standard manner:

$$x^0(t) \text{ arbitrary in } L^2, \quad x^{m+1}(t) = f(t) + \lambda \int_a^b k(t,s)x^m(s)\,ds, \quad m \ge 0.$$

With minor changes in comparison with the argument conducted in the preceding section, in the case of Volterra integral equations (1.3.22), we find that the unique solution of (1.4.12), under assumption (1.4.14), is given by the formula

$$x(t) = f(t) + \lambda \int_a^b \tilde{k}(t,s,\lambda)f(s)\,ds, \tag{1.4.15}$$

where the *resolvent kernel* $\tilde{k}(t,s,\lambda)$ is given by

$$\tilde{k}(t,s,\lambda) = \sum_{j=1}^{\infty} k_j(t,s)\lambda^{j-1}, \tag{1.4.16}$$

with

$$k_1(t,s) \equiv k(t,s), \quad k_m(t,s) = \int_a^b k_{m-1}(t,u)k_1(u,s)\,du, \quad m \ge 2. \tag{1.4.17}$$

From condition (1.4.13), all $k_m(t,s)$, $m \ge 1$, are in $L^2(Q, \mathscr{L}(R^n, R^n))$. Moreover, the series (1.4.16) is convergent in L^2, for those values of λ satisfying (1.4.14). Indeed, if we let

$$A^2(t) = \int_a^b |k(t,s)|^2\,ds, \quad B^2(s) = \int_a^b |k(t,s)|^2\,dt, \tag{1.4.18}$$

then both $A(t)$ and $B(s)$ are in $L^2(Q, R)$, and

$$K^2 = \int_a^b \int_a^b |k(t,s)|^2\,dt\,ds = \int_a^b A^2(t)\,dt = \int_a^b B^2(s)\,ds. \tag{1.4.19}$$

This immediately leads to the estimate

$$|k_m(t,s)| \le A(t)B(s)K^{m-2}, \quad m \ge 2, \tag{1.4.20}$$

almost everywhere on Q. Consequently, the terms of the series in the right-hand side of (1.4.16) are dominated by those of the series

$$A(t)B(s)|\lambda| \sum_{m=0}^{\infty} (|\lambda|K)^m, \tag{1.4.21}$$

almost everywhere in Q. Taking into account (1.4.19), we can see that the series (1.4.21) is convergent in $L^2(Q, R)$ for λ satisfying (1.4.14). Moreover, the series (1.4.16) converges absolutely, almost everywhere on Q.

What happens to (1.4.12) if condition (1.4.14) for λ is not satisfied, but k is such that (1.4.13) holds?

In other words, what can we say about the solvability of equation (1.4.12) when λ is an arbitrary complex number? Fredholm succeeded in providing a complete answer to the problem formulated above, known as the *Fredholm alternative*, and we must emphasize that this result constitutes one of the master-pieces of classical analysis, as well as the demiurge of linear functional analysis.

In order to prove the Fredholm alternative, we will limit our considerations to the space $L^2([a,b], \mathscr{C})$, where $\mathscr{C}$ stands for the complex number field. We will also allow λ to take complex values. It is obvious that the discussion carried out above for the linear equation (1.4.12) is still valid in the case of the space $L^2([a,b], \mathscr{C}^n)$, and, in particular, for the space $L^2([a,b], \mathscr{C})$.

Before we state and prove the Fredholm alternative, we shall define the so-called *adjoint equation* to equation (1.4.12):

$$y(t) = g(t) + \bar\lambda \int_a^b \overline{k(s,t)} y(s) \, ds. \tag{1.4.22}$$

Sometimes, instead of $\overline{k(s,t)}$ we will simply write $k^*(t,s)$. The kernel $k^*(t,s)$ is called the *adjoint kernel* associated with $k(t,s)$. The function $g(t)$ in (1.4.22) is any L^2-function.

It is obvious that the adjoint equation to (1.4.22) is equation (1.4.12).

The homogeneous equations associated with (1.4.12) and (1.4.22) are respectively

$$x(t) = \lambda \int_a^b k(t,s) x(s) \, ds, \tag{1.4.23}$$

and

$$y(t) = \bar\lambda \int_a^b k^*(t,s) y(s) \, ds. \tag{1.4.24}$$

Theorem 1.4.2 (Fredholm).   *Either equations (1.4.12) and (1.4.22) have a unique solution in $L^2([a,b], \mathscr{C})$ for every f, $g \in L^2([a,b], \mathscr{C})$, in which case the homogeneous equations have only a trivial solution, or the homogeneous equations have nonzero solutions. In this case, equation (1.4.12) has solutions if and only if f is orthogonal to every solution $y(t)$ of the adjoint homogeneous equation (1.4.24):*

$$\int_a^b f(t) \overline{y(t)} \, dt = 0. \tag{1.4.25}$$

Remark.   In some cases, the statement of Fredholm's theorem provides more information than the Fredholm alternative: for instance, the fact that the dimension of the linear space of solutions to (1.4.23) or (1.4.24) is the same. We shall later be concerned with these matters.

Proof of Theorem 1.4.2. The proof of Theorem 1.4.2 will consists of three steps. In the first step we will give the proof in a rather special case when the kernel $k(t, s)$ has finite rank:

$$k(t, s) = \sum_{i=1}^{n} a_i(t)\overline{b_i(s)}. \tag{1.4.26}$$

In (1.4.26) we can assume that the functions $a_i(t)$, $i = 1, 2, \ldots, n$, are linearly independent. In the contrary case, we can rewrite that formula with a smaller number of terms in the right-hand side. A similar remark can be made in relation to the functions $b_i(s)$, $i = 1, 2, \ldots, n$. Of course, a_i and b_i are L^2-functions.

In the second step we will show that any kernel $k(t, s)$ satisfying condition (1.4.13) can be represented in the form

$$k(t, s) = k_n(t, s) + R_n(t, s), \tag{1.4.27}$$

where $k_n(t, s)$ is a kernel of finite rank n (i.e., it can be represented in the form (1.4.26) with linearly independent functions $a_i(t)$ and $b_i(t)$, $i = 1, 2, \ldots, n$), and $R_n(t, s)$ is such that

$$\int_a^b \int_a^b |R_n(t, s)|^2 \, dt \, ds = r_n^2 \tag{1.4.28}$$

can be made arbitrarily small, provided n is chosen large enough.

The third step will contain the proof of the Fredholm alternative for kernels $k(t, s)$ satisfying (1.4.13), based on the possibility of representing any such kernel in the form (1.4.27), and on the Corollary of Theorem 1.4.1.

First step. We shall consider integral equations of the form

$$x(t) = f(t) + \lambda \int_a^b \left[\sum_{i=1}^{n} a_i(t)\overline{b_i(s)} \right] x(s) \, ds, \tag{1.4.29}$$

in which f is an arbitrary L^2-function. As noticed above, we can assume that the functions $a_i(t)$, $i = 1, 2, \ldots, n$, are linearly independent, as well as the functions $b_i(t), i = 1, 2, \ldots, n$. From equation (1.4.29) we see that a representation of the form

$$x(t) = f(t) + \lambda \sum_{i=1}^{n} x_i a_i(t), \tag{1.4.30}$$

is valid, where

$$x_i = \int_a^b x(s)\overline{b_i(s)} \, ds, \quad i = 1, 2, \ldots, n. \tag{1.4.31}$$

If we substitute $x(t)$ given by (1.4.30) in equation (1.4.29), we obtain, equating the coefficients of $a_i(t)$,

$$x_i = \lambda \sum_{j=1}^{n} \alpha_{ij} x_j + f_i, \quad i = 1, 2, \ldots, n, \tag{1.4.32}$$

where

$$\alpha_{ij} = \int_a^b a_j(s)\overline{b_i(s)}\,ds, \quad f_i = \int_a^b f(s)\overline{b_i(s)}\,ds.$$

The existence of a solution for (1.4.29) is, therefore, equivalent to the existence of a solution for the linear algebraic system (1.4.32). A unique solution for (1.4.32) will exist for any free terms $f_i = 1, 2, \ldots, n$, if and only if $\det A \neq 0$, where $A = (\delta_{ij} - \lambda\alpha_{ij})_{n \times n}$, and δ_{ij} is the Kronecker delta. On the other hand, for the adjoint equation to (1.4.29),

$$y(t) = g(t) + \bar{\lambda} \int_a^b \left[\sum_{i=1}^{n} b_i(t)a_i(s) \right] y(s)\,ds, \tag{1.4.33}$$

the system similar to (1.4.32) can be written as

$$y_i = \bar{\lambda} \sum_{i=1}^{n} \overline{\alpha_{ji}} y_j + g_i, \quad i = 1, 2, \ldots, n, \tag{1.4.34}$$

with obvious notations for y_i and g_i, $i = 1, 2, \ldots, n$. The matrix of the coefficients in the system (1.4.34) is obviously the adjoint of the matrix A, i.e., $A^* = (\delta_{ji} - \bar{\lambda}\overline{\alpha_{ji}})_{n \times n}$, while $\det A^* = \overline{\det A}$. Hence, (1.4.29) and (1.4.34) have simultaneously a unique solution, for any L^2-functions f and g. It also follows that the corresponding homogeneous equations have simultaneously nontrivial solutions. If both homogeneous equations have nontrivial solutions, the dimensions of the corresponding spaces of solutions will be the same (because rank A = rank A^*).

It only remains to discuss what happens when the homogeneous equation attached to (1.4.29) has nontrivial solutions. In other words, what condition must f satisfy to secure the existence of solutions to (1.4.29)?

Of course, we can look again at the system (1.4.32), which can be rewritten as $Ax = f$, where $x = \operatorname{col}(x_i)$, $f = \operatorname{col}(f_i)$ are complex n-vectors. Since $(Ax, y) = (x, A^*y)$, for any x, y in $\mathscr{C}^n$, one finds out that $(f, y) = (x, A^*y) = 0$, for any solution y of the homogeneous adjoint equation. On the other hand, if $(f, y) = 0$, it means that f belongs to the orthogonal complement (in $\mathscr{C}^n$) of the space of solutions of the homogeneous adjoint equation. This orthogonal complement is generated by the vectors $v_j = \operatorname{col}_j A, j = 1, 2, \ldots, n$, and consequently the vector f can be represented as a linear combination of the vectors v_j: $f = c_1 v_1 + \cdots + c_n v_n$, or $f = Ac$. This means that the system $Ax = f$ has the solution $x = c = \operatorname{col}(c_j)$.

We have only to notice the fact that, for any solution $y(t)$ of the homogeneous equation associated to (1.4.34), we have

$$\int_a^b f(t)\overline{y(t)}\,dt = \int_a^b f(t)\left[\sum_{j=1}^n y_j\overline{b_j(t)}\right]dt = \sum_{j=1}^n f_j\overline{y}_j,$$

to be able to establish the connection between the scalar product in L^2 (see condition (1.4.25)), and that in $\mathscr{C}^n$.

The proof of the alternative in case of kernels of finite rank is thereby complete.

Second step. We shall now prove that any L^2-kernel $k(t,s)$, i.e., such that (1.4.13) holds, can be represented in the form (1.4.27), with

$$k_n(t,s) = \sum_{i=1}^n a_i(t)b_i(s), \qquad (1.4.35)$$

where $a_i(t)$, $b_i(s)$, $i = 1, 2, \ldots, n$, are conveniently chosen L^2-functions.

In the space $L^2([a,b], \mathscr{C})$, let us consider the orthonormal complete system $\{b_i(s)\}$, $i = 1, 2, \ldots$, for instance the system of trigonometric functions

$$\frac{1}{\sqrt{2\pi}}, \frac{1}{\sqrt{\pi}}\cos s, \frac{1}{\sqrt{\pi}}\sin s, \ldots, \frac{1}{\sqrt{\pi}}\cos ns, \frac{1}{\sqrt{\pi}}\sin ns, \ldots$$

when $a = 0$, $b = 2\pi$. Let us denote by

$$a_i(t) = \int_a^b k(t,s)\overline{b_i(s)}\,ds, \quad i = 1, 2, \ldots \qquad (1.4.36)$$

the Fourier coefficients of $k(t,s)$, regarded as a function of s, with respect to the system $\{b_i\}$. Then Parseval's equation holds:

$$\sum_{i=1}^\infty |a_i(t)|^2 = \int_a^b |k(t,s)|^2\,ds, \qquad (1.4.37)$$

almost everywhere on $[a,b]$. If we again apply Parseval's equation for the function

$$k(t,s) - \sum_{i=1}^n a_i(t)b_i(s),$$

we obtain

$$\sum_{i=n+1}^\infty |a_i(t)|^2 = \int_a^b \left| k(t,s) - \sum_{i=1}^n a_i(t)b_i(s)\right|^2\,ds. \qquad (1.4.38)$$

But each $a_i(t)$, $i = 1, 2, \ldots$, is an L^2-function and, integrating both sides of (1.4.38), we obtain

$$\sum_{i=n+1}^n \int_a^b |a_i(t)|^2\,dt = \int_a^b \int_a^b \left| k(t,s) - \sum_{i=1}^n a_i(t)b_i(t)\right|^2\,dt\,ds. \qquad (1.4.39)$$

We have only to show that the sum of the series appearing in the left-hand side of (1.4.39) can be made arbitrarily small, provided n is chosen sufficiently large.

From (1.4.37) we obtain by integration

$$\sum_{i=1}^{\infty} \int_a^b |a_i(t)|^2 \, dt = \int_a^b \int_a^b |k(t,s)|^2 \, dt \, ds. \tag{1.4.40}$$

Therefore, in the left-hand side of (1.4.39) we have the remainder of a (numerical) convergent series. Hence, that quantity can be made arbitrarily small, provided we take n sufficiently large.

Letting

$$R_n(t,s) = k(t,s) - \sum_{i=1}^{n} a_i(t)b_i(s),$$

one can assert from (1.4.38) that

$$\lim_{n\to\infty} \int_a^b \int_a^b |R_n(t,s)|^2 \, dt \, ds = 0. \tag{1.4.41}$$

This completes the proof of the statement made in the second step of the proof of Theorem 1.4.1. Let us point out that the approximation property established above has various interesting consequences in relation to other problems in mathematical analysis.

Third step.   Based on the possibility of representing the kernel $k(t,s)$ in the form

$$k(t,s) = \sum_{i=1}^{n} a_i(t)b_i(s) + R_n(t,s), \tag{1.4.42}$$

for every natural number n, with $R_n(t,s)$ satisfying (1.4.41), we shall reduce the case of (general) kernels subject to (1.4.13) to the case of kernels of finite rank, for which the Fredholm alternative has been proved in the first step.

Notice that (1.4.12) can be written in the form

$$x(t) - \lambda \int_a^b R_n(t,s)x(s) \, ds = \lambda \int_a^b k_n(t,s)x(s) \, ds + f(t). \tag{1.4.43}$$

If we let

$$y(t) = x(t) - \lambda \int_a^b R_n(t,s)x(s) \, ds, \tag{1.4.44}$$

then for any n such that (see (1.4.14))

$$|\lambda| < \left\{ \int_a^b \int_a^b |R_n(t,s)|^2 \, dt \, ds \right\}^{-1/2}, \tag{1.4.45}$$

we can write (1.4.44) in the equivalent form

$$x(t) = y(t) + \lambda \int_a^b \tilde{R}_n(t,s,\lambda)y(s) \, ds, \tag{1.4.46}$$

where $\tilde{R}_n(t, s, \lambda)$ denotes the resolvent kernel corresponding to $R_n(t, s)\lambda$. It is worth while noticing that, according to (1.4.41), inequality (1.4.45) holds for any complex number λ, provided n is chosen sufficiently large.

If we now substitute $x(t)$, given by (1.4.46), into the equation

$$y(t) = f(t) + \lambda \int_a^b k_n(t, s) x(s) \, ds, \tag{1.4.47}$$

which is an immediate consequence of (1.4.43) and (1.4.44), we obtain for $y(t)$ the following integral equation:

$$y(t) = f(t) + \lambda \int_a^b \left[k_n(t, s) + \lambda \int_a^b k_n(t, u) \tilde{R}_n(u, s, \lambda) \, du \right] y(s) \, ds. \tag{1.4.48}$$

For any solution $x(t)$ of (1.4.12), there is a solution $y(t)$ of (1.4.48). The converse statement is also true because the relationship (1.4.44) between $x(t)$ and $y(t)$ is one-to-one, for sufficiently large n.

The kernel of the integral equation (1.4.48) is obviously of finite rank n, because it can be represented in the form

$$\sum_{i=1}^n a_i(t) \left[b_i(s) + \lambda \int_a^b b_i(u) \tilde{R}_n(u, s, \lambda) \, du \right].$$

We proved the validity of the Fredholm alternative for such kernels in the first step. Therefore, the validity of the alternative is also proved for equation (1.4.12).

It remains to show that the orthogonality condition (1.4.25) is the necessary and sufficient condition for the solvability of the nonhomogeneous equation (1.4.12), when the corresponding homogeneous equation has nontrivial solutions. We know that this is true in the case of kernels of finite rank (first step of the proof). In order to establish its validity in the general case, we shall prove that the adjoint homogeneous equation attached to (1.4.12) has the same solutions as the homogeneous equation attached to (1.4.48).

In order to be able to write the adjoint homogeneous equation for (1.4.48), we need to know how to obtain the adjoint kernel for

$$\int_a^b k_n(t, u) \tilde{R}_n(u, s, \lambda) \, ds,$$

which appears in (1.4.48). Of course, the particular meaning of k_n and $\tilde{R}_n$ does not matter for our purpose. We can easily check that the adjoint kernel of

$$k(t, s) = \int_a^b h(t, u) m(u, s) \, du, \tag{1.4.49}$$

where both h and m are L^2-kernels, is given by

$$k^*(t, s) = \int_a^b m^*(t, u)h^*(u, s)\,du. \tag{1.4.50}$$

Therefore, the adjoint homogeneous equation corresponding to (1.4.12) can be written as

$$z(t) = \bar{\lambda} \int_a^b \left[k_n^*(t, s) + \bar{\lambda} \int_a^b \tilde{R}_n^*(t, u, \lambda)k_n^*(u, s)\,du \right] z(s)\,ds. \tag{1.4.51}$$

Let us now proceed in proving the equation (1.4.24) has the same solutions as equation (1.4.51). In order to avoid some rather tedious calculations, we shall adopt an operator algebra approach to writing equations (there is some resemblance with the matrix case). Thus, equation (1.4.51) will be rewritten in the form

$$[I - \bar{\lambda}(I + \bar{\lambda}\tilde{R}_n^*)k_n^*]z = 0, \tag{1.4.52}$$

while equation (1.4.24) becomes

$$[I - \bar{\lambda}k_n^* - \bar{\lambda}R_n^*]y = 0, \tag{1.4.53}$$

after using the decomposition $k = k_n + R_n$ for the kernel.

It is useful to notice the validity of the following relationship:

$$(I - \bar{\lambda}R_n^*)(I + \bar{\lambda}\tilde{R}_n^*) = I. \tag{1.4.54}$$

This is true for any complex λ, provided we choose n sufficiently large. The proof of (1.4.54) comes easily if one takes into account the equations of the resolvent kernel, obtainable the same way in the Fredholm case as in the Volterra case. More exactly, from (1.4.16) one obtains for small values of λ

$$\tilde{k}(t, s, \lambda) = k(t, s) + \lambda \int_a^b k(t, u)\tilde{k}(u, s, \lambda)\,du, \tag{1.4.55}$$

as well as

$$\tilde{k}(t, s, \lambda) = k(t, s) + \lambda \int_a^b \tilde{k}(t, u, \lambda)k(u, s)\,du. \tag{1.4.56}$$

Of course, these equations must be adapted to the kernels appearing in equations (1.4.53) and (1.4.54).

If we now consider equation (1.4.52), and multiply on the left by $I - \bar{\lambda}R_n^*$, we obtain the equation

$$[I - \bar{\lambda}R_n^* - \bar{\lambda}(I - \bar{\lambda}R_n^*)(I + \bar{\lambda}\tilde{R}_n^*)k_n^*]z = 0,$$

which because of (1.4.54) reduces to

$$[I - \bar{\lambda}R_n^* - \bar{\lambda}k_n^*]z = 0. \tag{1.4.57}$$

Equation (1.4.57) coincides with equation (1.4.53). Therefore, any solution of equation (1.4.52), which is another form of equation (1.4.48), also satisfies equation (1.4.53), which is another form of (1.4.24).

Let us now start from equation (1.4.53), and obtain (1.4.52) as a consequence. If we notice that the factors in the left-hand side of (1.4.54) commute, and multiply both sides in (1.4.53) by $I + \bar{\lambda}\tilde{R}_n^*$, we obtain

$$[(I + \bar{\lambda}R_n^*)(I - \bar{\lambda}R_n^*) - \bar{\lambda}(I + \bar{\lambda}\tilde{R}_n^*)k_n^*]y = 0,$$

or, using (1.4.54),

$$[I - \bar{\lambda}(I + \bar{\lambda}R_n^*)k_n^*]y = 0,$$

which is exactly equation (1.4.52).

The proof of Theorem 1.4.1 is thereby complete.

Remark. One can prove that the homogeneous equations (1.4.23) and (1.4.24) have the same maximum number of linearly independent solutions.

Of course, when we are in the first case of the Fredholm's alternative, both equations (1.4.23) and (1.4.24) have only the zero solution. Therefore, this case is not of any interest. It remains to check the validity of the statement in the second case of the alternative.

As shown in the third step, equation (1.4.23) is equivalent to the equation obtained from (1.4.48) for $f(t) \equiv 0$, i.e.,

$$[I - \lambda k_n(I + \lambda\tilde{R}_n)]y = 0, \tag{1.4.58}$$

via the linear transformation $x \to y$ given by (1.4.44), or equivalently by (1.4.46). As noticed above, this transformation preserves linear dependence (independence) if condition (1.4.45) is satisfied.

On the other hand, equation (1.4.24) was found to be equivalent to equation (1.4.51), which is the adjoint equation of (1.4.58). Since both equations (1.4.51) and (1.4.58) are equations with kernels of finite rank and adjoint to each other, according to the first step of the proof they must have the same (maximum) number of linearly independent solutions ($n - $ rank A).

In view of further discussion related to Fredholm's alternative, let us define now the concepts of *eigenvalue* and *eigenfunction* related to a given L^2-kernel $k(t, s)$; or, if we prefer, related to the homogeneous equation (1.4.23) which is completely determined by the kernel $k(t, s)$.

According to Fredholm's alternative, equation (1.4.23) might have nontrivial solutions for some values of the complex parameter λ. Any such value of λ is by definition an *eigenvalue* of (1.4.23), or of the kernel $k(t, s)$. Any nonzero solution of (1.4.23) is called an *eigenfunction* of this equation, or of the kernel $k(t, s)$, corresponding to the eigenvalue λ.

Sometimes, one uses the term *characteristic value* for λ^{-1}, where λ is an eigenvalue ($\lambda = 0$ cannot be an eigenvalue!), and *characteristic function* for an eigenfunction.

A basic problem relating to eigenvalues and eigenfunctions is: what is the distribution of eigenvalues in the complex plane for a given L^2-kernel?

As we already know, the Volterra kernel ($k(t,s) = 0$ for $s > t$) does not have eigenvalues at all, even if it is a continuous function (because of the uniqueness of the solution, the homogeneous equation has only the zero solution).

In the case of Fredholm equations of finite rank, we have seen in the first step of the proof of Theorem 1.4.1 that the eigenvalues are the roots of an algebraic equation. Therefore, for kernels of finite rank, there are eigenvalues and their number is always finite (equal to the number of the functions $a_i(t)$, which are supposed to be linearly independent; as well as the number of the functions $b_i(s)$).

In the general case of an L^2-kernel, the discussion is somewhat more complicated. What we want to show is that the homogeneous equation (1.4.23) has at most a finite number of eigenvalues in any disk $|\lambda| \leq r$, $r > 0$ being arbitrary. In other words, equation (1.4.23) can have nonzero solutions only for finitely many values of λ that satisfy $|\lambda| \leq r$.

Indeed, as seen in the third step of the proof of Theorem 1.4.1, equation (1.4.23) is equivalent, as far as the existence of solutions is concerned, with the homogeneous equation corresponding to (1.4.48), i.e.,

$$y(t) = \lambda \int_a^b \left[k_n(t,s) + \lambda \int_a^b k_n(t,u)\tilde{R}_n(u,s,\lambda)\,du \right] y(s)\,ds, \qquad (1.4.59)$$

whenever λ satisfies (1.4.45); or, equivalently, whenever n is sufficiently large. Therefore, we have to examine the existence of nonzero solutions to equation (1.4.59), with λ arbitrary in $|\lambda| \leq r$. This problem is not so difficult because equation (1.4.59) has a kernel of finite rank (as noticed when we considered equation (1.4.48): the nonhomogeneous counterpart of (1.4.59)). Nevertheless, we face a different problem from before because λ appears in the kernel itself, and not only in front of the integral as is the case with (1.4.23), for instance.

If we proceed as in the first step of the proof of Theorem 1.4.1, then it is obvious that the solutions of (1.4.59) must be sought in the form

$$y(t,\lambda) = \lambda \sum_{i=1}^{n} c_i(\lambda)a_i(t). \qquad (1.4.60)$$

For $c_i(\lambda)$, $i = 1, 2, \ldots, n$, we obtain the homogeneous system $A(\lambda)c(\lambda) = 0$, where $c(\lambda) = \mathrm{col}(c_i(\lambda))$, $A(\lambda) = (\delta_{ij} - \lambda\alpha_{ij}(\lambda))$, $i, j = 1, 2, \ldots, n$, with

$$\alpha_{ij}(\lambda) = \int_a^b a_j(s)\left[b_i(s) + \lambda \int_a^b b_i(u)\tilde{R}_n(u,s,\lambda)\,du \right]ds.$$

Since $\tilde{R}_n(u,s,\lambda)$ is an analytic (holomorphic) function of λ (being represented as a convergent power series in λ, for λ satisfying (1.4.45)), it means that all $\alpha_{ij}(\lambda)$ are also analytic within the same disk. The necessary and sufficient condition for the existence of nonzero solutions to the equation $A(\lambda)c(\lambda) = 0$ is $\det A(\lambda) = 0$. But $\det A(\lambda)$ is obviously an analytic function of λ, for λ satisfying (1.4.45), and it cannot be identically zero because $\det A(0) = 1$. Since the zeros of an analytic function $\not\equiv 0$ have no accumulation point in the domain of analyticity, the set of those λ for which $\det A(\lambda) = 0$ within the disk $|\lambda| \leq r$ is a finite set. Of course, n has to be chosen such that (1.4.45) holds, which is always possible because of (1.4.41).

As a result of the discussion conducted above in relation to the distribution of the eigenvalues of an L^2-kernel, we can state the following conclusion: *the set of eigenvalues of an L^2-kernel ($\not\equiv 0$) is at most countable; if the set is infinite, the only accumulation point of this set is ∞.*

Consequently, if $k(t,s)$ is an L^2-kernel, then we can always arrange its eigenvalues $\lambda_j, j = 1, 2, \ldots$, in a sequence with increasing modules:

$$0 < |\lambda_1| \leq |\lambda_2| \leq \cdots \leq |\lambda_n| \leq \cdots \qquad (1.4.61)$$

where $|\lambda_n| \to \infty$ as $n \to \infty$. Of course, this last property has meaning only when there are infinitely many eigenvalues. It is usually agreed to repeat an eigenvalue in the sequence (1.4.61) as many times as its *multiplicity* indicates. By *multiplicity* of an eigenvalue we mean the dimension of the linear space of eigenfunctions corresponding to that eigenvalue. It is easily seen that this dimension is always finite (see, for instance, C. Corduneanu [6]).

In accordance with (1.4.61), we consider also the sequence of eigenfunctions of the kernel $k(t,s)$:

$$\phi_1(t), \quad \phi_2(t), \ldots, \quad \phi_n(t), \ldots \qquad (1.4.62)$$

where

$$\phi_j(t) = \lambda_j \int_a^b k(t,s)\phi_j(s)\,ds. \qquad (1.4.63)$$

While, as mentioned above, the sequence of eigenvalues might contain the same number repeated as many times as its multiplicity indicates, the sequence (1.4.62) contains only distinct elements. Moreover, if an eigenvalue has multiplicity m, then in the linear space of its eigenfunctions one chooses m linearly independent elements which will appear in the sequence (1.4.62) consecutively: $\phi_p(t), \phi_{p+1}(t), \ldots, \phi_{p+m-1}(t)$.

Of particular significance is the case of *Hermitian kernels*:

$$k^*(t,s) = k(t,s) \quad \text{a.e. in } Q. \qquad (1.4.64)$$

Such kernels were systematically investigated by Hilbert and his followers at the beginning of this century. One major result is:

Theorem 1.4.2 (Hilbert).   *Let $k(t, s)$ be a Hermitian L^2-kernel, nonvanishing identically in $Q = [a, b] \times [a, b]$. Then there exists at least one eigenvalue of $k(t, s)$.*

The proof of this basic result can be found in many books on the theory of integral equations. See, for instance, C. Corduneanu [6], F. Tricomi [1]. Since a Hermitian kernel generates a self-adjoint integral operator in $L^2([a, b], \mathscr{C})$, the eigenvalues are always real, and eigenfunctions corresponding to distinct eigenvalues are orthogonal.

Indeed, using the operator algebra notations as above, we easily find that $(\phi, k\phi) = (k^*\phi, \phi)$ is real, for every $\phi \in L^2$. If $\phi = \lambda k\phi$, $\phi \neq 0$, then we obtain $(\phi, \phi) = \lambda(\phi, k\phi)$, which shows that λ must be real.

For the orthogonality property of eigenfunctions we notice that $\phi = \lambda k\phi$ and $\psi = \mu k\psi$ imply $(\phi, \psi) = (\phi, \mu k\psi) = \mu(k\phi, \psi) = (\mu/\lambda)(\phi, \psi)$. This is possible only if $(\phi, \psi) = 0$.

If we apply these remarks to the sequence (1.4.61) of the eigenvalues, we find out that the eigenvalues of a Hermitian L^2-kernel can be arranged into a sequence $\{\lambda_j\}, j \in Z - \{0\}$, such that $\lambda_j < \lambda_{j+1}, \lambda_{-1} < 0 < \lambda_1$. Of course, this sequence may be infinite in both directions or only in one direction, if indeed it contains infinitely many terms.

With regard to the sequence of eigenfunctions (1.4.62), we can make the following important addition to what we said above, in the general case of L^2-kernels: if the kernel $k(t, s)$ is a Hermitian L^2-kernel, then the sequence of eigenfunctions is an orthogonal sequence, provided we agree that, for every multiple eigenvalue we choose an orthogonal basis for the space of its eigenfunctions (since a basis can certainly be chosen, it remains to apply the Gramm–Schmidt orthogonalization procedure to obtain an orthogonal basis). Therefore, if (1.4.62) constitutes the sequence of eigenfunctions associated with the Hermitian L^2-kernel $k(t, s)$, then $(\phi_i, \phi_j) = 0$, for $i \neq j$.

The theory of integral equations with Hermitian kernel (Hilbert's theory) is one of the most interesting parts of the theory of integral equations, which led to the development of such branches of modern analysis as the theory of orthogonal series/sequences, the variational methods and others.

Let us conclude this section with a result on the absence of eigenvalues for L^2-kernels. This result was proved first, under particular conditions, by Lalesco [1], and then obtained in the form given below by Krachkovskii [1].

Theorem 1.4.3.   *The L^2-kernel $k(t, s)$ has no eigenvalues if and only if the following conditions are satisfied by the iterated kernels*

$$\int_a^b k_n(t, t)\, dt = 0, \quad \text{for } n \geq 3, \tag{1.4.65}$$

where $k_n(t, s)$ is the iterated kernel defined by (1.4.17).

We omit the proof of this result, which clarifies the problem of the existence or absence of eigenvalues.

1.5 Nonlinear equations of Hammerstein type

A class of nonlinear integral equations which appear as a direct generalization of the linear ones was first investigated by Hammerstein [1]. The results included in this section are due to Dolph [1].

Unlike the preceding sections of this chapter, this section will assume some acquaintance with the basic facts of operator theory on Banach/Hilbert spaces. See, for instance, Gohberg and Goldberg [1].

The Hammerstein type equation we want to investigate in this section can be written in the form

$$x(t) = \int_a^b k(t,s)f(s,x(s))\,ds, \quad t \in [a,b], \tag{1.5.1}$$

where $k(t,s)$ is an L^2-kernel, and $f(t,x)$ is a (nonlinear) map from $[a,b] \times R$, into R. Further details will be specified below.

In general, $f(t,0) \neq 0$, so that $x = 0$ is not a solution of (1.5.1). It is obvious from (1.5.1) that, if there is any solution of this equation, it must belong to the range of the linear integral operator generated by the kernel $k(t,s)$, namely

$$(Kx)(t) = \int_a^b k(t,s)x(s)\,ds, \quad t \in [a,b]. \tag{1.5.2}$$

From the conclusion of the investigation conducted in the preceding section, if we choose $f(t,x) = \gamma x + g(t)$, with γ a complex number that does not coincide with an eigenvalue of the kernel $k(t,s)$, and $g \in L^2$, then (1.5.1) has unique solution in L^2.

We will assume in the sequel that $k(t,s)$ is a Hermitian operator, and therefore the eigenvalues associated with this kernel, or with the linear integral operator (1.5.2), are real.

In the special case considered above we have $f_x(t,x) = \gamma$, and according to the assumption that γ does not coincide with an eigenvalue of the kernel $k(t,s)$ we must find a pair of consecutive eigenvalues of $k(t,s)$ such that $\lambda_j < \gamma < \lambda_{j+1}$. A natural idea in connection with this remark is to assume that the nonlinearity $f(t,x)$ occurring in equation (1.5.1) is such that

$$\lambda_j < \mu_j \leq f_x \leq \mu_{j+1} < \lambda_{j+1}, \tag{1.5.3}$$

for some fixed j. The numbers μ_j and μ_{j+1} are also fixed, and their role is to prevent the nonlinearity of the equation being too close to the eigenvalues, for which the existence of a solution may not be assured.

Instead of (1.5.3), one can encounter in the literature several other conditions which are more or less equivalent to it. For instance, Hammerstein assumed a condition of the form

$$\int_0^x f(t, y)\, dy \le \frac{1}{2}\lambda_1 x^2 + C,$$

where λ_1 is the first positive eigenvalue of the kernel $k(t, s)$; while Dolph assumes that $f(t, x)$ is such that

$$\lambda_j < \mu_j \le \frac{f(t, x) - f(t, y)}{x - y} \le \mu_{j+1} < \lambda_{j+1}, \quad x \ne y.$$

It is obvious that condition (1.5.3) is basically the same as the condition used by Dolph. Hammerstein's condition, while not requiring anything about the derivative f_x, has been related only to the first eigenvalue.

To obtain solutions for equation (1.5.1) the basic idea is to use a rather simple linearization technique, which can be formulated as follows: we rewrite equation (1.5.1) in the equivalent form

$$x(t) = \gamma \int_a^b k(t, s)x(s)\, ds + \int_a^b k(t, s)[f(s, x(s)) - \gamma x(s)]\, ds, \qquad (1.5.4)$$

where the real number γ is chosen such that $\gamma \ne \lambda_j$, with λ_j an arbitrary eigenvalue of $k(t, s)$. In particular, one can take $\gamma = \frac{1}{2}(\mu_j + \mu_{j+1})$, when $f(t, x)$ satisfies (1.5.3). It is, of course, assumed that the eigenvalues λ_j and λ_{j+1} of $k(t, s)$ are consecutive.

Equation (1.5.4) can be discussed, and the existence of a solution proved, using the same method of successive approximations that we have already used in most of the cases considered earlier. Starting with an arbitrary $x^0(t)$ in L^2, we set

$$x_{n+1}(t) = \gamma \int_a^b k(t, s)x_{n+1}(s)\, ds + \int_a^b k(t, s)[f(s, x_n(s)) - \gamma x_n(s)]\, ds, \quad (1.5.5)$$

for any $n \ge 0$. It is important to notice that (1.5.5) has a unique solution at each step, because γ is not an eigenvalue of the kernel $k(t, s)$, which implies that we are in the first case described by Fredholm's alternative. It is certainly sufficient to assume that $f(t, x(t))$ is an L^2-function, for any $x(t)$ in the same space.

In order to prove the convergence of the successive approximations defined above, which is not necessarily an easy task, we shall again appeal to the operator notation that was used in Section 1.4.

If we take (1.5.2) into account, and let $F: x(t) \to f(t, x(t))$ be the operator on L^2, generated by the function $f(t, x)$, then equation (1.5.4) can be written as

$$x = \gamma Kx + K(Fx - \gamma x), \qquad (1.5.6)$$

or, in a slightly modified form,

$$(I - \gamma K)x = K(Fx - \gamma x). \qquad (1.5.7)$$

But the equation $(I - \gamma K)x = f \in L^2$ always has a unique solution since we are in the first case of the Fredholm's alternative. Therefore, the inverse of the operator $I - \gamma K$ does exist and, according to Banach's theorem about inverses of linear operators, $(I - \gamma K)^{-1}$ is continuous on L^2. If we let

$$H = (I - \gamma K)^{-1} K = K(I - \gamma K)^{-1}, \tag{1.5.8}$$

then equation (1.5.7) can be rewritten in the form

$$x = H(Fx - \gamma x), \quad x \in L^2. \tag{1.5.9}$$

Since the operator K given by (1.5.2) is a self-adjoint operator on L^2, it can easily be seen that $I - \gamma K$, its inverse, and the operator H are self-adjoint. Of course, all these operators are also bounded on L^2.

In order to prove the existence and uniqueness of the solution of the equation (1.5.4), which is equivalent to (1.5.1), it is sufficient to show that the operator appearing in the right-hand side of (1.5.9) is a contraction mapping on L^2.

If we denote $Tx = H(Fx - \gamma x)$, then the following relationship is obtained:

$$Tx - Ty = H[Fx - Fy - \gamma(x - y)], \quad x, y \in L^2. \tag{1.5.10}$$

If we notice that $Fx - Fy = f_x^*(x - y)$, where f_x^* denotes an intermediate value of the derivative f_x, then (1.5.10) leads immediately to the inequality

$$|Tx - Ty|_{L^2} \leq |H| \sup|f_x^* - \gamma| |x - y|_{L^2}, \tag{1.5.11}$$

where H denotes the operator norm of H. Since H is a bounded self-adjoint operator on the space L^2, its norm is given by the well-known formula

$$|H| = \sup\{|\sigma_i|^{-1}\}, \quad \sigma_i \text{ is an eigenvalue of } H. \tag{1.5.12}$$

It remains therefore to find out what the eigenvalues of the operator H are, in terms of the eigenvalues of the operator K (or kernel $k(t, s)$). This can be done relatively easily. Taking into account the first expression for H, in formula (1.5.8), we obtain from $\phi = \sigma H\phi$, with $\phi \in L^2$, $\phi \neq \theta$,

$$(I - \gamma K)\phi = \sigma K\phi, \quad \text{or } \phi = (\gamma + \sigma)K\phi,$$

which shows that ϕ is also an eigenfunction of K, corresponding to the eigenvalue $\gamma + \sigma$. Conversely, if ϕ is an eigenfunction of H, then it is also an eigenfunction of K, corresponding to the eigenvalue shown in the above formula. Hence, the operators K and H have the same eigenfunctions, and eigenvalues of these operators are related by the formula $\gamma + \sigma_i = \lambda_i$.

Taking into account that the eigenvalues of H are given by $\sigma_i = \lambda_i - \gamma$, and the formula (1.5.12) for $|H|$, we obtain

$$|H| = \sup \frac{1}{\lambda_i - \gamma}, \quad \lambda_i \text{ is an eigenvalue of } K, \tag{1.5.13}$$

which, with a suitable choice of γ, leads to the estimate

$$|H| < \frac{2}{\mu_{j+1} - \mu_j}. \tag{1.5.14}$$

On the other hand, from (1.5.3) we obtain the estimate

$$\sup|f_x^* - \gamma| \le (\mu_{j+1} - \mu_j)/2. \tag{1.5.15}$$

From (1.5.14) and (1.5.15) we obtain

$$|H|\sup|f_x^* - \gamma| < 1, \tag{1.5.16}$$

which combined with (1.5.11) proves that the operator $Tx = H(Fx - \gamma x)$ is a contraction mapping on the space L^2.

Summing up the discussion conducted in this section, we conclude:

Theorem 1.5.1. *Consider the Hammerstein equation* (1.5.1), *in which* $k(t,s)$ *is a symmetric* L^2-*kernel, and* $f(t,x)$ *is a continuous function on* $[a,b] \times R$, *with values in* R. *Assume that* f_x *does exist, and verifies the inequality* (1.5.3), *where* λ_j *and* λ_{j+1} *are two consecutive eigenvalues of the kernel* $k(t,s)$, *while* μ_j *and* μ_{j+1} *are two fixed numbers. Then choosing* $\gamma = (\mu_j + \mu_{j+1})/2$, *the successive approximations* (1.5.5) *converge in* L^2 *to the unique solution* $x(t)$ *of* (1.5.1) *belonging to this space.*

Remark. Under the assumptions of Theorem 1.5.1, the operator F defined above by $(Fx)(t) = f(t,x(t))$ is indeed acting from $L^2([a,b], R)$ into itself. It suffices to notice that $f(t,x(t)) = (f_x^*)x + f(t,0)$, and take (1.5.3) into account, in order to reach the stated property. In fact, the result applies to more general functions than continuous (for instance, the functions satisfying the so-called Carathéodory conditions).

The original approach of Hammerstein was based on what is usually known as the *variational method*. More results related to the Hammerstein type equations and their applications can be found in the above quoted paper by C. L. Dolph. More recent references on these matters can be also found in the books by Krasnoselskii [1] and Vainberg [1].

In the remaining part of this section we will present, following C. L. Dolph, a modified version of the process of successive approximations described above for equation (1.5.1), still convergent in L^2 to the solution of this equation. The interesting feature of this modified process is that each approximation is a finite linear combination of eigenfunctions of the operator K. Thus, we are dealing with a version of the Ritz–Galerkin procedure, which can be considered as a numerical or approximate method of solving equation (1.5.1).

To be more specific, the operator H will first be approximated by the terms of a sequence $\{H_n\}$ of operators whose range is finite-dimensional: $|H_n - H| \to 0$, as $n \to \infty$. Then, the successive approximations are constructed according to the formula $x_{n+1} = H_n(Fx_n - \gamma x_n)$, starting with an arbitrary element of L^2.

Instead of concentrating on the particular case specified above, we shall establish a rather general result related to the equation

$$x = HG(x), \quad x \in E, \tag{1.5.17}$$

where E stands for a Banach space, H is a linear continuous operator on E, and G is a map from E into E – in general nonlinear – satisfying a Lipschitz condition

$$|G(x) - G(y)|_E \leq \alpha |x - y|_E, \quad x, y \in E. \tag{1.5.18}$$

Since the operator occurring in the right hand side of (1.5.17) also satisfies a Lipschitz condition with constant $|H|\alpha$, it follows that equation (1.5.17) has a unique solution in E, provided

$$|H|\alpha < 1. \tag{1.5.19}$$

An immediate consequence of (1.5.19) is the convergence in E of the sequence

$$x_{n+1} = HG(x_n), \quad x_0 \in E \text{ being arbitrary.} \tag{1.5.20}$$

On the other hand, let us consider the sequence

$$\bar{x}_{n+1} = H_n G(\bar{x}_n), \quad \bar{x}_0 \in E \text{ being arbitrarily chosen.} \tag{1.5.21}$$

Since the sequence (1.5.20) converges in E to the unique solution of equation (1.5.17), it will suffice to show that

$$|x_n - \bar{x}_n|_E \to 0 \quad \text{as } n \to \infty, \tag{1.5.22}$$

in order to conclude that the sequence (1.5.21) also converges to the solution of equation (1.5.17).

From (1.5.20) and (1.5.21) we obtain

$$x_{n+1} - \bar{x}_{n+1} = (H - H_n)G(x_n) + H_n(G(x_n) - G(\bar{x}_n)),$$

and taking the norm in E we derive the inequality

$$|x_{n+1} - \bar{x}_{n+1}| \leq |H - H_n||G(x_n)| + |H_n||G(x_n) - G(\bar{x}_n)|, \tag{1.5.23}$$

Since $\{x_n\}$ is a convergent sequence in E, this means it is bounded. Moreover, taking into account the Lipschitz condition (1.5.18) for G we conclude that $\{G(x_n)\}$ is also a bounded sequence (even convergent) in E. Therefore, we can find a positive constant A, such that $|G(x_n)| \leq A$ for all natural n. If we apply the Lipschitz condition to the second term in the right hand side of (1.5.23), and notice that $|H_n|\alpha \leq \beta < 1$ for $n \geq N$, then (1.5.23) leads to the recurrent inequality

$$|x_{n+1} - \bar{x}_{n+1}| \leq A|H - H_n| + \beta|x_n - \bar{x}_n|, \quad n \geq N. \tag{1.5.24}$$

On the other hand, because $|H - H_n| \to 0$ as $n \to \infty$, we have

$$|x_{n+1} - \bar{x}_{n+1}| \le \beta|x_n - \bar{x}_n| + \varepsilon, \quad \text{for } n \ge m(\varepsilon). \tag{1.5.25}$$

The inequality (1.5.25) leads to the estimate

$$|x_{m+p} - \bar{x}_{m+p}| \le \beta^p|x_m - \bar{x}_m| + \varepsilon(1 - \beta)^{-1}, \quad p \ge 1. \tag{1.5.26}$$

Since $\beta^p \to 0$ as $p \to \infty$, we derive immediately (1.5.22) from (1.5.26). Hence, the scheme of successive approximations described by (1.5.21) is convergent in E to the unique solution of equation (1.5.17).

In view of the above discussion, we can state the following result.

Proposition 1.5.2. *Let E be a Banach space, $H: E \to E$ a continuous linear operator, and $G: E \to E$ a map satisfying the Lipschitz condition (1.5.18). If condition (1.5.19) is verified, then the unique solution $x \in E$ of equation (1.5.17) can be approximated with any degree of accuracy by means of the successive approximations (1.5.21), where $\{H_n\}$ is a sequence of linear continuous operators on E, such that $|H_n - H| \to 0$ as $n \to \infty$.*

Applying Proposition 1.5.2 to equation (1.5.1), in its equivalent form (1.5.9), is now an elementary exercise. Notice that condition (1.5.16) is actually condition (1.5.19) in Proposition 1.5.2.

Perhaps more attention should be given to the condition $|H - H_n| \to 0$ as $n \to \infty$. But in Section 1.4 we showed that any L^2-kernel can be approximated in $L^2(Q, R)$ by kernels of finite rank. In particular, if the system of eigenfunctions of $k(t, s)$ is complete in L^2, it can be used in the scheme.

Bibliographical notes

The material included in Section 1.1 is related to the work of Israilov [1] who obtained the integral equation (1.1.16) starting from the differential system (1.1.8), with boundary value conditions (1.1.9); C. Corduneanu [15] who considered the integral equation (1.1.23) in connection with the first-order partial differential equation (1.1.18), motivated by the problem of constructing Liapunov functions for ordinary differential systems of the form (1.1.8); Saaty [1] for the nonlinear Volterra singular equation (1.1.35); Corduneanu and Poorkarimi [1] for the integral equation (1.1.43) for the hyperbolic equation (1.1.36) with data on characteristics (1.1.37); and Pazy [1] for the integral equation (1.1.45) which provides the mild solution of differential equations of the form (1.1.44), with $-A$ an infinitesimal generator of a C_0-semigroup. With regard to the integral equations generated by inverse problems for partial differential equations, besides the monographs quoted in the text, we also mention the paper of Kabanikhin [1]. Further references are included in the sources mentioned above.

In Section 1.2, the examples of both Volterra and Fredholm integral equations (1.2.8) and (1.2.9), for the motion of a sliding chain, are due to Myller [1]. The integral inclusion for the temperature control by means of a thermostat (1.2.23) was obtained by Glashoff and Sprekels in their papers [1], [2]. See also Prüss [2]. For integral equations occurring in the theory of engineering systems see C. Corduneanu [4], Desoer and Vidyasagar [1], Popov [1]. The integrodifferential equation (1.2.37), describing the motion of a viscoelastic body, is taken from the paper by Renno [1], which contains further references for the linear case. For the nonlinear case, see Nohel [1], [2]. For an example of an integral equation occurring in population dynamics we used the model presented by Thieme [1], [2]. At least two monographs have been dedicated to the mathematical aspects of the dynamical problems of population growth: Cushing [3] and Webb [4]. A good deal of examples regarding the use of integral or integrodifferential equations in the modelling of various physical systems can be found in the recent book by Jerri [1]. In particular, Chapter 2 of Jerri's book contains examples of integral equations occurring in mechanics and control theory. Another useful reference for those interested in the use of integral equations in mathematical physics is the book by Guenther and Lee [1].

Section 1.3 contains a few basic results on the theory of Volterra equations in which the unknown function depends on a single variable. By means of the method of successive approximations (see, for instance, C. Corduneanu [6]) one obtains the existence and uniqueness results. The method of successive approximations is also applied under the abstract form of the contraction mapping principle. Some results that go beyond the strict existence part, hinting at the asymptotic behavior, are also included. Most classical books on integral equations, such as Bocher [1], Lalesco [1], Volterra [1], [2], Petrovskii [1], include basic results of the nature of those discussed in Section 1.3. For contributions to the theory of Volterra equations, during the period 1960–75, see the survey paper of Caljuk [2]. This paper is mostly based on the list of those contributions to the theory of Volterra integral equations, which were reviewed in *Referativnyi Zurnal* (Mathematics Series).

The literature concerning the Volterra type inequalities is particularly rich, and we refer the reader to the book by Martinjuk and Gutowski [1] which contains a good many references on these topics. A recent paper by Beesack [1], surveys most of the work conducted in this area, including inequalities of Volterra type for functions dependent on several variables. See also references under the names of Pachpatte, A. Corduneanu, Yeh, Turinici, as well as the references contained in these papers. Also, for integral inequalities of Volterra type, see the standard references Làkshmikantham and Leela [1], and Walter [1].

In Section 1.4 we tried to present the basic facts of classical Fredholm theory, without making appeal to functional analysis. It is likely that a certain number of readers would have preferred the more concise presentation of this theory, based on the use of the concept of a completely continuous operator on a Hilbert or a Banach space. While there is no doubt that the abstract approach is more economical and more general, I considered it adequate to follow the classical approach, which actually preceded the functional analytic apparatus and terminology, and had a decisive influence on the development of the ideas that have finally contributed to the foundation of modern abstract analysis (including the theory of linear spaces and operators which has become classical itself). My presentation is, perhaps, close to that adopted in Petrovskii's textbook [1]. I have experience of using this approach in teaching students who had completed a minimal course in real analysis. Another approach, which could be used with students who have no knowledge of real analysis, but a good knowledge of classical analysis and some elements of complex analysis, has been adopted in C. Corduneanu [6]. The original approach of Fredholm was based on substituting an integral sum for the integral thus reducing the equation to an algebraic linear system. This system suggests the construction of the Fredholm entire functions which play a significant role in the discussion of the equation.

The theory of Fredholm, including the alternative presented in Theorem 1.4.1, can be found in many books on the theory of integral equations. I shall mention here Lalesco's book [1], which seems to be the first to include a complete presentation of what we now understand by classical Fredholm theory; the book by Lovitt [1], which has known a wide circulation and has been translated into several languages; the third volume of the treatise by Goursat [1]; the books by Hamel [1], Hoheisel [1]; those of Cochran [1], Fenyo and Stolle [1], Tricomi [1], Vivanti [1], Schmeidler [1], Heywood and Fréchet [1], Kneser [1], Wiarda [1], Yosida [2], Smithies [1].

The modern development of Fredholm theory is presented in a book edited by Ruston [1]. Besides results pertaining to the classical heritage, such as those due to Riesz (see also Riesz and Sz.-Nagy [1]), one finds more recent contributions by A. Grothendieck, A. Pietsch, and many other authors. The list of references includes over 700 items, and covers practically all significant contributions to Fredholm theory.

The material included in Section 1.5, dedicated to what we may call an elementary introduction to the theory of Hammerstein equations, has been selected – as acknowledged in the text – from the classical paper by Dolph [1]. This reference also contains several interesting results based on the variational approach, as used by Hammerstein in his paper [1]. More recent contributions related to the theory of Hammerstein integral equations in various function spaces will be discussed in Chapter 4. In all the references mentioned above with respect to Hammerstein equations one can also find applications of the basic results (particularly) to the theory of partial differential equations. A good deal of results concerning Hammerstein equations can be found in the book by Krasnoselskii [1], as well as other suitable references.

Further useful references relating to the topics discussed in Chapter 1 can be found in the survey paper by Imanaliev et al. [1], in which the main contributions made by Soviet mathematicians during the 1960s and the 1970s are briefly examined. Only the names of the authors and the particular field in which they obtained results are mentioned.

For the theory of integral equations with Hermitian kernel, which we did not cover in this book, the following references are significant: Hilbert [1], Hellinger and Toeplitz [1], Mikhlin [1], Zabreyko et al. [1]. Most of the references listed above (Section 1.4) also contain the basic facts related to equations with Hermitian kernel.

2

Function spaces, operators, fixed points and monotone mappings

In this chapter we shall present various topics related to function spaces, operators on function spaces (including, of course, integral operators), as well as certain auxiliary results from nonlinear functional analysis. All these topics will be needed in the subsequent chapters of the book, and we think it is useful to review them in a separate chapter. Sometimes, we might go beyond the strict needs of the book, and reach deeper into the theory of function spaces or integral operators. This may be motivated by the fact that most topics discussed in this book are susceptible of further investigation and improvement.

2.1 Spaces of continuous functions

The space of continuous functions that is most often encountered is the space usually denoted by $C([t_0, T], R^n)$, which consists of all continuous maps of the closed interval $[t_0, T]$ into R^n. The norm is the usual supremum norm, i.e.,

$$|x|_C = \sup\{|x(t)|; t \in [t_0, T]\},$$

where $|\cdot|$ stands for the Euclidean norm in the space R^n. It is commonly known that $C([t_0, T], R^n)$ is a Banach space.

If the interval $[t_0, T]$ is substituted by the semi-open interval $[t_0, T)$, making the case $T = +\infty$ possible, then the set of all continuous maps from $[t_0, T)$ into R^n is no longer a Banach space. It is a linear metric space, with translation-invariant metric, which is also complete. Such linear spaces are usually called Fréchet spaces.

More precisely, if $x \in C([t_0, T), R^n)$, then for every natural number m we define the seminorm

$$|x|_m = \sup\{|x(t)|; t \in [t_0, t_m]\},$$

where $t_m \uparrow T$. Then the metric is defined by

$$\rho(x,y) = \sum_{m=1}^{\infty} \frac{1}{2^m} \frac{|x-y|_m}{1+|x-y|_m},$$

and it can be seen that the topology induced by this metric does not depend on the particular choice of the sequence $\{t_n\}$, with $t_m \uparrow T$.

The topology of the space $C([t_0, T), R^n)$, induced by the metric ρ defined above, is the topology of uniform convergence on every compact interval belonging to $[t_0, T)$.

The set $M \subset C([t_0, T), R^n)$ is compact in this space, if and only if M is uniformly bounded and equicontinuous (i.e., Arzelà's criterion conditions are satisfied) on each compact interval of $[t_0, T)$. In particular, we have to check these conditions on any $[t_0, t_m]$, with $t_m \uparrow T$.

The case of the space $C((t_0, T), R^n)$ is similar to the case discussed above, the topology being defined by the family of seminorms

$$|x|_m = \sup\{|x(t)|; t \in (\tau_m, t_m)\},$$

where $\tau_m \downarrow t_0$ and $t_m \uparrow T$. Of course, a translation-invariant metric can be defined as above, and compactness conditions can be stated similarly.

The space $C([t_0, T), R^n)$ contains all continuous maps from $[t_0, T)$ into R^n, and therefore we cannot derive too much information from the fact that $x \in C([t_0, T), R^n)$. We need to specialize somewhat the space of functions, to be able to get some information about the (asymptotic) behavior of its elements. This can be achieved in many ways, and we are led to various classes (spaces) of continuous functions, with conspicuous applications in the theory of integral operators/equations.

Let us indicate first a procedure for defining spaces of continuous functions, all continuously imbedded in the space $C([t_0, T), R^n)$. Actually, we have already dealt with such spaces in Section 1.3.

Assume $g: [t_0, T) \to (0, \infty)$ is a continuous map. Denote by $C_g([t_0, T), R^n)$ the subset of $C([t_0, T), R^n)$ consisting of those maps $x: [t_0, T) \to R^n$, such that

$$\sup\{|x(t)|/g(t); t \in [t_0, T)\} < \infty. \qquad (2.1.1)$$

It is easily seen that the set $C_g([t_0, T), R^n)$ is a Banach space, in which the norm is defined by

$$|x|_g = \sup\{|x(t)|/g(t); t \in [t_0, T)\}. \qquad (2.1.2)$$

This space was used in Chapter 1, and more details can be found in C. Corduneanu [4].

Since (2.1.1) means that for each $x \in C_g$ there exists a constant $A_x > 0$, such that $|x(t)| \le A_x g(t)$ on $[t_0, T)$, one can view C_g as the space of continuous functions whose growth (as $t \to T$) is of the same nature as the growth (or decay!) of the function g. This property may seem too restrictive in some circumstances, and in order to be able to describe more nuanced behavior (not necessarily the same for each component of x), we shall introduce (see Gollwitzer [1]) the spaces $C_g([t_0, T), R^n)$ for which g is not necessarily a scalar function. More precisely, we shall see that for a matrix g, whose entries are real-valued functions on $[t_0, T)$, the space $C_g([t_0, T), R^n)$ can be naturally defined, and it satisfies some basic properties similar to those corresponding to the case when g is a scalar function.

If $\mathscr{L}(R^n, R^n)$ is the linear space of all linear maps from R^n into itself and, if an orthogonal basis of R^n is fixed, then $\mathscr{L}(R^n, R^n)$ coincides with the class of all n by n matrices with real entries. For any matrix $A \in \mathscr{L}(R^n, R^n)$, let us denote by N_A and R_A respectively, the kernel (null space) and the range of A. By P_A we denote the orthogonal projector of R^n onto R_A, and let $\tilde{A}$ be the restriction of A to the orthogonal complement (in R^n) of N_A, say $N_A^\perp$. Since $\tilde{A}$ is obviously invertible on $R_A = R_{\tilde{A}}$, we can introduce the map

$$A_{-1} = \tilde{A}^{-1} P_A, \quad \forall A \in \mathscr{L}(R^n, R^n). \tag{2.1.3}$$

The above definition of A_{-1} implies $A_{-1} \colon R^n \to N_A^\perp$, and $A_{-1} A x = x$ for any $x \in N_A^\perp$. In case A is nonsingular, one has $N_A = \{0\}$, $N_A^\perp = R^n$, which means that $A_{-1} = A^{-1}$. Therefore, A_{-1} given by (2.1.3) should be regarded as a generalized inverse of A. For details on generalized inverses see, for instance, Groetsch [1] or Nashed [1]. The generalized inverse (Moore–Penrose) is completely determined by the relations $A_{-1} A x = x$ for any $x \in N_A^\perp$, and $A_{-1} y = \theta$ for any $y \in N_A^\perp$.

Assume now that $g \colon [t_0, T) \to \mathscr{L}(R^n, R^n)$ is a map, not necessarily continuous, but such that $|g(t)|$ is bounded on any compact interval in $[t_0, T)$, and define $C_g([t_0, T), R^n)$ by means of the following properties:

(1) $x \in C([t_0, T), R^n)$, i.e., it is continuous;
(2) $P_{g(t)} x(t) = x(t)$, $\forall t \in [t_0, T)$;
(3) $g_{-1}(t) x(t)$ is bounded on $[t_0, T)$.

The norm in C_g is defined by

$$|x|_g = \sup\{|g_{-1}(t) x(t)|; t \in [t_0, T)\}. \tag{2.1.4}$$

We shall prove now that $C_g([t_0, T), R^n)$, with g as described above, is a Banach space in which convergence is stronger than the convergence in $C([t_0, T), R^n)$.

It is elementary to see that $|\cdot|_g$, as defined by (2.1.4), defines a norm. Now let $\{u_m(t)\}$ be a Cauchy sequence in the space $C_g([t_0, T), R^n)$. In this case, the sequence $\{v_m(t)\}$, $v_m(t) = g_{-1}(t) u_m(t)$, $m \geq 1$, is a Cauchy sequence in the space of all bounded functions on $[t_0, T)$, with values in R^n, with respect to the sup norm (uniform convergence on $[t_0, T)$). Hence $\{v_m(t)\}$ converges uniformly on $[t_0, T)$ to a bounded function $v_0(t)$, such that $v_0(t) \in N_{g(t)}^\perp$ for every $t \in [t_0, T)$. If we let $u_0(t) = g(t) v_0(t)$, then $u_0(t) \in C_g([t_0, T), R^n)$, because: (1) it is continuous on

$[t_0, T)$, since, for each $[t_0, t_1] \subset [t_0, T)$, one has

$$|u_m(t) - u_0(t)| = |g(t)(v_m(t) - v_0(t))| \leq \bar{g}|v_m(t) - v_0(t)|, \qquad (2.1.5)$$

where $\bar{g} \geq |g(t)|$ on $[t_0, t_1]$; (2) $P_{g(t)}u_0(t) = u_0(t)$ for any $t \in [t_0, T)$, according to the definition of $U_0(t) = g(t)V_0(t)$; (3) $g_{-1}(t)u_0(t) = g_{-1}(t)g(t)v_0(t) = v_0(t)$, which is bounded on $[t_0, T)$ by construction.

The above discussion shows that the space $C_g([t_0, T), R^n)$ is complete with respect to the norm $|\cdot|_g$, and hence it is a Banach space.

Remark. When $g(t) = g_0(t)I$, where $g_0(t)$ is a scalar positive function on $[t_0, T)$, uniformly bounded on each compact interval of $[t_0, T)$, and I is the unit matrix of dimension n, one obtains a function space of vector-valued continuous functions whose components have the same order of growth (as $t \to T$) as $g_0(t)$. A more general situation corresponds to $g(t) = \text{diag}(g_1(t), g_2(t), \ldots, g_n(t))$, with $g_i(t)$, $i = 1, 2, \ldots, n$, positive and uniformly bounded on any compact of $[t_0, T)$. Then, the order of growth of the components of a function $x \in C_g$ can be different for each of the components (for instance, some tend to zero, while the remaining ones tend to infinity as $t \to T$; such a situation takes place in the case of conditional stability).

A basic question regarding the space $C_g([t_0, T), R^n)$ is the following: under what conditions on the matrix-valued maps $g(t)$ and $h(t)$ can we assert the identity of the spaces C_g and C_h?

This question can be satisfactorily answered, under fairly general assumptions on $g(t)$ and $h(t)$ (see Pandolfi [1]).

Theorem 2.1.1. *The following conditions are necessary and sufficient for the identity (as sets) of the spaces* $C_g([t_0, T), R^n)$ *and* $C_h([t_0, T), R^n)$, *with both* $g, h: [t_0, T) \to \mathscr{L}(R^n, R^n)$ *continuous:*

(a) $R_{g(t)} = R_{h(t)} \neq \{\theta\}$, *for any* $t \in [t_0, T)$;
(b) $\alpha = \sup\{|h_{-1}(t)g(t)|; t \in [t_0, T)\} < +\infty$;
(c) $\beta = \sup\{|g_{-1}(t)h(t)|; t \in (t_0, T)\} < +\infty$.

It follows also that $|\cdot|_g$ *and* $|\cdot|_h$ *are equivalent norms.*

Proof. Let us denote by $g^i(t)$, $i = 1, 2, \ldots, n$, and $h^i(t)$, $i = 1, 2, \ldots, n$, the vectors which represents the columns of the matrix $g(t)$ and $h(t)$ respectively. Since the vector-valued functions $g^i(t)$, $i = 1, 2, \ldots, n$, are continuous from $[t_0, T)$ into R^n, and furthermore $P_{g(t)}g^i(t) = g^i(t)$, $i = 1, 2, \ldots, n$ (because $R_{g(t)}$ is spanned by the vectors $g^i(t)$, $i = 1, 2, \ldots, n$), we have only to show that $g_{-1}(t)g^i(t)$, $i = 1, 2, \ldots, n$, are bounded on $[t_0, T)$, in order to conclude that $g^i(t) \in C_g([t_0, T], R^n)$, $i = 1, 2, \ldots, n$. Indeed, if $\{e_i\}$, $1 \leq i \leq n$, is the basis in R^n, with respect to which

$g(t)$ is defined, then $g^i(t) = g(t)e_i$, $i = 1, 2, \ldots, n$, and hence

$$|g_{-1}(t)g^i(t)| = |g_{-1}(t)g(t)e^i| \leq |g_{-1}(t)g(t)| = 1,$$

taking also into account that $x \in N_g$ implies $g_{-1}(t)g(t)x = x$, for any $t \in [t_0, T)$.

If we admit now that C_g and C_h have the same elements, it means that $g^i(t) \in C_h$, $i = 1, 2, \ldots, n$. Consequently, $R_{g(t)} \in R_{h(t)}$, and if we keep in mind the symmetric role of g and h we obtain $R_g \equiv R_h$, i.e., condition (a). But $g^i(t) \in C_h$, $i = 1, 2, \ldots, n$, imply the boundedness on $[t_0, T)$ of each function $h_{-1}(t)g^i(t)$, $i = 1, 2, \ldots, n$. This obviously implies the boundedness of $|h_{-1}(t)g(t)|$ on $[t_0, T)$, which is condition (b) in Theorem 2.1.1. Similarly, one obtains condition (c) as a necessary condition for the identity of the spaces C_g and C_h.

Let us now assume the validity of conditions (a), (b), (c) in Theorem 2.1.1, and prove their sufficiency, i.e., the identity (as sets!) of the spaces C_g and C_h. Assume $x \in C_g([t_0, T], R^n)$. From condition (a), we have $P_{h(t)}x(t) = x(t)$. Furthermore, for each $t \in [t_0, T)$ we have $g(t)g_{-1}(t) = P_{g(t)}$, and therefore for $x \in C_g$ we obtain

$$|h_{-1}(t)x(t)| = |h_{-1}(t)g(t)g_{-1}(t)x(t)| \leq |h_{-1}(t)g(t)| \cdot |g_{-1}(t)x(t)|.$$

By taking the supremum with respect to $t \in [t_0, T)$ and keeping in mind condition (b) and $x \in C_g$, we have

$$\sup|h_{-1}(t)x(t)| \leq \alpha|x|_g < +\infty. \qquad (2.1.6)$$

Consequently, $x \in C_h$, and we have

$$|x|_h \leq \alpha|x|_g. \qquad (2.1.7)$$

A similar argument, based on conditions (a) and (c) in Theorem 2.1.1 leads to the conclusions $x \in C_h$ implies $x \in C_g$, and

$$|x|_g \leq \beta|x|_h. \qquad (2.1.8)$$

Therefore, the conditions of Theorem 2.1.1 imply $C_g = C_h$, and from (2.1.7), (2.1.8) one derives

$$\alpha^{-1}|x|_h \leq |x|_g \leq \beta|x|_h, \qquad (2.1.9)$$

which shows that the norms $|\cdot|_g$ and $|\cdot|_h$ are equivalent. Hence, the spaces C_g and C_h are isomorphic.

This ends the proof of Theorem 2.1.1.

Remark. If we assume $g(t)$ and $h(t)$ are continuous, conditions (b) and (c) of Theorem 2.1.1 can be replaced by the following ones:

$$0 < \inf\{|g_{-1}(t)h(t)|; t \in [t_0, T)\}, \qquad (2.1.10)$$

$$0 < \inf\{|h_{-1}(t)g(t)|; t \in [t_0, T)\}. \qquad (2.1.11)$$

In order to derive (2.1.10) from condition (b), we need to notice that, under assumption (a), one has $(h_{-1}g)_{-1} = g_{-1}h$, and also $|g| \geq |g_{-1}|^{-1}$. While the first formula can be obtained from the definition of g_{-1}, the second follows from the considerations below:

$$|g| = \sup\{|gx|; |x| = 1\} = \sup\{|gx|; |x| = 1, x \in N_g^{\perp}\}$$

$$= |\tilde{g}| \geq |\tilde{g}^{-1}|^{-1} = (\sup\{|\tilde{g}^{-1}x|; |x| = 1, x \in R_g\})^{-1}$$

$$= (\sup\{|\tilde{g}^{-1}P_g x|; |x| = 1)^{-1} = (\sup\{|g_{-1}x|; |x| = 1\})^{-1}.$$

Therefore, we have

$$|h_{-1}(t)g(t)| \geq |g_{-1}(t)h(t)|^{-1},$$

for any $t \in [t_0, T)$. Taking condition (b) into account, we obtain condition (2.1.10). Similarly for condition (c), which implies condition (2.1.11).

The problem of the identity of the spaces C_g and C_h when both g and h are continuous positive scalar functions can be answered in the following simple manner: $C_g \equiv C_h$ if and only if there are two positive constants λ and μ, such that

$$\lambda g(t) \leq h(t) \leq \mu g(t), \quad t \in [t_0, T).$$

This follows immediately from the Remark to Theorem 2.1.1, more precisely from (2.1.10) and (2.1.11). It is enough to notice that $R_g = R_h = R^n$ and $g_{-1} = g^{-1}$.

The problem of the identity of the spaces C_g and C_h, when g and h are matrix-valued maps not necessarily continuous, is still open.

Let us discuss now a few other spaces of continuous functions, which will occur in later chapters.

First, we take the subspace of $C([t_0, T), R^n)$ consisting of all bounded functions on $[t_0, T)$, which coincides with the space $C_g([t_0, T), R^n)$ for $g \equiv 1$, with the usual supremum norm:

$$|x| = \sup\{|x(t)|; t \in [t_0, T)\}.$$

Sometimes, this space is denoted by $BC([t_0, T), R^n)$, for obvious reasons.

Another interesting function space is the space $C_\ell([t_0, T), R^n)$, which is the subspace of $BC([t_0, T), R^n)$, containing all the continuous maps from $[t_0, T)$ into R^n, such that

$$\lim_{t \to T} x(t) = x_T \in R^n. \tag{2.1.12}$$

It can easily be seen that the space $C_\ell([t_0, T), R^n)$ is isomorphic to the space $C([t_0, T], R^n)$, when $T < +\infty$. It suffices to identify the element $x(t)$ of $C_\ell([t_0, T), R^n)$, with the element $\bar{x}(t) \in C([t_0, T], R^n)$ defined as follows: $\bar{x}(t) = x(t)$ for $t \in [t_0, T)$, and $\bar{x}(T) = x_T$, with x_T given by (2.1.12). When $T = +\infty$, a similar argument holds.

The above possibility of regarding C_ℓ as a space of continuous functions on a compact interval leads to a rather simple, but useful, criterion of compactness in $C_\ell([t_0, T), R^n)$.

The following conditions are necessary and sufficient for a set $M \subset C_\ell([t_0, T), R^n)$ to be compact:

(a) M is bounded in C_ℓ;
(b) M is equicontinuous on any interval $[t_0, t_1] \subset [t_0, T)$;
(c) $\lim x(t) = x_T$, as $t \to T$, exists uniformly with respect to $x \in M$.

For a proof of this criterion see, for instance, C. Corduneanu [4].

A remarkable subspace of the space $C_\ell([t_0, T), R^n)$ is $C_0([t_0, T), R^n)$, consisting of those $x \in C_\ell$ for which $x_T = 0$. If $T = +\infty$, this is the natural space for investigating asymptotic stability.

The following relationship

$$C_\ell = C_0 \oplus R^n \qquad (2.1.13)$$

can be obtained without difficulty. It provides better insight into the structure of the space C_ℓ.

Another remarkable space of continuous functions that will be encountered in subsequent chapters is the space $AP(R, \mathscr{C}^n)$ of almost periodic functions on R, with values in the n-dimensional complex space $\mathscr{C}^n$.

The easiest way to define the space $AP(R, \mathscr{C}^n)$ is as follows. We consider the space $BC(R, \mathscr{C}^n)$ of all continuous bounded maps from R into $\mathscr{C}^n$, with the topology induced by the supremum norm; it is obvious that any trigonometric polynomial

$$\sum_{k=1}^{n} a_k \exp(i\lambda_k t), \qquad (2.1.14)$$

with $a_k \in \mathscr{C}^n$ and $\lambda_k \in R, k = 1, 2, \ldots, n$, belongs to $BC(R, \mathscr{C}^n)$; the space $AP(R, \mathscr{C}^n)$ is the closure in $BC(R, \mathscr{C}^n)$ of all trigonometric polynomials of the form (2.1.14).

Another definition of the function space $AP(R, \mathscr{C}^n)$ is the classical one, given by H. Bohr: the continuous map $x: R \to \mathscr{C}^n$ is almost periodic if for every $\varepsilon > 0$ there exists $l(\varepsilon) > 0$, such that any interval $[a, a + l] \subset R$ contains a number τ with the property

$$|x(t + \tau) - x(t)| < \varepsilon, \quad \forall t \in R. \qquad (2.1.15)$$

The number τ is called an ε-almost-period of x, and the definition requires that the set of all ε-almost-periods be relatively dense in R.

For details on the theory of almost periodic functions see Corduneanu [3].

In concluding this section, we notice that other spaces of continuous functions will be useful in the sequel. For instance, the space $C^{(1)}([t_0, T], R^n)$ consisting of all continuously differentiable functions on $[t_0, T]$, with values in R^n. The derivative is understood to be the right derivative at t_0, and the left derivative at T. A convenient norm on $C^{(1)}([t_0, T], R^n)$ is given by

$$|x|_1 = \sup(|x(t)| + |x'(t)|; t \in [t_0, T]). \tag{2.1.16}$$

Other function spaces, necessary in the subsequent development, will be defined when needed.

2.2 Spaces of measurable functions

Most of the spaces of measurable functions we shall deal with in this book are function spaces whose elements are defined on measurable subsets of R or R^n, with values in the same spaces. The measure will always be the Lebesgue measure on R or R^n, unless otherwise specified.

The most popular spaces of measurable functions are the Lebesgue spaces $L^p(I, R^n)$, $1 \le p \le \infty$, where I stands for an interval of R (in particular, I could be the positive semi-axis R_+, or even the whole real axis).

These spaces are well investigated, and their basic theory can be found in many books. See, for instance, Kantorovich and Akilov [1], Dunford and Schwartz [1].

Among various properties of the spaces L^p, we mention here some compactness criteria.

Theorem 2.2.1. *Let $M \subset L^p([t_0, t_1], R^n)$, $1 < p < \infty$. Necessary and sufficient conditions for the relative compactness of M in L^p are:*

(1) *M is bounded in L^p;*
(2) *$\int_{t_0}^{t_1} |x(t + h) - x(t)|^p \, dt \to 0$ as $h \to 0$,*

uniformly with respect to $x \in M$.

Theorem 2.2.1 is known as the Riesz compactness criterion.

Another criterion is due to A. N. Kolmogorov, and relies on the use of mean functions associated with the functions in $M \subset L^p$, given by

$$x_h(t) = \frac{1}{h} \int_t^{t+h} x(u) \, du. \tag{2.2.1}$$

Theorem 2.2.2. *The following conditions are both necessary and sufficient for the relative compactness of the set $M \subset L^p([t_0, t_1], R^n)$, $1 \le p < \infty$:*

(1) *M is bounded in L^p;*
(2) *$x_h \to x$ as $h \to 0$, uniformly with respect to $x \in M$.*

In both Theorems 2.2.1 and 2.2.2, it is agreed that $x \in M$ is extended to an interval $(a, b) \supset [t_0, t_1]$, by letting $x(t) = 0$ outside $[t_0, t_1]$.

Sometimes, the weak compactness plays an important role in the theory of functions spaces and operators defined on such spaces.

Theorem 2.2.3. *The following conditions are necessary and sufficient for a set $M \subset L^1([t_0, t_1], R^n)$ to be weakly sequentially compact:*

(1) *M is bounded in L^1;*
(2) *The functions of M are equi-absolutely continuous, i.e., for every $\varepsilon > 0$, there exists $\delta(\varepsilon) > 0$ such that*

$$\int_E |x(t)|\, dt < \varepsilon \qquad (2.22)$$

for any $x \in M$, provided $E \subset [t_0, t_1]$ is a measurable set with meas $E < \delta(\varepsilon)$.

This classical result is due to N. Dunford and B. J. Pettis. It is remarkable that the result remains true when R^n is substituted by a reflexive Banach space. This case has been discussed by Brooks and Dinculeanu [1].

The spaces $L^p(\Omega, R^n)$, where $\Omega \subset R^m$ is a measurable set, usually of finite Lebesgue measure, have many properties in common with the spaces $L^p([t_0, t_1], R^n)$. We omit details about these spaces, and send the interested reader to classical references such as Dunford and Schwartz [1], Kantorovich and Akilov [1], Edwards [1].

Let us now consider other spaces of measurable functions which will occur in the subsequent chapters.

We will first define the spaces L^p with respect to a weight g. As in the case of spaces of continuous functions, we could deal with the case when g is scalar valued, or with the case when g is matrix valued.

The case when $g = g(t) > 0$ is a measurable function on $[t_0, T)$ is straightforward, and the space $L_g^p([t_0, T], R^n)$, $1 \leq p \leq \infty$, can be defined as consisting of those maps from $[t_0, T)$ into R^n, such that $g^{-1}(t)x(t) \in L^p([t_0, T), R^n)$, the norm being given by

$$|x|_{L_g^p} = |g^{-1}(t)x(t)|_{L^p}, \quad 1 \leq p \leq \infty. \qquad (2.2.3)$$

This case was considered in Dotseth [1], and more details and applications of these spaces are given in that reference.

Let us now consider the case when $g: [t_0, T) \to \mathscr{L}(R^n, R^n)$ is a measurable map. In this case, $g_{-1}(t)$ is defined for every $t \in [t_0, T)$, but in order to apply Gollwitzer's weighting scheme to construct the spaces $L_g^p([t_0, T), R^n)$ we need the measurability of the map $t \to g_{-1}(t)$.

The following representation of the generalized inverse $g_{-1}(t)$ of $g(t)$ can be found in Groetsch [1]:

$$g_{-1}(t) = \int_0^\infty \exp(-g^*(t)g(t)u)g^*(t)\, du. \qquad (2.2.4)$$

The integral converges in the uniform topology of the space $\mathscr{L}(R^n, R^n)$, i.e., with respect to the topology induced in the linear space of matrices by the norm $|A| = \sup\{|Ax|: x \in R^n, |x| \le 1\}$. From (2.2.4) we derive the measurability of the map $t \to g_{-1}(t)$, if we assume $g(t)$ is measurable. Similarly, $g_{-1}(t)$ is continuous when $g(t)$ is.

Now, we can proceed to the definition of the spaces $L_g^p([t_0, T), R^n)$, $1 \le p \le \infty$, in the following manner: first, we require the measurability of each $x \in L_g^p$; second, we assume the condition $P_{g(t)}x(t) = x(t)$, a.e. on $[t_0, T)$, holds; third, we assume $g_{-1}(t)x(t) \in L^p([t_0, T), R^n)$. The norm in $L_g^p([t_0, T), R^n)$ is defined by a formula similar to (2.2.3):

$$|x|_{L_g^p} = |g_{-1}(t)x(t)|_{L^p}, \quad 1 \le p \le \infty. \tag{2.2.5}$$

It can easily be shown that the space $L_g^p([t_0, T), R^n)$ is a Banach space.

Milman [2], has further extended the application of Gollwitzer's [1] weighting scheme, replacing the spaces $BC([t_0, T), R^n)$ or $L^p([t_0, T), R^n)$, $1 \le p \le \infty$, with a general function space E, consisting of measurable functions.

Let us assume first that we have a normed space $E([t_0, T), R)$ of locally integrable functions, such that the following conditions hold:

(1) if $g \in E$, f is measurable and $|f| \le |g|$ a.e., then $f \in E$, and $|f|_E \le |g|_E$;

(2) if $S \subset [t_0, T)$ is a measurable set of finite measure (meas $S = +\infty$ is possible when $T = +\infty$) then $\chi_s \in E$;

(3) if $0 \le f_m \uparrow f$ a.e., and $f_m \in E$ for every $m \ge 1$, then in case $\{|f_m|_E\}$ is bounded on R, one has $f \in E$ and

$$\lim_{m \to \infty} |f_m|_E = |f|_E.$$

In order to define $E([t_0, T), R^n)$, we must take the direct product of n identical factors of the space $E([t_0, T), R)$.

The function spaces $E([t_0, T), R)$ defined above by means of the axioms (1), (2), and (3) are indeed Banach spaces. Condition (3), which is in fact Fatou's lemma, enables us to prove the completeness.

The associate space $E'([t_0, T), R)$ is defined by means of

$$E'([t_0, T), R) = \left\{ y; y \text{ measurable and } \sup \int_{t_0}^T |x(t)y(t)| \, dt < \infty \text{ for } |x|_E \le 1 \right\}.$$

The norm in E' is the supremum appearing in the definition of E'.

The theory of Banach function spaces, as defined above, is mainly due to W. A. J. Luxemburg and A. C. Zaanen. See Zaanen [1].

Starting from an arbitrary Banach function space $E([t_0, T), R^n)$, as defined above, M. Milman has defined the space $E_g([t_0, T), R^n)$ by means of the weighting scheme proposed by H. Gollwitzer, as follows:

(1) $x \in E_g$ is measurable on $[t_0, T)$;
(2) $P_{g(t)}x(t) = x(t)$ a.e. on $[t_0, T)$, for any $x \in E_g$;
(3) $g_{-1}(t)x(t) \in E([t_0, T), R^n)$, for any $x \in E_g$.

It is easily seen that the norm

$$|x|_{E_g} = |g_{-1}(t)x(t)|_E \qquad (2.2.6)$$

satisfies the usual conditions. For instance, if $|x|_{E_g} = 0$, this means $g_{-1}(t)x(t) = 0$ a.e. on $[t_0, T)$. This implies $g(t)g_{-1}(t)x(t) = 0$ a.e. on $[t_0, T)$. But according to the definition of $g_{-1}(t)$ we have $g(t)g_{-1}(t)x(t) = x(t)$, which implies $x(t) = 0$ a.e. on $[t_0, T)$.

To show that $E_g([t_0, T), R^n)$ is a Banach space, it remains to check the completeness condition. Let $\{x^m\} \subset E_g$ be a Cauchy sequence. This implies that the sequence $\{y^m(t)\}$, $y^m(t) = g_{-1}(t)x^m(t)$, is a Cauchy sequence in E. Therefore, there exists $y \in E$ such that $|y^m - y|_E \to 0$ as $m \to \infty$. Consider now the map $x(t) = g(t)y(t)$ a.e. on $[t_0, T)$. Since $g^m(t) \in N_{g(t)}^{\perp}$ a.e. on $[t_0, T)$, we obtain $y(t) \in N_{g(t)}^{\perp}$ a.e. on $[t_0, T)$, which means $P_{g(t)}x(t) = x(t)$ a.e. on $[t_0, T)$. Moreover, from $|x^m - x|_{E_g} = |y^m - y|_E$ and $y^m \to y$ in E, we obtain $x^m \to x$ in E_g.

The construction of the space E_g is also possible when E is a Banach function space whose elements are maps from a given interval $I \subset R$, or from a domain $\Omega \subset R^m$, into a Banach space. A basic assumption that has to be made, which is not necessary in the finite-dimensional case, is the closedness of the range of $g(t)$ for almost all $t \in [t_0, T)$. See M. Milman [1] and C. W. Groetsch [1] for the necessary details and background.

In Chapter 4 we shall need some properties related to measure spaces, usually denoted by (Ω, Σ, μ). Ω stands for an arbitrary set; Σ represents a non-empty collection of subsets of Ω which constitutes a σ-algebra (i.e., any countable union of elements of Σ belongs to Σ, the complementary set of any set in Σ is also in Σ, and $\varnothing \in \Sigma$); and μ is a countably additive set function on the σ-algebra Σ with values in $\bar{R}_+ = [0, \infty]$.

The spaces $L^p(\Omega, \mu)$, $1 \le p \le \infty$, are defined in the usual way, and the norm is given by

$$|x|_{L^p} = \left\{ \int_\Omega |x(s)|^p \, d\mu(s) \right\}^{1/p}, \quad 1 \le p < \infty, \qquad (2.2.7)$$

and

$$|x|_{L^\infty} = \mu\text{-ess-sup}_{t \in \Omega} |x(t)|, \qquad (2.2.8)$$

where x is assumed, for instance, a real-valued function.

For basic properties of these spaces we send the reader to Dunford and Schwartz [1], Kantorovich and Akilov [1], Yosida [1]. Function spaces on measure spaces will be considered in the next section.

We shall now consider some function spaces, consisting of measurable functions defined on the whole real axis R, or on the positive half-axis $R_+ = [0, \infty)$. Such spaces are related to the L^p-spaces, but are more complex. Moreover, they can be taken as underlying spaces in many problems related to functional equations, including almost periodicity problems in a broader sense than the one encountered in Section 2.1.

Let us start with the space of locally integrable functions on R_+, with values in R^n. This space consists of those maps from R_+ into R^n, $x\colon R_+ \to R^n$, such that, on any compact interval of R_+, x is (Lebesgue) integrable. It cannot be organized as a Banach space, but it can be organized as a Fréchet space if we use the seminorms

$$|x|_m = \int_0^m |x(t)|\,dt, \quad m = 1, 2, \ldots, \tag{2.2.9}$$

and define the linearly invariant metric by means of

$$\rho(x, y) = \sum_{m=1}^{\infty} \frac{1}{2^m} \frac{|x - y|_m}{1 + |x - y|_m}. \tag{2.2.10}$$

The space described above, with the topology induced by the complete family of seminorms (2.2.9), or equivalently by the metric ρ given by (2.2.10), is complete. It is usually denoted by $L_{loc}(R_+, R^n)$.

The meaning of the notations $L_{loc}^p(R_+, R^n)$, $1 \le p \le \infty$, or $L_{loc}(R, R^n)$ is obvious. The space $L_{loc}(R_+, R^n)$ is the space $L_{loc}^p(R_+, R^n)$, with $p = 1$.

Of course, $L_{loc}(R_+, R^n)$ contains all spaces $L^p(R_+, R^n)$, $1 \le p \le \infty$, and the topology of each L^p is stronger than the topology of $L_{loc}(R_+, R^n)$.

It is interesting to point out that a Banach function space can be constructed, continuously imbedded in $L_{loc}(R_+, R^n)$, and such that all L^p spaces on R_+ are its subspaces. More precisely, let

$$M(R_+, R^n) = \left\{ x; x \in L_{loc}(R_+, R^n), \sup \int_t^{t+1} |x(s)|\,ds < \infty, t \in R_+ \right\}, \tag{2.2.11}$$

and take as a norm on M

$$|x|_M = \sup \int_t^{t+1} |x(s)|\,ds, \quad t \in R_+. \tag{2.2.12}$$

The spaces $M^p(R_+, R^n)$, $1 \le p < \infty$, are defined similarly, and the one corresponding to $p = 1$ is exactly the space $M(R_+, R^n)$ defined above.

The meaning of the notation $M(R, R^n)$ is obvious.

Let us also define a subspace of $M(R_+, R^n)$, consisting of those elements for which

$$\int_t^{t+1} |x(s)|\,ds \to 0 \quad \text{as } t \to \infty. \tag{2.2.13}$$

This space is denoted by $M_0(R_+, R^n)$, and it is worth noticing that it also contains all L^p-spaces, with $1 \leq p < \infty$. Indeed, we have for such values of p

$$\int_t^{t+1} |x(s)| \, ds \leq \left\{ \int_t^{t+1} |x(s)|^p \, ds \right\}^{1/p}, \quad t \in R_+, \qquad (2.2.14)$$

which shows that (2.2.13) holds for any $x \in L^p$, $1 \leq p < \infty$. Hence, $L^p(R_+, R^n) \subset M_0(R_+, R^n)$, $1 \leq p < \infty$. From (2.2.14) one also derives that the imbedding of L^p into M_0 is continuous.

A noteworthy subspace of the space $M(R, R^n)$ is the space $S(R, R^n)$ of *almost periodic functions* in the sense of Stepanov. We shall mention two different definitions of this space.

The first definition can be formulated as follows: $S(R, R^n)$ is the closure in $M(R, R^n)$ of the set of all (real) trigonometric polynomials of the form

$$\text{Re} \left\{ \sum_{j=1}^N a_j \exp(i\lambda_j t) \right\}, \qquad (2.2.15)$$

where $a_j \in \mathscr{C}^n, j = 1, 2, \ldots, N$, and $\lambda_j \in R, j = 1, 2, \ldots, N$.

The second definition is similar to Bohr's definition (see Section 2.1): $x \in M(R, R^n)$ is Stepanov almost periodic if, for every $\varepsilon > 0$, there exists $l(\varepsilon) > 0$, such that any interval $(a, a + l) \subset R$ contains a point τ with the property

$$|x(t + \tau) - x(t)|_M < \varepsilon, \quad \forall t \in R. \qquad (2.2.16)$$

Similar definitions can be formulated for the spaces $M(R, \mathscr{C}^n)$ and $S(R, \mathscr{C}^n)$ where $\mathscr{C}^n$ stands for the n-dimensional complex space.

The connection between the spaces of almost periodic functions $S(R, \mathscr{C}^n)$ and $AP(R, \mathscr{C}^n)$ is very simple. First, it is obvious that $AP(R, \mathscr{C}^n) \subset S(R, \mathscr{C}^n)$ because for any $x \in AP(R, \mathscr{C}^n)$ one has for each $t \in R$

$$\int_t^{t+1} |x(s + \tau) - x(s)| \, ds \leq \sup_{t \in R} |x(t + \tau) - x(t)|. \qquad (2.2.17)$$

Second, it can be shown (see C. Corduneanu [3], for instance) that $x \in S(R, \mathscr{C}^n)$ is also in $AP(R, \mathscr{C}^n)$ if and only if x is uniformly continuous on R.

We shall need in Chapter 4 a property related to convolution, for functions belonging to the spaces M or S.

It is known (see, for instance, Massera and Schäffer [2]) that for $x \in M$

$$\int_{-\infty}^t |x(s)| \exp\{-\alpha(t - s)\} \, ds, \quad \int_t^\infty |x(s)| \exp\{\alpha(t - s)\} \, ds \in BC(R, R), \qquad (2.2.18)$$

for any $\alpha > 0$. In particular, the norm in $BC(R, R)$ of any function in the left-hand side of (2.2.18) is a norm on $M(R, R^n)$, equivalent to the norm $|\cdot|_M$.

A similar property to (2.2.18), corresponding to the case $x \in S$, is

$$\int_{-\infty}^{t} |x(s)| \exp\{-\alpha(t-s)\} \, ds, \quad \int_{t}^{\infty} |x(s)| \exp\{\alpha(t-s)\} \, ds \in AP(R,R), \quad (2.2.19)$$

where α is a positive number. It states that any of the integral transforms appearing in the left side of (2.2.19) takes the space of Stepanov almost periodic functions into the space of Bohr's almost periodic functions.

2.3 Operators on function spaces

Most operators we shall encounter in the subsequent chapters of this book are classical integral operators such as the Volterra linear operator

$$(Kx)(t) = \int_{t_0}^{t} k(t,s)x(s) \, ds, \quad t \in [t_0, T), \quad (2.3.1)$$

or the Fredholm linear operator

$$(Fx)(t) = \int_{t_0}^{T} f(t,s)x(s) \, ds, \quad t \in [t_0, T]. \quad (2.3.2)$$

As standard examples of nonlinear integral operators we can quote the Hammerstein integral operator

$$(Hx)(t) = \int_{t_0}^{T} h(t,s)g(s,x(s)) \, ds, \quad t \in [t_0, T], \quad (2.3.3)$$

the Volterra nonlinear operator

$$(Vx)(t) = \int_{t_0}^{t} k(t,s,x(s)) \, ds, \quad t \in [t_0, T), \quad (2.3.4)$$

the Urysohn (nonlinear) operator

$$(Ux)(t) = \int_{t_0}^{T} k(t,s,x(s)) \, ds, \quad t \in [t_0, T], \quad (2.3.5)$$

or operators of the form

$$(Ax)(t) = f\left(t, x(t), \int_{t_0}^{t} a(t,s)x(s) \, ds\right), \quad (2.3.6)$$

which are encountered in many papers.

Of course, in order to be precise about such operators, we should make clear what function spaces are involved and what conditions are to be imposed on such functions as $k(t,s)$, $f(t,s)$, $h(t,s)$, $g(s,x)$, $k(t,s,x)$, etc., appearing in the above formulas. The properties of the operators, like those listed above, intervening in the associated integral equations are basically continuity and compactness.

We will provide results that guarantee such properties for various classes of operators, with the main emphasis on integral operators (i.e., those operators having representations of the form indicated above in this section).

Another type of operator which is involved in (2.3.3), for instance, is the so-called *Niemytzki operator*

$$(Gx)(t) = g(t, x(t)), \quad t \in [t_0, T), \tag{2.3.7}$$

which is generated by a certain map $(t, x) \to g(t, x)$ between finite-dimensional spaces. Some properties of this type of operator are very useful in the study of integral equations and we shall provide some related results.

We shall also consider *abstract Volterra* operators, in the sense that is usually associated with the work of Tonelli [2] and Tychonoff [1].

Another important problem which we would like to discuss here is the integral representation of linear operators acting between two function spaces. Unlike the case of integral representation of linear functionals, the problem of integral representation of operators involves some difficulties. Nevertheless, several satisfactory results in this direction are available in the literature, and we shall include them in our presentation, in view of their application in subsequent chapters.

We shall start the series of results about operators acting on function spaces with a result concerning the Volterra linear operator (2.3.1), in which we assume $k(t, s)$ to be continuous on $\Delta = \{(t, s); t_0 \le s \le t < T\}$, with values in $\mathcal{L}(R^n, R^n)$.

The pair (C_g, C_G) is called *admissible* with respect to the operator K given by (2.3.1) if $KC_g \in C_G$. In this case, the closed graph theorem enables us to conclude that K is continuous.

The following theorem provides necessary and sufficient conditions for the admissibility of the pair (C_g, C_G), with respect to the operator K given by (2.3.1):

Theorem 2.3.1. *The pair of spaces $(C_g([t_0, T), R^n), C_G([t_0, T), R^n))$ is admissible with respect to the Volterra operator K given by (2.3.1) where $k(t, s)$ is continuous, if and only if the following conditions are satisfied:*

(a) $$P_{G(t)} \left\{ \int_{t_0}^t k(t, s) x(s) \, ds \right\} = \int_{t_0}^t k(t, s) x(s) \, ds, \quad t \in [t_0, T),$$

for any $x \in C_g([t_0, T), R^n)$;
(b) *there exists $M > 0$, such that*

$$\int_{t_0}^t |G_{-1}(t) k(t, s) g(s)| \, ds \le M, \quad t \in [t_0, T). \tag{2.3.8}$$

Proof. To prove the sufficiency of conditions (a), (b), we need only prove that

$$\sup \left| G_{-1}(t) \int_{t_0}^t k(t, s) x(s) \, ds \right| \le M |x|_g, \quad t \in [t_0, T), \tag{2.3.9}$$

where $|x|_g$ is given by (2.1.4). If $x \in C_g$, one has $x(t) = g(t)u(t)$, with $u \in BC([t_0, T), R^n)$. Therefore,

$$\sup \left| G_{-1}(t) \int_{t_0}^t k(t,s)x(s)\,ds \right| \le \sup \left| \int_{t_0}^t G_{-1}(t)k(t,s)g(s)u(s)\,ds \right|$$

$$\le M \sup |u(t)|, \quad t \in [t_0, T). \tag{2.3.10}$$

But $g_{-1}(t)x(t) = u(t)$, and according to (2.1.4) $\sup |u(t)|$, $t \in [t_0, T)$, is exactly $|x|_g$. Therefore, (2.3.10) implies (2.3.9). The sufficiency of conditions (a), (b) is thus proven. Of course, (2.3.9) also shows the continuity of K from C_g into C_G.

To prove the necessity of conditions (a), (b), we will assume that (C_g, C_G) is admissible with respect to the operator K, given by (2.3.1), in which $k(t,s)$ is continuous from $\Delta = \{(t,s); t_0 \le s \le t < T\}$, into $\mathscr{L}(R^n, R^n)$. Then condition (a) must hold, obviously. For condition (b), we need a more elaborate argument. First, we notice that the admissibility of the pair (C_g, C_G) with respect to the operator K implies the continuity of K from C_g into C_G. As noticed above, the closed graph theorem applies directly. We have to keep in mind that K is continuous from $C([t_0, T), R^n)$ into itself, and that the topologies of C_g and C_G are stronger than the topology of $C([t_0, T), R^n)$. Let us denote by $(e^1, e^2, \ldots, e^n)$ an orthonormal basis of R^n and define for every pair i, j, $1 \le i, j \le n$, a linear operator

$$(S_{ij}v)(t) = \int_{t_0}^t \langle G_{-1}(t)k(t,s)g(s)e^j, e^i \rangle v(s)\,ds, \tag{2.3.11}$$

for any $v \in C([t_0, T), R^n)$. The inner product under the integral is exactly the (i,j)-element of the matrix $G_{-1}(t)k(t,s)g(s)$ in the basis $(e^1, e^2, \ldots, e^n)$. It is now obvious that if we can show the existence of a constant $M_{ij} > 0$, such that for $t \in [t_0, T)$

$$\int_{t_0}^t |\langle G_{-1}(t)k(t,s)g(s)e^j, e^i \rangle|\,ds \le M_{ij}, \tag{2.3.12}$$

then condition (2.3.8) of the theorem will be satisfied with $M = \sum M_{ij}$.

Therefore, we need show only (2.3.12), $i, j = 1, 2, \ldots, n$, which means that the operator S_{ij} is continuous from $BC([t_0, T), R)$ into itself. This fact is almost common knowledge. Nevertheless, a proof can be found in C. Corduneanu [4], where the case $T = +\infty$ is considered.

Theorem 2.3.1 is thus proven.

Remark. A condition equivalent to condition (a) in Theorem 2.3.1 can be formulated as follows:

$$P_{G(t)}k(t,s)g(s) = k(t,s)g(s), \tag{2.3.13}$$

for $(t,s) \in \Delta = \{(t,s); t_0 \le s \le t < T\}$.

Indeed, condition (2.3.13) implies condition (a) of Theorem 2.3.1, since the integral can be interchanged with the projector $P_{G(t)}$, for every fixed $t \in (t_0, T)$. Conversely, if condition (a) holds, and the pair (C_g, C_G) is admissible with respect to the operator K given by (2.3.1), then we can write

$$P_{G(t)} \int_{t_0}^t k(t,s)g(s)u(s)\,ds = \int_{t_0}^t k(t,s)g(s)u(s)\,ds, \qquad (2.3.14)$$

for every $u \in BC([t_0, T), R^n)$. Since $BC([t_0, T), R^n)$ contains only continuous maps which are bounded on each compact interval that belongs to $[t_0, T)$, it is possible to extend the validity of equation (2.3.14) to any u which is piecewise continuous on compact intervals belonging to $[t_0, T)$. Indeed, this can easily be seen if we take into account the fact that each continuous map on a compact interval can be uniformly approximated by means of step functions, and then apply the Lebesgue-dominated criterion of convergence of integrals. One must also take into account the fact that the projector $P_{G(t)}$ is a continuous operation (we have also assumed $G(t)$ to be continuous).

Let us fix now $(t, \tau) \in \Delta$, and denote by χ_δ the characteristic function of the interval $[\tau - \delta, \tau + \delta]$, $\delta > 0$. In (2.3.14) we choose successively $u = (2\delta)^{-1} e^i \chi_\delta$, $i = 1, 2, \ldots, n$. We obtain

$$P_{G(t)}(2\delta)^{-1} \int_{t_0}^t k(t,s)g(s)e^i\chi_\delta(s)\,ds = (2\delta)^{-1} \int_{\tau-\delta}^{\tau+\delta} k(t,s)g(s)e^i\chi_\delta(s)\,ds, \quad (2.3.15)$$

for $i = 1, 2, \ldots, n$. If we let $\delta \to 0$ in (2.3.15), the following equations are obtained:

$$P_{G(t)}k(t,\tau)g(\tau)e^i = k(t,\tau)g(\tau)e^i, \quad i = 1, 2, \ldots, n. \qquad (2.3.16)$$

Since $(e^1, e^2, \ldots, e^n)$ is a basis in R^n, one derives from (2.3.16) the condition (2.3.13).

Another admissibility result, which is due to Milman [1], relates to a triplet (E_g, C_G, K), with K given by

$$(Kx)(t) = \int_{t_0}^T k(t,s)x(s)\,ds, \quad t \in [t_0, T). \qquad (2.3.17)$$

The space E_g is constructed as shown in Section 2.2, starting with a Banach function space E, and applying the weighting scheme described in that section.

In order to formulate the admissibility result related to the triplet (E_g, C_G, K), with K as in (2.3.17), we need one more condition on the norm of the space E.

We say that $E = E([t_0, T), R)$ has an *absolutely continuous norm* if the following property holds: for every $f \in E$, and any sequence $\{A_m\}$ of measurable subsets of $[t_0, T]$ such that $A_m \downarrow \emptyset$, one has

$$|f\chi_{A_m}| \to 0 \quad \text{as } m \to \infty. \qquad (2.3.18)$$

Motivated by the needs of the theory of integral equations, it is useful to impose one more condition on the operator K, besides the simple admissibility condition $KE_g \subset C_G$. We shall say that the triplet (E_g, C_G, K) is *strongly admissible*, if the following conditions hold:

(1) $KE_g \subset C_G$;
(2) for every fixed $\tau \in [t_0, T)$, the set KB, with $B \subset E_g$ bounded, is equicontinuous at τ.

Theorem 2.3.2. *Consider the triplet (E_g, C_G, K) in which E_g has absolutely continuous norm, and $k(t, s)$ is measurable on $[t_0, T) \times [t_0, T)$. Then the following conditions are necessary and sufficient for strong admissibility:*

(a)
$$P_{G(t)} \int_{t_0}^{T} k(t, s)x(s)\,ds = \int_{t_0}^{T} k(t, s)x(s)\,ds,$$

for any $t \in [t_0, T)$, and any $x \in E_g$;

(b)
$$|G_{-1}(t)k(t, \cdot)g(\cdot)|_{E'} \in L^{\infty}([t_0, T), R);$$

(c)
$$\lim_{t \to \tau} |[k(t, \cdot) - k(\tau, \cdot)]g(\cdot)|_{E'} = 0,$$

for any $\tau \in [t_0, T)$.

The proof of this theorem can be carried out following the same pattern as in the proof of Theorem 2.3.1. In particular, the absolute continuity of the norm is required in order to establish the boundedness of the operators S_{ij}, similar to those defined by (2.3.11). It is also useful to take a look at the proof of Theorem 5.1 in C. Corduneanu [4], where only scalar weight functions are considered.

Remark. Under the conditions of Theorem 2.3.2, the operator K is obviously a compact operator from E_g into $C([t_0, T), R^n)$. It may not be compact from E_g into C_G.

Similar results to those given in the last two theorems can be found in Milman [2], Dotseth [1]. See also C. Corduneanu [4], [7].

Among various classes of operators we can consider on function spaces, the class of *integral operators* is certainly the most significant with respect to the theory of integral equations.

We shall provide here a few definitions and results related to integral operators defined on spaces of measurable functions. Although we shall refer only to the Lebesgue measure of sets in Euclidean spaces, most of the following considerations can easily be formulated for general measure spaces (Ω, Σ, μ).

Let $S = S(\Omega, R)$ be a linear space of real-valued measurable functions, on a measurable set $\Omega \subset R^m$. The following two properties will be useful in what follows.

(1) If $x \in S$ and y is measurable, then $|y(u)| \leq |x(u)|$, a.e. on Ω implies $y \in S$.

(2) If $\operatorname{supp} x = \{u \in \Omega; x(u) \neq 0\}$, then $\bigcup_{x \in S} \operatorname{supp} x = \Omega$.

Property (1) is almost the same as the corresponding property in the definition of a Banach function space (Section 2.2). Here, one does not require the norm $|\cdot|_s$ to be monotone.

Property (2) states, in fact, that there is no proper subset of Ω containing all $\operatorname{supp} x$, for $x \in S$. This remark is necessary because $\operatorname{supp} x$ is determined up to a part of measure zero.

A linear operator $T: S \to S$ is called an *integral operator*, if there exists a real-valued measurable function $k(t, s)$ on $\Omega \times \Omega$, such that T admits the representation

$$(Tx)(t) = \int_\Omega k(t, s) x(s) \, ds, \quad \forall x \in S. \tag{2.3.19}$$

Of course, (2.3.19) should be regarded as valid almost everwhere for $t \in \Omega$. The exceptional set will depend, in general, on x.

The definitions formulated above with respect to $S = S(\Omega, R)$ and T can be carried over to the case of the space $S = S(\Omega, R^n)$, with T an operator on this space. In this case, $k(t, s) \in \mathscr{L}(R^n, R^n)$, for almost all $(t, s) \in \Omega \times \Omega$.

Theorem 2.3.3. *Let $S = S(\Omega, R^n)$ be a linear space of measurable maps, such that conditions (1) and (2) formulated above hold, and let T be a linear operator on S, with values in the space of all R^n-valued measurable functions on Ω. Then the following conditions with respect to T are equivalent:*

(1) *T is an integral operator on S;*

(2) *if $x^p \to 0$ in measure, and the sequence is dominated in S, say $|x^p| \leq |x|$, with $x \in S$, $p > 1$, then $(Tx^p u)(t) \to 0$ a.e. on Ω;*

(3) *if $x^p \to 0$ a.e. on Ω, and the sequence is dominated in S, then $(Tx^p)(t) \to 0$ a.e. on Ω; moveover, if $\Omega_p \subset \Omega$, $p \geq 1$, are such that $\operatorname{meas} \Omega_p \to 0$ as $p \to \infty$, and $\{\chi_\Omega\}$ is dominated in $S(\Omega, R)$, then $T\chi_{\Omega_p} \to 0$ a.e. on Ω.*

The proof of Theorem 2.3.3 is given in Kantorovich and Akilov [1], and the result is credited to Bukhvalov [1].

By means of Theorem 2.3.3, Lebesgue's dominated convergence theorem, and elementary measure theory, one can prove the following useful (see Section 3.3) result:

Theorem 2.3.4. *Let S be a function space as in Theorem 2.3.3, and assume that $\phi(t, s)$ is a map from $\Omega \times \Omega$ into $\mathscr{L}(R^n, R^n)$, not necessarily measurable, such that* *the function*

$$y(t) = \int_\Omega \phi(t,s)x(s)\,ds, \quad a.e. \text{ for } t \in \Omega, \qquad (2.3.20)$$

is measurable and a.e. finite, for every $x \in S$. *If we define* $y = Tx$, $x \in S$, *then there exists a measurable kenel* $K(t,s)$, *with the property that*

$$(Tx)(t) = \int_\Omega K(t,s)x(s)\,ds \qquad (2.3.21)$$

for almost all $t \in \Omega$, *and any* $x \in S$.

Under the conditions of Theorem 2.3.4, one can see that property (2) in the statement of Theorem 2.3.3 is verified.

Based on Theorem 2.3.4, one can obtain an integral representation for abstract operators acting on functions spaces whose elements are measurable functions. We will illustrate here such a possibility, in view of the application to be indicated in Section 3.3.

Assume that we are dealing with the space $L^2([t_0, T], R)$, on which a linear continuous abstract operator is given: $K: L^2 \to L^2$. We also assume that K is a nonanticipative operator, i.e., $x(s) = y(s)$ a.e. on $[t_0, t]$, implies $(Kx)(s) = (Ky)(s)$ a.e. on $[t_0, t]$.

Let us now consider the operator

$$(Lx)(t) = \int_{t_0}^t (Kx)(s)\,ds, \quad t \in [t_0, T], \qquad (2.3.22)$$

which takes the space $L^2([t_0, T], R)$ into itself (actually, into a subset consisting of absolutely continuous functions). For every $t \in (t_0, T]$, the map $x \to (Lx)(t)$, from $L^2([t_0 t], R)$ into R, is obviously linear and continuous. Hence, for every $t \in (t_0, T)$, there exists a function $k(t, \cdot)$ which belongs to $L^2([t_0, t], R)$ in respect to the second argument, such that (Riesz representation theorem) for any $x \in L^2([t_0, t], R)$

$$(Lx)(t) = \int_{t_0}^t k(t,s)x(s)\,ds. \qquad (2.3.23)$$

Of course, $k(t,s)$ is measurable in s for every fixed $t \in [t_0, T]$. One can set $k(t_0, s) = 0$, $s \in [t_0, T]$. But we are not sure about the measurability of $k(t,s)$ in (t,s), which is a requirement for deriving the fact that L is indeed an integral operator (of Volterra type, or nonanticipative). At this point, Theorem 2.3.4 can be applied, and we obtain for the operator L an integral representation of the form (2.3.23), where $k(t,s)$ is measurable.

Of course, the above considerations hold in the case $L^2([t_0, T], R^n)$. Carrying over to the space $L^2_{loc}([t_0, T), R^n)$ is then straight-forward.

We shall now discuss a few basic facts concerning *abstract Volterra operators*, which have been mentioned above in this section. In Sections 3.3 and 3.4, we shall deal in more detail with such operators, in relation to various classes of functional equations. Such equations appear as a natural generalization of classical Volterra equations, and their properties are strong enough to construct a theory that has many common features with the classical theory of Volterra integral equations. In particular, the so-called 'hereditary mechanics' has been based on the use of these kinds of operators and equations.

Let us consider the case of operators acting from a function space $E = E([t_0, T], R^n)$, into another function space $F = F([t_0, T], R^n)$, $V: E \to F$.

We shall say that V is an *abstract Volterra operator* (as opposed to classical Volterra integral operator), if the following condition holds: for any functions $x, y \in E$, such that $x(s) = y(s)$ for $s \in [t_0, t]$, $t \leq T$, one has $(Vx)(t) = (Vy)(t)$.

In other words, the value of Vx at t is determined by the values of x for $s \leq t$. For this reason, the Volterra operators are also called *nonanticipative*. Such operators are of great interest in investigating certain evolutionary processes in which the heredity is to be taken into consideration. Sometimes, for Volterra operators one uses the term *causal*.

The definition formulated above does not involve other properties of the operator V, such as linearity, continuity, closedness, compactness, etc. Such properties will be useful, and we shall impose them in order to derive existence, uniqueness and other facts related to abstract Volterra equations of the form $x(t) = (Vx)(t)$, with $V: E \to E$, or the differential functional equation of Volterra type $\dot{x}(t) = (Vx)(t)$.

Of course, the definition formulated above has meaning when we deal with spaces of continuous functions on $[t_0, T]$. For the case of functions spaces consisting of measurable functions, the definition has to be reformulated as follows: $V: E \to F$ is an *abstract Volterra operator* if for any functions $x, y \in E$, such that $x(s) = y(s)$ a.e. on $[t_0, t]$, $t \leq T$, one has $(Vx)(s) = (Vy)(s)$ a.e. on $[t_0, t]$. As usual, when we speak of an operator V defined on a function space E, consisting of measurable functions, we have in mind such operators that take classes of equivalent functions (i.e., a.e. equal), into classes consisting also of equivalent functions. So, we really have in mind operators defined in E, with values in F.

Certainly, one can consider various modifications of the above definitions for Volterra operators. For instance, in case E is a space of measurable functions and F consists of continuous functions, the definition should be adapted as follows. $V: E \to F$ is of *Volterra type* if $x(s) = y(s)$ a.e. on $[t_0, t]$, $t \leq T$, implies $(Vx)(t) = (Vy)(t)$. Obviously, $(Vx)(s) = (Vy)(s)$, $s \in [t_0, t]$.

There are several elementary properties of abstract Volterra operators, such as: (a) if $V_1, V_2: E \to F$ are Volterra operators, then $V_1 + V_2$ is also a Volterra operator; (b) if $V: E \to F$ is of Volterra type, and $a(t)$ is a scalar continuous/measurable function on $[t_0, T]$, then the operator aV given by $(aVx)(t) = a(t)(Vx)(t)$, $x \in E$, is also of Volterra type; (c) if $V_1, V_2: E \to E$ are Volterra operators, then $W = V_1 V_2$, defined by $(Wx)(t) = (V_1(\dot{V_2}x))(t)$ is also a Volterra operator.

From the properties stated above we can see that Volterra operators can be organized as algebras. Restricting our considerations, for instance, to the case of linear Volterra operators on a given space E, we can easily see that they constitute an algebra (over a scalar field, or even over a function field). It seems that such a property of Volterra operators has not been yet exploited, in relation to their applications to functional equations.

Another kind of property for Volterra type operators is related to the possibility of 'localizing' such operators. More precisely, let $V: E \to F$ be a Volterra operator, where $E = E([t_0, T], R^n)$, and $F = F([t_0, T], R^m)$. We shall fix a number $t_1, t_0 < t_1 < T$. Then we can consider the restrictions of the elements of E and F to $[t_0, t_1]$. It is obvious that the operator V generates another Volterra operator from $E_1([t_0, t_1], R^n)$, into $F_1([t_0, t_1], R^m)$, where E_1 and F_1 are the spaces obtained from E and F, taking the restrictions of their elements to the interval $[t_0, t_1]$.

We shall now mention a property of Volterra type operators related to the possibility of 'localizing' them on an interval $[t_1, t_2]$, with $t_0 < t_1$ and $t_2 \leq T$. Therefore, we assume that V is acting from $E([t_0, T], R^n)$ into $F([t_0, T], R^m)$. This can be achieved in different ways in accordance with our purpose. Let us illustrate this matter in connection with the continuation (in the future) of solutions that are locally defined, for the abstract Volterra equation $(Vx)(t) = f(t)$, with $x \in E$ as the unknown function, and $f \in F$. If we know a solution $x = x^0(t)$, defined on some interval $[t_0, t_1]$, $t_1 < T$, the problem that naturally arises is whether we can construct another local solution of the same equation, defined on the interval $[t_1, t_2]$, $t_2 < T$, which can be regarded as a continuation of the solution $x^0(t)$ defined on $[t_0, t_1]$. Let us point out the fact that if E is a space of measurable functions, we may not necessarily try to match the values of the solutions defined on $[t_0, t_1]$ and $[t_1, t_2]$, at t_1. If E consists of continuous functions these solutions must coincide at t_1. But what is the meaning of the 'restriction' of the operator V to the space of function defined on $[t_1, t_2]$, as restrictions of the elements of E? It seems that the natural way to define this 'restriction' is in the following manner. We agree to extend any function given on $[t_1, t_2]$ to $[t_0, t_2]$, by letting it equal the local solution on $[t_0, t_1]$. Then, on the space $\tilde{E}([t_1, t_2], R^n)$ consisting of the restrictions of the functions in $E([t_0, T], R^n)$ to $[t_1, t_2]$, we define the operator $(\tilde{V}x)(t) = (V\bar{x})(t)$, $t \in [t_1, t_2]$, where $\bar{x}: [t_0, t_2] \to R^n$ is defined by $\bar{x}(t) = x^0(t)$, for $t \in [t_0, t_1)$, and $\bar{x}(t) = \tilde{x}(t)$ for $t \in [t_1, t_2)$. The problem of continuing the local solution $x^0(t)$ reduces now to the construction of another local solution on $[t_1, t_2)$ for the equation $(\tilde{V}\tilde{x})(t) = f(t)$. In such circumstances, the 'localized' operator $\tilde{V}$ is useful, and enables us to define solutions on larger intervals, by piecing together several local solutions. With an operator which is not of Volterra type, such constructions are meaningless.

A remarkable class of Volterra type operators is the so-called *Niemytzki operators* which – roughly speaking – are operators representable in the form

$$(Gx)(t) = g(t, x(t)), \quad t \in [t_0, T), \tag{2.3.7}$$

where $(t, x) \to g(t, x)$ is a map from the set $[t_0, T) \times D \subset R^{n+1}$, into R^n (assuming $x \in R^n$). These superposition operators occur in many types of integral equations, especially Hammerstein equations, which will be considered in several sections of Chapter 4. It turns out that rather general operators, acting between function spaces, can be represented in the form (2.3.7), with $g(t, x)$ satisfying the Carathéodory conditions (i.e., measurability with respect to t, for fixed x, and continuity in x for almost all t).

The Niemytzki operator G, given by (2.3.7), is easily recognized as verifying the following property.

(α) If $x(t)$, $y(t)$ are measurable functions, defined on $[t_0, T]$ and taking values in $D \subset R^n$, such that $x(t) = y(t)$ a.e. on some set $A \subset [t_0, T)$, then $(Gx)(t) = (Gy)(t)$ a.e. on A.

It is obvious that condition (α) is stronger than the definition of Volterra operators (in which A can only be an interval $[t_0, t]$, $t < T$). On the other hand, property (α) enables us to 'localize' the operator G on any interval (or subset) of $[t_0, T)$. Operators verifying condition (α) are called *locally defined*.

If $g(t, x)$ satisfies the Carathéodory conditions, the Niemytzki operator (2.3.7) is continuous in measure. In other words, if $x^p \to x$ in measure on $[t_0, T)$, then $Gx^p \to Gx$ in measure. For the proof of this statement, under rather general assumptions, see Krasnoselskii *et al.* [1; p. 355].

Various converse theorems have been obtained for the above statement. We mention here a recent result obtained by A. V. Ponosov [1], which we particularize to the case of operators defined on function spaces with elements measurable functions on $[t_0, T)$, whose values are in R^n.

Theorem 2.3.5. *Let G be an operator taking the space of all measurable functions on $[t_0, T)$, with values in R^n, into itself. If G is locally defined and continuous in measure, then there exists a map $g: [t_0, T) \times R^n \to R^n$, satisfying Carathéodory conditions, such that G can be represented by (2.3.7).*

Theorem 2.3.5 allows to answer in a positive manner a conjecture of Niemytzki. Namely, assuming that G is given by (2.3.7), in which $g(t, x)$ does not necessarily satisfy the Carathéodory conditions, there exists a function $\bar{g}(t, x)$ which satisfies those conditions and generates the same Niemytzki operator G.

The answer obtained by Ponosov [1] is based on Theorem 2.3.5, and can be formulated as follows.

Corollary. *If $g(t, x)$ is a map from $[t_0, T) \times R^n$, such that the Niemytzki operator G, generated by g according to formula (2.3.7), is continuous in measure, then there* *exists a map* $\bar{g}: [t_0, T) \times R^n \to R^n$, *satisfying Carathéodory conditions, and such that*

$$(Gx)(t) = \bar{g}(t, x(t)) \quad a.e. \text{ on } [t_0, T), \qquad (2.3.24)$$

for any measurable x from $[t_0, T)$ *into* R^n.

Generally speaking, the map $\bar{g}$ is (essentially) different from g. Krasnoselskii and Pokrovskii [1] have constructed an example of a map $g(t, x)$ which does not satisfy the Carathéodory conditions, but takes (continuously in measure!) measurable functions into measurable ones. They make use of the continum hypothesis.

The last property we want to discuss in relation to Niemytzki operators is continuity. The following result is due to Krasnoselskii, and can be found in many books (see, for instance, Krasnoselskii *et al.* [1]).

Theorem 2.3.6. *Let* $g: [t_0, T] \times R^n \to R^n$ *satisfy the Carathéodory conditions, and consider the Niemytzki operator G defined by* (2.3.7). *Then G is continuous from* $L^p([t_0, T), R^n)$ *into* $L^q([t_0, T], R^n)$, $0 \le p, q \le \infty$, *if and only if there exists a function* $a(t) \in L^q([t_0, T], R)$ *and a positive number b, such that*

$$|g(t, x)| \le a(t) + b|x|^{p/q}, \qquad (2.3.25)$$

a.e. for $t \in [t_0, T]$, *and for all* $x \in R^n$.

Let us notice that for Hammerstein operators (2.3.3), which can be represented as the product of a linear integral operator and a Niemytzki operator, $H = LG$, where G is given by (2.3.7) and L is given by

$$(Lx)(t) = \int_{t_0}^{T} h(t, s) x(s) \, ds,$$

continuity or compactness conditions can be obtained from the corresponding conditions for linear integral operators and Niemytzki operators. For instance, if L is continuous only but G is compact, then $H = LG$ is compact.

To conclude this section we will give some properties for convolution operators. These properties are encountered in subsequent chapters, and they are not always found in standard references.

We shall consider operators that can be (formally) represented by the formula

$$(Lx)(t) = \int_{R} x(t - s) \, d\mu(s), \qquad (2.3.26)$$

where μ is a finite signed measure on R. (In other words, μ is a function with bounded variation on R, taking values in R, or may be in $\mathscr{C}$.) We limit our considerations to the scalar case, but the conclusions reached are obviously valid when x takes values in R^n, and $\mu: R \to \mathscr{L}(R^n, R^n)$ is a matrix-valued map, whose entries are functions with bounded variation on R.

Basically, we are interested in two properties of the convolution operator L:

(1) L is continuous on the space $L^1(R, R)$, and

$$|Lx|_{L^1} \leq (\operatorname{Var} \mu)|x|_{L^1}, \quad x \in L^1; \tag{2.3.27}$$

(2) L is continuous on the space $M(R, R)$, and

$$|Lx|_M \leq (\operatorname{Var} \mu)|x|_M, \quad x \in M. \tag{2.3.28}$$

The proofs of these properties are rather similar, and they are based on Fubini's theorem for double integrals. It is also useful to notice that, if $x(t)$ is measurable on R, then $x(t - s)$ is measurable on R^2.

Since $\mu(s) = \mu_1(s) - \mu_2(s)$, $s \in R$, where $\mu_1(s)$ and $\mu_2(s)$ are positive finite measures on R, and $v = |\mu| = \mu_1 + \mu_2$ is a positive finite measure on R, with $v(R) = \operatorname{Var} \mu$, we can write for any $\alpha \in R$

$$\int_\alpha^{\alpha+1} |(Lx)(t)|\, dt = \int_\alpha^{\alpha+1} \left| \int_R x(t - s)\, d\mu(s) \right| dt$$

$$= \int_\alpha^{\alpha+1} \left| \int_R x(t - s)\, d\mu_1(s) - \int_R x(t - s)\, d\mu_2(s) \right| dt$$

$$\leq \int_\alpha^{\alpha+1} \left(\int_R |x(t - s)|\, d\mu_1(s) + \int_R |x(t - s)|\, d\mu_2(s) \right) dt$$

$$= \int_\alpha^{\alpha+1} \int_R |x(t - s)|\, dv(s)\, dt \leq \int_R \int_\alpha^{\alpha+1} |x(t - s)|\, dt\, dv(s)$$

$$\leq |x|_M \int_R dv(s) = (\operatorname{Var} \mu)|x|_M.$$

Then the supremum has to be taken with respect to $\alpha \in R$, in order to obtain (2.3.28). This proves property (2) for the convolution operator L. Property (1) can be obtained similarly, taking the integral on R, instead of the integral on $[\alpha, \alpha + 1]$.

In the special case $d\mu(s) = k(s)\, ds$, where $k \in L^1(R, R)$, it is commonly known that

$$|x * k|_E \leq |x|_E |k|_{L^1}, \quad x \in E, \tag{2.3.29}$$

where E stands for any of the spaces $L^p(R, R)$, $1 \leq p \leq \infty$, or M.

See for instance, C. Corduneanu [4], or Edwards [2].

2.4   Fixed point theorems; monotone operators

This last section is devoted to the topics indicated in the title. These topics pertain to nonlinear functional analysis, and their use in the theory of functional equations (ordinary differential equations, partial differential equations, integral equations, integrodifferential equations, delay equations) is probably the most powerful method in obtaining existence theorems.

We shall start with the Schauder–Tychonoff fixed point theorem, which we choose to state in the framework of a locally convex space.

Theorem 2.4.1. *Let E be a locally convex Hausdorff space, and assume T: K → E is a continuous map, with K ⊂ E convex, and*

$$TK \subset A \subset K,$$

where A is compact. Then there exists at least one fixed point for T, i.e. Tx = x for some x ∈ K.

The proof of Theorem 2.4.1 is rather lengthy. It can be found, for instance, in C. Corduneanu [4], Edwards [2]. The following corollary is known as the Schauder fixed point theorem, and it can be stated as follows.

Corollary. *Let E be a real Banach space, and K ⊂ E a closed, bounded and convex set. If T: K → K is compact, then T has at least one fixed point.*

Proofs of the Corollary can be found in Deimling [2], Kantorovich and Akilov [1].

We shall repeatedly apply Theorem 2.4.1 or its corollary to obtain existence theorems for various classes of equations. Here, we want to indicate a quick illustration of the corollary (Schauder fixed point theorem).

Let us consider the classical Volterra equation

$$x(t) = f(t) + \int_0^t k(t, s, x(s)) \, ds, \tag{2.4.1}$$

where f and k satisfy the following assumptions:

(1) $f: [0, a] \to R$ is continuous;
(2) $k(t, s, x)$ is continuous on the set $\Delta_a \times R$, where $\Delta_a = \{(t, s); 0 \le s \le t \le a\}$ into R, and is bounded: $|k(t, s, x)| \le M$.

Under conditions (1) and (2), there exists at least one continuous solution $x(t)$ of (2.4.1), defined on the whole interval $[0, a]$.

Indeed, from (2.4.1) we see that if a solution does exist on $[0, a]$, one must have $|x(t)| \le |f(t)| + Ma$, $t \in [0, a]$.

In the space $C([0, a], R)$ we consider the convex set K defined by the inequality

$$|x(t)| \le |f(t)| + Ma, \quad t \in [0, a]. \tag{2.4.2}$$

K is obviously closed and bounded in the space $C([0, a], R)$.

Let us define the operator T by

$$(Tx)(t) = f(t) + \int_0^t k(t, s, x(s)) \, ds. \tag{2.4.3}$$

This operator is defined and continuous on the whole space $C([0, a], R)$. It remains to show that T takes K into a compact subset of K.

Based on our assumptions, it is obvious that (2.4.3) implies

$$TK \subset K. \tag{2.4.4}$$

That TK is relatively compact in $C([0, a], R)$ follows from Arzela's test of compactness. First, TK is uniformly bounded, being part of K which is bounded. Second, $k(t, s, x)$ is uniformly continuous on $\Delta_a \times [-A, A]$, with $A \geq |f(t)| + Ma$, $t \in [0, a]$. Hence

$$|(Tx)(t) - (Tx)(u)| \leq |f(t) - f(u)| + \left| \int_u^t k \, ds \right| + \int_0^a |k(t, s, x(s)) - k(u, s, x(s))| \, ds$$

$$\leq |f(t) - f(u)| + M|t - u| + a \sup |k(t, s, x) - k(u, s, x)|,$$

where the supremum is taken with respect to $(t, x, s), (u, s, x) \in \Delta_a \times [-A, A]$. One sees that $|(Tx)(t) - (Tx)(u)| < \varepsilon$ for any $x \in K$, t, $u \in [0, a]$, provided $|t - u| < \delta(\varepsilon)$, where δ is chosen in such a way that $|f(t) - f(u)| < \varepsilon/3$, $M|t - u| < \varepsilon/3$ and $|k(t, s, x) - k(u, s, x)| < \varepsilon/3$.

The corollary yields the existence of at least one solution to (2.4.1) in K.

Another basic result in the theory of fixed point methods is the famous Leray–Schauder theorem. The following formulation is taken from Gilbarg and Trudinger [1], where a proof is provided.

Theorem 2.4.2. *Let E be a Banach space, and let $T: E \times [0, 1] \to E$ be a compact mapping, such that $T(x, 0) = x$ for all $x \in E$. If there exists a constant $M > 0$, such that*

$$|x|_E \leq M \tag{2.4.5}$$

for all $(x, \lambda) \in E \times [0, 1]$ satisfying $x = T(x, \lambda)$, then the mapping $T_1(x) = T(x, 1)$ of E into itself has a fixed point.

Notice that compact mapping means continuous mapping taking bounded sets in relatively compact sets in E.

A special case of Theorem 2.4.2, which is also known as the Leray–Schauder fixed point theorem, can be stated:

Corollary. *Let $T: E \to E$ be a compact mapping of the Banach space E into itself, and suppose there exists a constant $M > 0$ with the property $|x|_B \leq M$ for all $x \in E$ and $\lambda \in [0, 1]$ satisfying $x = \lambda Tx$. Then T has a fixed point in E.*

For the proof, we refer the reader to Gilbarg and Trudinger [1]. See also Deimling [2].

Applications of the Leray–Schauder fixed point theorem will be considered in Chapter 4.

The next topic we wish to discuss in view of applications to integral equations is the concept of *monotone operators*. This concept, introduced in the 1960s, has proven to be very effective in obtaining existence results in nonlinear problems. One of the reasons is certainly the lack of compactness among the basic requirements. Compactness is not always easy to check and it does represent a rather severe restriction on the operator.

Let E be a real Banach space, and denote by E^* its dual space.

We shall say that the map $A: E \to E^*$ is *monotone*, if the following condition is verified:

$$(Ax - Ay, x - y) \geq 0, \qquad (2.4.6)$$

for any $x, y \in E$, where $(\cdot, \cdot)$ is the usual bilinear pairing of E and E^*, $(x, x^*) = (x^*, x) = x^*(x)$, for $x \in E$ and $x^* \in E^*$.

Another useful concept that we need in stating some basic results related to monotone operators is that of *hemicontinuity*. The operator $A: E \to E^*$ is called *hemicontinuous* on E, if $\lim_{t \to 0} A(x + ty) = Ax$ in the weak topology, for any $x, y \in E$.

Finally, the concept of *local boundedness* of the operator $A: E \to E^*$ is defined as follows: A is locally bounded at $x_0 \in E$ if there exists a neighborhood V_{x_0}, such that $A(V_{x_0})$ is bounded in E^*.

The following theorem will be necessary in investigating nonlinear integral equations in Section 4.3.

Theorem 2.4.3. *Let E be a reflexive Banach space and $A: E \to E^*$ a monotone hemicontinuous operator with A^{-1} locally bounded. Then A is onto E^*, i.e. $R(A) = E$.*

For a proof of Theorem 2.4.3, we refer the reader to Browder [1]. For results equivalent to Theorem 2.4.3, or closely related to it, see Barbu [4], Deimling [2], or Brézis and Browder [3]. Of course, A^{-1} is, in general, a multi-valued map.

Let us now give another result on monotone operators in which the condition of local boundedness is replaced by the so-called condition of coercivity. $A: E \to E^*$ is called *coercive* if

$$\frac{(x, Ax)}{|x|} \to \infty \quad \text{as } |x| \to \infty. \qquad (2.4.7)$$

Theorem 2.4.4. *Let E be a reflexive Banach space, and $A: E \to E^*$ a map which is monotone, hemicontinuous and coercive. Then $R(A) = E^*$.*

The proof of this theorem is given, for instance, in Deimling [2], Barbu and Precupanu [1].

It is interesting that when hemicontinuity is replaced by weak continuity (i.e., for every fixed $x_0 \in E$, $(Ay, x_0) \to (Ax, x_0)$ for $y \to x$ in the weak topology of E), the monotonocity condition is no longer necessary. More precisely, the following result can be found in Kačurovskii's paper [1]:

Theorem 2.4.5.   *Let E be a reflexive Banach space, and $A: E \to E^*$ a map which is weakly continuous and coercive. Then $R(A) = E^*$.*

This result is basically a consequence (apparently, not immediate) of the Schauder–Tychonoff fixed point Theorem (see Theorem 2.4.1).

A consequence of Theorem 2.4.5 will be needed in Section 4.5, in connection with boundary value problems for ordinary differential equations. Its basic idea consists in extending (continuing) an operator given on a ball, to an operator which is defined on the whole space and is coercive. We shall assume that E is a real Hilbert space, because this case covers the situation to be discussed in Section 4.5.

Corollary 2.4.6.   *Let E be a real Hilbert space, and $A: E \to E$ a weakly continuous map. Denote by E_s the space whose elements are the same as those of E, but endowed with a different norm $|\cdot|_s$. If $B: E \to E$ is a weakly continuous coercive operator such that $Ax = Bx$ for $|x|_s \le R$, and $\{y; By = f, |f| \le F\} \subset \{x; |x|_s \le R\}$ for some positive constants R and F, then the equation $Ax = f$ has at least one solution for those $f \in E$ such that $|f| \le F$.*

Proof.   Since B is weakly continuous and coercive, the equation $Bx = f$ has at least one solution, for every $f \in E$ (according to Theorem 2.4.5). If $f \in E$ is such that $|f|_s \le F$, then any solution of $Bx = f$ is in the ball $|x|_s \le R$. But A and B coincide on this ball, which means that $Ax = f$.

Remark.   The procedure described in Corollary 2.4.6 is called the coercive-continuation method. Corollary 2.4.6 is due to Gaponenko [1].

Bibliographical notes

Most of the material in Section 2.1 is common knowledge for readers interested in the theory of integral equations. The spaces C_g, where g is a matrix-valued function, were introduced and investigated by Gollwitzer [1]. The equivalence of these spaces was obtained, under special conditions, by Pandolfi [1]. Contributions to the theory of these spaces when g is a scalar function can be found in C. Corduneanu [4].

The results related to spaces of measurable functions are also common knowledge, and adequate references have been given in the text. A special mention is due to the paper by Milman [1] which contains a very good synthesis of various necessary results. The paper by Brooks and Dinculeanu [1] contains the infinite-dimensional variant of the Dunford–Pettis result on weak compactness of families of measurable functions and is instrumental in generalizing some of the results to Banach spaces. Further basic references are indicated in the text.

Section 2.3 contains only a few available results concerning integral operators on function spaces, as well as some generalities about abstract Volterra operators (also known as causal or nonanticipative operators). In no way should this section be regarded as an attempt to present a general picture of the theory of integral or related operators on function spaces. As already mentioned, we are interested only in providing some background and orientation for the discussion of the topics related to integral (both classical and abstract) equations in the subsequent chapters of this book.

Basic facts about the theory of integral operators are included in most books on functional analysis. The following references at least, devote substantial attention to the theory of integral operators: Dunford and Schwartz [1], Kantorovich and Akilov [1], Edwards [1], Fenyo and Stolle [1], Schwabik *et al.* [1], C. Corduneanu [4].

The mathematical literature of the last 20 years offers several books entirely dedicated to the theory of integral/related operators on various function spaces. Most of the results included in these books are certainly related to the theory of integral equations in a direct way, but there are significant results obtained for integral operators which have not yet found expected applications to the theory of integral equations. It seems that the theory of integral operators, which has now been successfully expanded, remains – together with the methods of the functional analysis (mostly nonlinear) – the main source of new results for the theory of integral equations.

There are several books available which deal exclusively with the theory of integral operators: The books by Okikiolu [1], Krasnoselskii *et al.* [1], which include the best available results on nonlinear integral operators; and those by Halmos and Sunder [1], Jörgens [1], and Korotkov [1]. The book by Korotkov contains the latest available results in the literature, and a list of selected references which provide a very sound orientation to the field. The results are exclusively related to linear operators in integral form but, unlike most of the results obtained for this kind of operators, there are significant results related to unbounded operators (Carleman type operators, the theory of which was initiated by Carleman [1]). Another reference relevant to the theory of linear integral operators is the monograph by Gohberg and Krein [1] on Volterra operators (the definition of a Volterra operator adopted in this book is as follows: a completely continuous linear operator whose spectrum contains only the number 0).

The journal literature related to the theory of integral operators is very rich, and cannot be covered adequately here. Nevertheless, I shall mention some valuable references which should stimulate the interest of the readers in pursuing the development of this branch of analysis, and provide a few hints about the new directions that seem to be becoming dominant in the theory of integral operators/equations.

A classical paper is due to von Neumann [1], in which the problem of integral representation of abstract operators is considered. Theorem 2.3.3 constitutes an answer to this, which was obtained by Bukhvalov [1] 40 years after it was raised by von Neumann. During this time, partial results were obtained by several authors. A good deal of references are listed in the book by Korotkov [1], as well as in the survey paper by Bukhvalov [1]. von Neumann's paper is also relevant to Carleman operators, which represent one of the first classes of unbounded linear operators to be investigated. The condition on the kernel $k(t,s)$ which leads to the concept of a Carleman type operator on the L^2-space can be simply written as $\int |k(t,s)|^2 \, ds < \infty$ for almost all t. Instead of the Lebesgue measure, one may consider a general measure on a measure space. Another classical paper about integral operators on measure spaces is due to Aronszajn and Szeptycki [1]. It was continued by Szeptycki in a long series of papers, and by many researchers in this field. A recent contribution is that of Labuda [1]. Other pertinent references are: Sunder [1]; Sourour [1]; Gretsky and Uhl, Jr [1]; Schep [1]; Weis [1]; Schachermayer [1]; Victory, Jr [1]; Dodds and Schep [1]; Aliprantis, Burkinshaw, and Duhoux [1]. A recent paper by Rubio de Francia, Ruiz and Torrea is dedicated to convolution operators with operator-valued kernel, and contains a good many references related to the Calderon–Zygmund theory of singular integral operators.

Developing some classical results due to Carleman and Weyl, Pietsch [1] and Elstner and Pietsch [1] have obtained significant results in the problem of the distribution of eigenvalues of integral operators. The underlying function spaces are the so-called Besov spaces (very useful in the theory of partial differential operators). The apparatus is quite sophisticated, which indicates the complexity of the problem. Even Besov spaces with respect to a weight function had to be considered in order to deal with the eigenvalue problem and its discussion. It is interesting to note that, under appropriate conditions for the kernel, the sequence of eigenvalues is shown to belong to a convenient sequence space (which provides, of course, some information about their distribution in the complex plane).

Further useful references on the theory of integral operators on various function spaces are: Novitskii [1], Maksimova [1], Kalton [1], Pavlov [1], Dijkstra [1], Milman [2], and Colton [1]. Colton's paper shows how integral operators occur in problems involving parabolic type equations, and their equivalence (i.e., by means of an integral operator, one establishes a correspondence between the space of solutions of a rather general parabolic equation and the space of solutions of the classical heat equation).

On the theory of abstract Volterra operators briefly sketched in Section 2.3, the literature is rather sparse, which shows that this theory is in some sense in its early stage. A basic problem in relation to abstract nonlinear Volterra operators, which can be traced back to Volterra and Fréchet, and which concerned Wiener [1], is the problem of representing analytically such operators by means of the so-called Volterra series. More precisely, one looks for representations of the form

$$(Tu)(t) = \sum_{i=1}^{\infty} \int_0^t \int_0^t \cdots \int_0^t k_i(t_1, t_2, \ldots, t_i) u(t - t_1) \ldots u(t - t_i) \, dt_1 \ldots dt_i,$$

which need to hold for $t \geq 0$. Of course, the convergence of the series can be understood in various senses. In many cases (see for instance Sandberg [7] or Boyd and Chua [1]) one speaks about approximations of the Volterra (causal) operators by means of (finite) Volterra series.

The most valuable results on the use of Volterra series in representing general causal operators were obtained relatively recently by Sandberg [1]–[7]. Applications to integral equations are also dealt with in most of Sandberg's papers quoted above. There are also two books dedicated to the Volterra–Wiener approach to describing linear systems: Schetzen [1] and Rugh [1].

For the topics discussed in Section 2.4, an additional reference to those already in the text is Zeidler [1].

3

Basic theory of Volterra equations:
integral and abstract

In this chapter we shall discuss in depth the problem of (local) existence
for Volterra equations in spaces of continuous/measurable functions. Besides
classical Volterra equations involving integral operators, we shall also deal with
general Volterra equations that involve causal (or nonanticipative) operators,
not necessarily of integral type. Applications to the existence (and related) problems for various classes of functional-differential equations, including equations
with infinite delay, will also be considered.

Certain global problems related to the existence of solutions to Volterra
equations, such as continuability of solutions, escape time, and other related
concepts, will also be discussed in this chapter. Occasionally, some problems
concerning asymptotic behavior will be considered, even though we focus our
attention on this kind of problem in Chapter 4.

3.1 Continuous solutions for Volterra integral equations

We shall discuss in this section the existence of at least one solution for the
Volterra nonlinear equation

$$x(t) = f(t) + \int_0^t k(t, s, x(s)) \, ds, \qquad (3.1.1)$$

in which f and k are assumed to be continuous on their domains of definition:
$f: [0, a] \to R$, and $k: \bar{D} \to R$, where $\bar{D}$ is defined by the inequalities

$$0 \le s \le t \le a, \quad |x| \le r, \qquad (3.1.2)$$

with both a and r positive numbers.

From the beginning, we want to stress that the results we shall establish here
for scalar equations can be extended without any difficulty to the case of vector-
valued functions, i.e., $f \in R^n$, $k \in R^n$, at each point of their domains of definition.

As we shall notice immediately, it is necessary in general to restrict the interval on which a solution to (3.1.1) can be defined. This feature was actually present in the case we discussed in Section 1.3, where k was also assumed to satisfy a Lipschitz type condition of the form

$$|k(t, s, x) - k(t, s, y)| \leq L|x - y|, \tag{3.1.3}$$

with $L > 0$ a constant, and any $(t, s, x), (t, s, y) \in \bar{D}$.

Let us now consider a sequence of continuous functions on $\bar{D}$, $\{k_n(t, s, x)\}$, $n = 1, 2, \ldots$, such that

$$\lim_{n \to \infty} k_n(t, s, x) = k(t, s, x), \tag{3.1.4}$$

uniformly on $\bar{D}$, and such that

$$|k_n(t, s, x) - k_n(t, s, y)| \leq L_n|x - y|, \tag{3.1.5}$$

with $\{L_n\}$ a sequence of positive numbers, generally unbounded. Of course, we can choose k_n to be polynomials in all three variables, according to the Weierstrass approximation theorem (see for instance, C. Corduneanu [3]).

Another way of choosing $k_n(t, s, x)$ can be described as follows: one considers a partition of the interval $[-r, r]$ into $2n$ equal subintervals by means of the points $x_j = jr/n$, $|j| = 0, 1, 2, \ldots, n$. We then define $k_n(t, s, x_j) = k(t, s, x_j)$, and extend k_n to the whole interval $|x| \leq r$ (as far as x is concerned) by letting

$$k_n(t, s, x) = k_n(t, s, x_j) + \frac{n}{r}[k_n(t, s, x_{j+1}) - k_n(t, s, x_j)](x - x_j) \quad \text{for } x_j < x < x_{j+1}, \tag{3.1.6}$$

$|j| = 0, 1, 2, \ldots, n$. In other words, $k_n(t, s, x)$ is a piecewise linear function in x, from which it follows that $k_n(t, s, x)$ is Lipschitz in x. It is obviously continuous in $\bar{D}$. Moreover, we see that $k_n(t, s, x)$ is a *monotone function* in x, whenever $k(t, s, x)$ possesses this property. Since

$$k(t, s, x) - k_n(t, s, x) = k(t, s, x) - k(t, s, x_j) - \frac{n}{t}[k(t, s, x_{j+1}) - k(t, s, x_j)](x - x_j)$$

for a convenient j (such that $x_j \leq x < x_{j+1}$), the uniform continuity of k in $\bar{D}$ leads immediately to (3.1.4). One has also to take into account that $n(x - x_j) \leq r$.

If we consider now the auxiliary equations

$$x_n(t) = f(t) + \int_0^t k_n(t, s, x_n(s)) \, ds, \tag{3.1.7}$$

$n = 1, 2, \ldots$, then the result established in Theorem 1.3.1 is applicable, provided we also assume

$$\sup|f(t)| = b < r, \quad t \in [0, a]. \tag{3.1.8}$$

Therefore, equation (3.1.7) has a unique continuous solution $x_n(t)$, defined on the interval $[0, \delta]$, with

$$\delta = \min\left\{a, \frac{r-b}{\tilde{M}}\right\}, \tag{3.1.9}$$

where $\tilde{M}$ is such that

$$|k_n(t, s, x)| \leq \tilde{M} \text{ in } \bar{D}. \tag{3.1.10}$$

Of course, such an $\tilde{M}$ does exist, because of (3.1.4).

If we consider now the sequence $\{x_n(t)\}$, $n = 1, 2, \ldots$, in $C([0, \delta], R)$, we notice:

(1) $|x_n(t)| \leq r$, for $n = 1, 2, \ldots$, $t \in [0, \delta]$, which implies the uniform boundedness property for the sequence (or simply, the boundedness of the sequence in $C([0, \delta], R)$);
(2) for every $n = 1, 2, \ldots$, and $t, u \in [0, \delta]$,

$$|x_n(t) - x_n(u)| \leq \left| f(t) - f(u) + \left| \int_u^t k_n(t, s, x_n(s))\, ds \right| \right.$$

$$+ \left| \int_0^u [k_n(t, s, x_n(s)) - k_n(u, s, x_n(s))]\, ds \right|$$

$$\leq |f(t) - f(a)| + \tilde{M}|t - u|$$

$$+ \int_0^u |k_n(t, s, x_n(s)) - k_n(u, s, x_n(s))|\, ds$$

which means that the sequence is equicontinuous in $C([0, \delta], R)$.

By the Ascoli–Arzela condition of compactness in $C([0, \delta], R)$, we can state the existence of a subsequence $\{x_{n_j}\}$, $j = 1, 2, 3, \ldots$, such that

$$\lim_{j \to \infty} x_{n_j}(t) = x(t) \text{ in } C([0, \delta], R), \tag{3.1.11}$$

which immediately leads, from (3.1.7) with $n = n_j$, to (3.1.1) for the continuous function $x(t)$ defined by (3.1.11).

Therefore, the following result has been proven:

Theorem 3.1.1. *Consider equation (3.1.1) under the following assumptions:*

(1) $f \in C([0, a], R)$;
(2) $k \in C(\bar{D}, R)$, *with $\bar{D}$ defined by (3.1.2)*;
(3) *condition (3.1.8) holds.*

Then there exists at least one solution of (3.1.1) in $C([0, \delta], R)$, where δ is given by (3.1.9), with $\tilde{M}$ an arbitrary positive number such that $\tilde{M} > M = \sup|k(t, s, x)|$ in $\bar{D}$.

Remark. Since $\tilde{M}$ can be chosen as close to M as we like, because of (3.1.10), we can say that the interval of existence in Theorem 3.1.1 is 'almost' the same as in Theorem 1.3.1. Actually, one can prove that the theorem holds with M instead of $\tilde{M}$, which we shall see using Tonelli's procedure [1].

A second proof of Theorem 3.1.1. The method of proof used above is based on the construction of a sequence of approximate solutions. At the limit of this sequence (more precisely, a convenient subsequence) provides a solution of the equation (3.1.1).

Tonelli has provided another method for constructing approximate solutions, with arbitrary degree of accuracy. Namely, consider the partition of the interval $[0, \delta]$, $\delta = \min\{a, (r - b)/M\}$ given by the points $t = t_j = j/n$, $j = 0, 1, 2, \ldots, n$, where $n > 1$ is an arbitrary natural number. For this fixed n, we construct the approximate solution $x_n(t)$ of (3.1.1) by letting:

$$\left.\begin{aligned}
x_n(t) &= f(t), && 0 \le t \le t_1, \\
x_n(t) &= f(t) + \int_0^{t-(\delta/n)} k(t, s, x_n(s))\, ds, && t_1 \le t \le t_n = \delta.
\end{aligned}\right\} \tag{3.1.12}$$

The first formula in (3.1.12) helps us to find $x_n(t)$ on $[\delta/n, 2\delta/n]$, using the second formula. Then, we use the values of $x_n(t)$ on $[\delta/n, 2\delta/n]$ to obtain those on $[2\delta/n, 3\delta/n]$, etc. All functions $x_n(t), n = 1, 2, \ldots$, are continuous on $[0, \delta]$ because we obtain from the second formula in (3.1.12) the estimate

$$|x_n(t)| \le b + M\left(t - \frac{\delta}{n}\right) < b + M\delta \le r. \tag{3.1.13}$$

Since (3.1.13) shows the uniform boundedness of the sequence $\{x_n(t)\}$ in $C([0, \delta], R)$, we need to prove that the sequence is also equicontinuous. Indeed, one can easily see that

$$|x_n(t) - x_n(u)| \le |f(t) - f(u)| + M|t - u| + \int_0^u |k(t, s, x_n(s)) - k(u, s, x_n(s))|\, ds,$$

$$\tag{3.1.14}$$

for any $n = 1, 2, \ldots$, and $t, u \in [0, \delta]$.

The existence of the sequence $\{x_{n_j}(t)\}$ satisfying (3.1.11) is thereby proven.

In order to derive the existence of the continuous solution for (3.1.1) we can proceed as follows: let $t \in [0, \delta]$ be fixed; then $(\delta/n) < t$ for large enough n, so we have to look at the second formula in (3.1.12); at the limit, when $n \to \infty$, this formula leads to equation (3.1.1). This ends the second proof of Theorem 3.1.1. Another version of Theorem 3.1.1 can be stated as follows:

Let us consider equation (3.1.1), *under the following hypotheses*:

(1) $f \in C([0, a], R)$;
(2) $k \in C(\bar{\Delta}, R)$, *where* $\bar{\Delta}$ *is defined by*

$$0 \leq s \leq t \leq a, \quad |x - f(t)| \leq b. \tag{3.1.15}$$

Then there exists at least one continuous solution $x = x(t)$ *of equation* (3.1.1), *defined for* $0 \leq t \leq \delta$, *where*

$$\delta = \min\left\{a, \frac{b}{M}\right\}, \qquad M = \sup_{\bar{\Delta}} |k(t, s, x)| \tag{3.1.16}$$

The proof is obtained, for instance, using the Tonelli's method, in the same manner as in the second proof of Theorem 3.1.1. See for details C. Corduneanu [6].

This variant of the basic existence theorem for Volterra integral equations is more useful than Theorem 3.1.1 if we want to discuss such topics as *continuability* of solutions, and related problems.

The next result concerning equation (3.1.1), under the assumption of continuity for both $f(t)$ and $k(t, s, x)$, is related to the (nonlinear) *integral inequalities*, and the existence of a *maximal (minimal) solution*. We shall again restrict our considerations to the scalar case. Various results for both vector and scalar cases are given in Azbelev and Caljuk [1], Lakshmikantham and Leela [1], Martinjuk and Gutowski [1].

Let $x(t)$ be a continuous function on some interval $[0, \delta'] \subset [0, \delta]$, with δ given by (3.1.9), and $\delta' < \delta$ (δ' can be as close to δ as we like). Assume that the integral inequality

$$x(t) \leq f(t) + \int_0^t k(t, s, x(s)) \, ds \tag{3.1.17}$$

holds true on $0 \leq x \leq \delta'$. Of course, this implies $|x(t)| \leq r$ on $[0, \delta']$. In particular, (3.1.17) is verified as an equality by the solution $x(t)$ whose existence is guaranteed by Theorem 3.1.1.

We shall again make use of the approximating functions $k_n(t, s, x)$, given by (3.1.6), modifying somewhat these functions, such that

$$k(t, s, x) \leq k_n(t, s, x) \text{ in } \bar{D}, \tag{3.1.18}$$

while (3.1.4) remains true. This can be achieved if we define

$$\varepsilon_n = \sup|k(t, s, x) - k_n(t, s, x)| \text{ in } \bar{D}, \tag{3.1.19}$$

and consider $k_n(t, s, x) + \varepsilon_n$, instead of $k_n(t, s, x)$, as approximating functions for $k(t, s, x)$. Of course, $k_n(t, s, x) + \varepsilon_n$ are Lipschitz continuous (because k_n are), and if $k_n(t, s, x)$ is monotone in x, so is $k_n(t, s, x) + \varepsilon_n$.

Therefore, without losing generality, we can assume that $k_n(t, s, x)$ verifies both (3.1.4) and (3.1.18). This means that

$$x(t) \le f(t) + \int_0^t k_n(t, s, x(s)) \, ds, \tag{3.1.20}$$

for any $x(t)$ satisfying (3.1.17).

Let us now consider the integral equations

$$x_n(t) = f(t) + \delta_n + \int_0^t k_n(t, s, x_n(s)) \, ds, \tag{3.1.21}$$

where $\{\delta_n\}$ is a sequence of positive numbers such that $\delta_n \to 0$ as $n \to \infty$. We notice that the (unique) solution of (3.1.21) is defined on $[0, \delta']$, provided n is chosen sufficiently large. Of course, we can assume without loss of generality that all solutions $x_n(t)$, $n = 1, 2, \ldots$, are defined on $[0, \delta']$. This is an immediate consequence of (3.1.9), if we notice that $|k_n(t, s, x)| \le M'$ in $\bar{D}$, $n = 1, 2, \ldots$, with some $M' > M$.

From (3.1.17) and (3.1.21) we obtain

$$x(t) - x_n(t) \le -\delta_n + \int_0^t [k_n(t, s, x(s)) - k_n(t, s, x_n(s))] \, ds \tag{3.1.22}$$

on $[0, \delta']$. Since $x(0) - x_n(0) \le -\delta_n < 0$ for every n, we can find for fixed n a largest number τ, $\tau \le \delta'$, such that $x(t) - x_n(t) \le 0$ on $[0, \tau]$, with $x(\tau) - x_n(\tau) = 0$. If $\tau = \delta'$, then we can assert

$$x(t) \le x_n(t), \quad t \in [0, \delta']. \tag{3.1.23}$$

If $\tau < \delta'$, then in any right neighborhood of τ there are values of t such that $x(t) - x_n(t) > 0$. This means that we can find a pair of numbers τ_1, t_1, with $\tau < \tau_1 < t_1 < \delta'$, where τ_1 could be as close to τ as we want, such that $x(t) - x_n(t) \ge 0$ on $[\tau_1, t_1]$. From (3.1.22) we derive

$$x(t) - x_n(t) \le -\delta_n + \int_0^\tau + \int_\tau^{\tau_1} + \int_{\tau_1}^t, \tag{3.1.24}$$

the integrands for each of the three integrals in (3.1.24) being the same as (3.1.22). But on $[0, \tau]$ we have $x(s) - x_n(s) \le 0$, and because of the monotonicity of k_n we find the first integral in the right-hand side of (3.1.24) is nonpositive. The second one can be taken arbitrarily small if we choose τ_1 sufficiently close to τ:

$$\left| \int_\tau^{\tau_1} [k_n(t, s, x(s)) - k_n(t, s, x_n(s))] \, ds \right| < \frac{1}{2} \delta_n.$$

Therefore, from (3.1.24) we obtain if we apply the Lipschitz condition to k_n:

$$x(t) - x_n(t) \leq -\frac{1}{2}\delta_n + \int_{\tau_1}^t L_n[x(s) - x_n(s)]\,ds \qquad (3.1.25)$$

for $t \in [\tau_1, t_1]$, where $L_n > 0$.

The inequality (3.1.25) is of Gronwall type, and leads to

$$x(t) - x_n(t) \leq -\tfrac{1}{2}\delta_n \exp\{L_n(t - \tau_1)\},$$

which contradicts the inequality $x(t) - x_n(t) \geq 0$ on $[\tau_1, t_1]$.

Consequently, $\tau < \delta'$ is not acceptable, and (3.1.23) is thus established for every n.

We have already noticed above that $x_n(t)$, $n = 1, 2, \ldots$, are uniformly bounded on $[0, \delta']$: $|x_n(t)| \leq r$. From (3.1.21) we derive

$$|x_n(t) - x_n(u)| \leq |f(t) - f(u)| + M'|t - u| + \int_0^u |k_n(t, s, x_n(s)) - k_n(u, s, x_n(s))|\,ds$$

$$(3.1.26)$$

on $[0, \delta']$, which proves that the sequence $\{x_n(t)\}$, $n = 1, 2, \ldots$, is equicontinuous on $[0, \delta']$. Therefore, the sequence $\{x_n(t)\}$ contains a subsequence, say $\{x_{n_k}(t)\}$, $k = 1, 2, \ldots$, such that

$$\lim_{k \to \infty} x_{n_k}(t) = x_M(t), \quad t \in [0, \delta'], \qquad (3.1.27)$$

uniformly on $[0, \delta']$. Of course, we can assume that the sequence $\{x_n(t)\}$ itself is uniformly convergent on $[0, \delta']$ to the (continuous) function $x_M(t)$. From (3.1.23) we then obtain

$$x(t) \leq x_M(t), \quad t \in [0, \delta']. \qquad (3.1.28)$$

But $x_M(t)$ is a solution of (3.1.1), as we can find out immediately from (3.1.21), letting $n \to \infty$ and taking (3.1.4) into account. Consequently, any solution of the inequality (3.1.17) satisfies (3.1.28). In particular, any solution of (3.1.1) satisfies (3.1.28), which means that $x_M(t)$ is the *maximal solution* of (3.1.1) on $[0, \delta']$.

The maximal solution is unique, as follows easily from the definition. Indeed, if $\bar{x}_M(t)$ is another maximal solution of (3.1.1) on $[0, \delta']$, then as seen above $\bar{x}_M(t) \leq x_M(t)$. On the other hand, if $\bar{x}_M(t)$ is maximal, one must have $x_M(t) \leq \bar{x}_M(t)$, which shows that $x_M(t) \equiv \bar{x}_M(t)$.

We now summarize the above discussion in the following result:

Theorem 3.1.2. *Let us consider equation (3.1.1) under the same conditions as in Theorem 3.1.1. Moreover, we assume $k(t, s, x)$ to be monotonically nondecreasing in x. Then, there exists a (unique) maximal solution $x_M(t)$ of (3.1.1), and any continuous solution of (3.1.17) on $[0, \delta'] \subset [0, \delta]$ satisfies (3.1.28).*

Remark 1.   The maximal solution can be defined on each $[0, \delta'] \subset [0, \delta]$, as seen above. Therefore, we can say it is defined on $[0, \delta]$. But from

$$|x_M(t) - x_M(u)| \leq |f(t) - f(u)| + M'|t - u| + \int_0^u |k(t, s, x_M(s)) - k(u, s, x_M(s))|\, ds,$$

one sees that $\lim x_M(t)$ does exist as $t \to \delta$. Hence, $x_M(t)$ is actually defined on $[0, \delta]$.

Of course, we have to keep in mind that two solutions which are maximal, must coincide on their common interval of existence.

Remark 2.   By using completely similar arguments to those used above for the maximal solution, one can show that, under the same monotonicity condition for $k(t, s, x)$, there exists a *minimal solution* for (3.1.1), say $x_m(t)$, such that any continuous solution of the integral inequality

$$x(t) \geq f(t) + \int_0^t k(t, s, x(s))\, ds, \tag{3.1.29}$$

on any interval $[0, \delta'] \subset [0, \delta]$, satisfies

$$x(t) \geq x_m(t). \tag{3.1.30}$$

It is interesting to point out that uniqueness for (3.1.1) is equivalent to $x_M(t) \equiv x_m(t)$. The comparison technique, which will be sketched in this section, can be applied in obtaining uniqueness results.

In order to discuss the existence problem for the equation (3.1.1) in a *global* framework, we shall assume that the following conditions are satisfied:

(1) $f \in C([0, a], R)$, with $0 < a \leq \infty$;
(2) $k \in C(\Delta_a \times R, R)$, where

$$\Delta_a = \{(t, s); 0 \leq s \leq t < a\}. \tag{3.1.31}$$

It is easily seen that conditions (1), (2) above guarantee the existence of a solution $x(t)$ to (3.1.1), defined in some interval $[0, \delta] \subset [0, a)$.

Indeed, let us fix a', $0 < a' < a$, and choose $r > 0$ such that $r = \sup |f(t)|$, $t \in [0, a']$. Then $k(t, s, x)$ is continuous on $\bar{\Delta}_{a'} \times [-r, r]$, and the conditions of Theorem 3.1.1 are verified. Hence, there exists $\delta > 0$, $\delta \leq a'$ and a continuous solution of (3.1.1) defined on $[0, \delta]$.

Moreover, if we are given a continuous solution $x(t)$ of (3.1.1), on a closed interval $[0, \delta]$, $\delta < a$, then under the above stated assumptions (1) and (2) for (3.1.1) we can *extend* $x(t)$ to a larger interval $[0, \delta']$, with $\delta' > \delta$.

Indeed, we can rewrite equation (3.1.1) in the form

$$x(t) = \bar{f}(t) + \int_\delta^t k(t, s, x(s))\, ds, \quad t > \delta, \tag{3.1.32}$$

where

$$\bar{f}(t) = f(t) + \int_0^\delta k(t, s, x(s)) \, ds, \quad t \in [0, a]. \tag{3.1.33}$$

Since $x(t)$ is known on $[0, \delta]$, (3.1.33) shows that $\bar{f}(t)$ is known. If we can show the existence of a solution to (3.1.32) in an interval $[\delta, \delta']$, $\delta' > \delta$, say $\bar{x}(t)$, then

$$X(t) = \begin{cases} x(t), & t \in [0, \delta], \\ \bar{x}(t), & t \in [\delta, \delta'], \end{cases}$$

constitutes an extension (continuation) of $x(t)$. $X(t)$ is continuous on $[0, \delta']$ because

$$x(\delta) = \bar{x}(\delta) = f(\delta) + \int_0^\delta k(\delta, s, x(s)) \, ds. \tag{3.1.34}$$

The existence of $\bar{x}(t)$ on $[\delta, \delta']$ can be proven immediately based on Theorem 3.1.1, or on the variant of that theorem which was mentioned after the second proof of Theorem 3.1.1.

Indeed, let us consider a number $a' < a$, and notice that $k(t, s, x)$ is continuous for $\delta \leq s \leq t \leq a'$, $|x - \bar{f}(t)| \leq r$, where $r > 0$ is a positive number. Let $M = \sup |k(t, s, x)|$ in the above set. Then a solution $\bar{x}(t)$ of (3.1.32) does exist on $[\delta, \delta']$, with $\delta' = \min\{a' - \delta, rM^{-1}\}$.

Let us now consider a solution $x(t)$ of (3.1.1) which is defined on a semi-open interval $[0, a')$, $a' \leq a$. If $a' = a$, then $x(t)$ is *maximally defined*, i.e., under conditions (1), (2) formulated above we cannot extend $x(t)$ to an interval which is larger than $[0, a)$. When $a' < a$, then two distinct situations can occur: first, the graph of $x(t)$ on $[0, a')$ does belong to some compact subset of the strip $[0, a') \times R$; second, the graph of $x(t)$ on $[0, a')$ does not belong to any compact subset of the strip $[0, a')$.

In the first case, we shall prove that the solution $x(t)$ is *continuable* to an interval which is larger than $[0, a')$. In fact, it is enough to show that $x(t)$ can be extended to become continuous on the closed interval $[0, a']$, and then apply the remark made above regarding the continuability of solutions of (3.1.1) which are defined on a closed interval $[0, \delta]$.

Indeed, if we take into account that the graph of $x(t)$ belongs to a compact set, we obtain

$$|x(t) - x(u)| \leq |f(t) - f(u)| + \left| \int_u^t k(t, s, x(s)) \, ds \right|$$

$$+ \left| \int_0^u [k(t, s, x(s)) - k(u, s, x(s))] \, ds \right|$$

$$\leq |f(t) - f(u)| + M|t - u| + \int_0^{a'} |k(t, s, x(s)) - k(u, s, x(s))| \, ds \tag{3.1.35}$$

for $t, u \in [0, a')$. Cauchy's criterion leads to the conclusion that

$$\lim x(t) \text{ exists as } t \to a' -. \tag{3.1.36}$$

If we let $x(a') = \lim x(t)$ as $t \to a' -$, then we realize that $x(t)$ is continuous on $[0, a']$, and from (3.1.1) we obtain immediately that it is a solution on $[0, a']$.

The second case emphasized in the above discussion of the solutions to (3.1.1) defined on semi-open intervals generates a solution which is *maximally defined*, under our assumptions (1) and (2).

Indeed, if we assume that $x(t)$ can be continued to a solution defined on $[0, a'')$, $a'' > a'$, then the graph of $x(t)$ on $[0, a']$ is itself a compact subset in $[0, a'] \times R$. Hence, the graph of $x(t)$ on $[0, a'] \times R$ belongs to a compact subset, which contradicts the property characterizing the second case, namely, that the graph should not belong to any compact subset of $[0, a) \times R$.

The discussion constructed above illustrates the fact that the solutions of (3.1.1), in $[0, a] \times R$, can be either *continuable* (unsaturated), or *maximally defined* (saturated).

Of course, the discussion does not guarantee the existence of *saturated* solutions. This problem constitutes the objective of the next theorem.

Theorem 3.1.3. *Consider the integral equation (3.1.1), under the continuity assumptions for f in $[0, a)$, $0 < a \leq \infty$, and for k in $\Delta_a \times R$.*

Then any continuable solution of (3.1.1) can be extended up to a maximally defined solution of the same equation.

Proof. As seen above, a solution of (3.1.1), defined in $[0, a')$, is continuable if and only if its graph belongs to a compact subset of $[0, a) \times R$. Therefore, what we need to prove is the fact that any continuable solution of (3.1.1) can be continued until the graph of the extended solution cannot be contained by any compact subset of $[0, a) \times R$.

Let us notice first that $[0, a) \times R$ can be represented as the union of the compact subsets $K_n = [0, a - 1/n] \times [-n, n]$, $n = 1, 2, 3, \ldots$, when a is finite, or as the union of the compact subsets $K_n = [0, n] \times [-n, n]$, $n = 1, 2, \ldots$, when $a = +\infty$. For any compact $K \subset [0, a) \times R$, there is a natural number m such that $K \subset K_m$. Therefore, it is sufficient to restrict our considerations to the compact sets K_m, $m = 1, 2, \ldots$.

We shall now prove that each solution of (3.1.1), say $x(t)$, $0 \leq t < \delta$, whose graph belongs to some K_n, can be continued until its graph leaves K_n. Indeed, since $K_n \subset K_{n+1}$, we can find positive numbers α and β, such that for each $(\tau, \zeta) \in K_n$, the set $\{(t, x); \tau \leq t \leq \tau + \alpha, |x - \zeta| \leq \beta\}$ belongs to K_{n+1}. Notice that $|k(t, s, x)|$ is bounded by some $M_n > 0$, when $(t, x) \in K_{n+1}$ and $0 \leq s \leq t$, because (t, s, x) belongs to a compact set. As seen above in this section, we can assume

$x(t)$ is defined on $[0, \delta]$, because of the compactness of K_n. Let $\tau = \delta$ and $\xi = x(\tau) = f(\tau) + \int_0^\tau k(\tau, s, x(s)) \, ds$. Since we want to prove the existence of a solution to the equation

$$x(t) = \bar{f}(t) + \int_\tau^t k(t, s, x(s)) \, ds, \quad t > \tau, \tag{3.1.37}$$

where $\bar{f}(t)$ is given by

$$\bar{f}(t) = f(t) + \int_0^\tau k(t, s, x(s)) \, ds, \tag{3.1.38}$$

it is sufficient to show (see the variant of Theorem 3.1.1) that $k(t, s, x)$ is continuous on a set of the form

$$\tau \leq s \leq t \leq \tau + \bar{\alpha}, \quad |x - \bar{f}(t)| \leq \beta, \tag{3.1.39}$$

with $\bar{\alpha} \leq \alpha$, and such that it does not depend on (τ, ξ). If we take into account that $\bar{f}(\tau) = x(\tau) = \xi$, then we obtain

$$|x - \xi| \leq |x - \bar{f}(t)| + |\bar{f}(t) - \bar{f}(\tau)| \leq \beta + \beta = 2\beta,$$

provided $|\bar{f}(t) - \bar{f}(\tau)| \leq \beta$, i.e., $\tau \leq t \leq \tau + \alpha'$, and x verifies (3.1.39). If we let $\bar{\alpha} = \min\{\alpha, \alpha'\}$, then it is obvious that the set defined by $\tau \leq t \leq \tau + \bar{\alpha}, |x - \bar{f}(t)| \leq \beta$ belongs to K_{n+1}. Therefore, if we let $\bar{\delta} = \min\{\bar{\alpha}, \beta M_n^{-1}\}$, we can assert the existence of a solution $\bar{x}(t)$ to (3.1.37), defined on $\tau \leq t \leq \tau + \bar{\delta}$. Then defining $X(t)$ by $X(t) = x(t), 0 \leq t \leq \tau = \delta, X(t) = \bar{x}(t), \delta \leq t \leq \delta + \bar{\delta}$, it is obvious we have obtained a continuation of $x(t)$.

At this point, it is important to notice that $\bar{\delta}$ is independent of (τ, ξ), and depends only on K_n. Indeed, if we show that α' can be chosen independently of (τ, ξ), this will obviously suffice to obtain $\bar{\delta}$ independent of $(\tau, \xi) \in K_n$. But α' has been determined from: $\tau \leq t \leq \tau + \alpha'$ implies $|\bar{f}(t) - \bar{f}(\tau)| \leq \beta$. Since $\bar{f}(t)$ is uniformly continuous on any $[0, a'] \subset [0, a)$, the independence of α' with respect to (τ, ξ) is proven.

Going back to the extended solution $X(t)$, we have to distinguish two cases: first, when the graph of this solution leaves K_n (but will remain in K_{n+1}!), and second, when the graph of $X(t)$ is still contained in K_n. In the second case, we can again extend $X(t)$, and therefore $x(t)$, such that the new solution is defined in $[0, \delta + 2\bar{\delta}]$. Of course, after a finite number of steps we have to leave K_n, because in K_n t remains finite, while $\delta + m\bar{\delta} \to \infty$ as $m \to \infty$.

In both cases emphasized above, we can extend $x(t)$, whose graph is contained in K_n, to a solution which has the graph in K_{n+1} (but not in K_n!). The above procedure, repeated indefinitely, will lead to a solution whose graph leaves any K_n, and therefore it must be a saturated solution of (3.1.1) which is defined on some interval $[0, a'), a' \leq a$.

Remark 1. If $x(t)$ is a saturated solution of (3.1.1) defined on $[0, a')$, with $a' < a$, then $x(t)$ must be unbounded in any left neighborhood of a': $a' - \varepsilon \leq t < a'$, $\varepsilon > 0$.

Indeed, if $x(t)$ would be bounded on $[a' - \varepsilon, a')$, then its graph would belong to some compact part of $[0, a) \times R$.

Of course, this fact can be expressed as

$$\lim \sup |x(t)| = \infty \text{ as } t \to a' -. \tag{3.1.40}$$

In this case, we say that $x(t)$ has a finite escape time at a'.

Remark 2. It is interesting to notice that we can continue the maximal solution so that it remains maximal, provided we assume the monotonicity of $k(t, s, x)$ in x. More precisely, in Theorem 3.1.2 we have shown the *local existence* of the maximal solution, under the assumption that $k(t, s, x)$ is monotonically non-decreasing in x. If $x_M(t)$ denotes the maximal solution of the equation (3.1.1) on the interval $[0, \delta]$, then it can be extended to some larger interval $[0, \delta_1]$, $\delta_1 > \delta$, by choosing as extension the maximal solution of (3.1.1) on $[\delta, \delta_1]$. This procedure raises the question whether the extended solution is the maximal solution of (3.1.1) on $[0, \delta_1]$.

Let $x_M(t)$ be the solution defined on $[0, \delta_1]$, obtained by means of the above procedure. If we assume that $\bar{x}_M(t)$ is the maximal solution of (3.1.1) on $[0, \delta_1]$, then $x_M(t) \equiv \bar{x}_M(t)$ on $[0, \delta]$ as an immediate consequence of the definition of the maximal solution. On $[\delta, \delta_1]$ one must have $\bar{x}_M(t) \leq x_M(t)$ because $x_M(t)$ is the maximal solution of the equation

$$x(t) = \bar{f}(t) + \int_\delta^t k(t, s, x(s)) \, ds, \tag{3.1.41}$$

with

$$\bar{f}(t) = f(t) + \int_0^\delta k(t, s, x_M(s)) \, ds, \tag{3.1.42}$$

and $\bar{x}_M(t)$ is a solution of (3.1.41) on $[\delta, \delta_1]$. On the other hand, if we assume $\bar{x}_M(t_1) < x_M(t_1)$ at some $t_1 > \delta$, then we contradict the definition of the maximal solution on $[0, \delta_1]$. Hence, $x_M(t) \equiv \bar{x}_M(t)$ on $[0, \delta_1]$, and the extension method based on taking the maximal solution at each step leads to the maximal solution which is maximally defined (saturated).

The results already obtained in this section enable us to establish a theorem of existence for the Volterra equation (3.1.1) in the vector case, and simultaneously find a connection between vector equations and scalar equations. Such results are usually called comparison results. See, for instance, Lakshmikantham and Leela [1], Miller [1], Martinjuk and Gutowski [1].

Theorem 3.1.4. *Consider the equation* (3.1.1) *under the following assumptions*:

(1) $f \in C([0, a), R^n), 0 < a \leq \infty$;
(2) $k \in C(\Delta_a \times R^n, R^n)$, *where* Δ_a *is defined by* (3.1.31);
(3) *there exists* $K \in C(\Delta_a \times R, R)$, *such that*

$$|k(t, s, x)| \leq K(t, s, |x|) \text{ on } \Delta_a \times R^n, \tag{3.1.43}$$

with $K(t, s, r)$ *monotonically nondecreasing in* r.

Then any saturated solution of (3.1.1) *does exist on the interval* $[0, a')$, *on which the (saturated) maximal solution of the comparison equation*

$$X(t) = F(t) + \int_0^t K(t, s, X(s)) \, ds \tag{3.1.44}$$

is defined, with $F(t) = |f(t)|$. *In particular, if the comparison equation* (3.1.44) *has its maximal solution defined on* $[0, a)$, *the saturated solutions of* (3.1.1) *are all defined on the same interval.*

Remark. Before we provide the proof of Theorem 3.1.4, let us comment on condition (3). This condition does not represent a severe restriction on the function k. Indeed, one can define

$$K(t, s, r) = \sup|k(t, s, x)|, \quad |x| \leq r, r > 0 \tag{3.1.45}$$

for any fixed $(t, s) \in \Delta_a$. This function may not be continuous, but it is nondecreasing in r. A smoothing procedure could be applied in order to replace K given by (3.1.45) by a continuous function which dominates K. Of course, such a procedure might reduce the interval of existence for solutions.

Proof of Theorem 3.1.4. Let $x(t)$ be a saturated solution of the equation (3.1.1). As we know already, this solution is defined on a certain interval $[0, a'')$, $a'' \leq a$. This fact was established in Theorem 3.1.3. Moreover, it was noticed in Remark 1 to that theorem that, if $a'' < a$, one has

$$\limsup|x(t)| = \infty \quad \text{as } t \to a''. \tag{3.1.46}$$

If we denote $Y(t) = |x(t)|$, $t \in [0, a'')$, then we obtain from (3.1.1)

$$Y(t) \leq F(t) + \int_0^t K(t, s, Y(s)) \, ds \tag{3.1.47}$$

on the whole interval $[0, a'')$.

On the other hand, the maximal solution of equation (3.1.44), say $X_M(t)$, does exist on some interval $[0, a')$, $a' \leq a$, as seen in Remark 2 to Theorem 3.1.3. The key point of the proof consists in showing that $a'' \geq a'$.

Assume, on the contrary, that $a'' < a'$. In such a case, the integral inequality (3.1.47) which is valid on $[0, a'')$, implies

$$|x(t)| = Y(t) \leq X_M(t), \quad t \in [0, a''). \tag{3.1.48}$$

Indeed, this inequality holds locally, i.e., in an interval $[0, \delta]$, $\delta < a''$. This was shown in Theorem 3.1.2. The local theory is going to help us in the present global case. Let $\tau = \sup\{t; 0 < t < a'', Y(t) \leq X_M(s), 0 \leq s \leq t\}$. If $\tau < a''$, then any right neighborhood of τ must contain points t such that $Y(t) > X_M(t)$. On the other hand, inequality (3.1.47) leads to

$$Y(t) \leq F(t) + \int_0^\tau K(t, s, Y(s)) \, ds + \int_\tau^t K(t, s, Y(s)) \, ds$$

$$\leq F(t) + \int_0^\tau K(t, s, X_M(s)) \, ds + \int_\tau^t K(t, s, Y(s)) \, ds$$

on $[\tau, a'')$, or

$$Y(t) \leq \bar{F}(t) + \int_\tau^t K(t, s, Y(s)) \, ds, \tag{3.1.49}$$

with

$$\bar{F}(t) = F(t) + \int_0^\tau K(t, s, X_M(s)) \, ds.$$

Since (3.1.44) can be rewritten for $X_M(t)$ in the form

$$X_M(t) = \bar{F}(t) + \int_\tau^t K(t, s, X_M(s)) \, ds, \tag{3.1.50}$$

Theorem 3.1.2 applies to (3.1.49) and (3.1.50): $Y(t) \leq X_M(t)$ on $[\tau, a'')$. This contradicts the definition of τ. Hence, we must have $\tau = a''$, which shows that (3.1.48) holds.

Taking (3.1.46) into account, (3.1.48) leads to a contradiction. Consequently, our hypothesis $a'' < a'$ cannot be accepted, and we must always have $a' \leq a''$.

This ends the proof of Theorem 3.1.4.

Remark 1. An immediate application of Theorem 3.1.4 can be obtained if we assume the following growth condition on $k(t, s, x)$:

$$|k(t, s, x)| \leq \lambda(t)\rho(|x|) \text{ in } \Delta_a \times R^n, \tag{3.1.51}$$

with $\lambda(t)$ continuous and positive on $[0, a)$, and $\rho(r) > 0$ continuous, nondecreasing on R_+, and such that

$$\int^\infty \frac{dr}{\rho(r)} = \infty. \tag{3.1.52}$$

Under conditions (3.1.51) and (3.1.52), the comparison equation (3.1.44) is

$$X(t) = F(t) + \lambda(t) \int_0^t \rho(X(t)) \, ds, \qquad (3.1.53)$$

where we let $X(t)$ be the maximal solution.

In order to estimate the interval of existence for $X(t)$, we shall reduce (3.1.53) to an ordinary differential equation. Let us denote

$$Y(t) = \int_0^t \rho(X(s)) \, ds, \qquad (3.1.54)$$

for those $t \geq 0$ for which $X(t)$ is defined. We obtain for $Y(t)$ the following ordinary differential equation:

$$Y'(t) = \rho(F(t) + \lambda(t) Y(t)), \qquad (3.1.55)$$

and notice that $X(t)$ and $Y(t)$ either have a common escape time $t_1 < a$, or are both defined on $[0, a)$. This fact follows from $X(t) = F(t) + \lambda(t) Y(t)$, if we take into account that $\lambda(t)$ is continuous and positive on $[0, a)$.

Therefore, it suffices to show that (3.1.55) cannot have a finite escape time $t_1 < a$. Indeed, if we assume $t_1 < a$ is an escape time for a solution of (3.1.55), then on $[0, t_1]$ we have $Y'(t) \leq \rho(F_0 + \lambda_0 y(t))$, with $F_0 = \sup F(t)$, $\lambda_0 = \sup \lambda(t)$ on $[0, t_1]$. Hence, for $t \in [t_1 - \varepsilon, t_1]$, where $\varepsilon > 0$ is fixed, one has

$$\int_{t_1 - \varepsilon}^t \frac{\lambda_0 Y'(s)}{\rho(F_0 + \lambda_0 Y(s))} \, ds \leq \lambda_0(t - t_1 + \varepsilon) \leq \lambda_0 \varepsilon$$

or

$$\int_{F_0 + \lambda_0 Y(t_1 - \varepsilon)}^{F_0 + \lambda_0 Y(t)} \frac{du}{\rho(u)} \leq \lambda_0 \varepsilon.$$

Since $Y(t) \to \infty$ as $t \to t_1$, the above inequality contradicts condition (3.1.52).

Therefore, under conditions (3.1.51) and (3.1.52), all (saturated) solutions of (3.1.1) are defined on $[0, a)$.

The reader who is familiar with the theory of ordinary differential equations will recognize above a direct generalization of Wintner's criterion of global existence.

In particular, conditions (3.1.51) and (3.1.52) are obviously satisfied in the case of linear (vector or scalar) equations of the form

$$x(t) = f(t) + \int_0^t k(t, s) x(s) \, ds.$$

One has $\rho(r) \equiv r$, while $\lambda(t)$ can be chosen as $\sup|k(t, s)|$ for $0 \leq s \leq t$. Of course, this result is known from Section 1.3.

Remark 2. Theorem 3.1.4 is only one example of what is usually called the *comparison technique*. Without trying to achieve now the greatest generality in this respect, we shall provide one more example in which as a 'measure' for the solution $x(t)$ we use a certain (Liapunov) function $V(t, x(t))$.

Assume first that $V(t, x)$ is continuous on $\Delta_a \times R^n$, with values in R, and is of class $C^{(1)}$ with respect to x. Moreover, we assume that the gradient of V is uniformly bounded on $\Delta_a \times R^n$.

Since $V(t, x(t)) = V(t, f(t) + \int_0^t k(t, s, x(s)) \, ds)$, we can obviously write

$$V(t, x(t)) = V(t, f(t)) + \nabla_x \tilde{V} \cdot \int_0^t k(t, s, x(s)) \, ds, \qquad (3.1.56)$$

where the tilde on V indicates the fact that the gradient is taken at an intermediate value (in between $f(t)$ and $x(t)$). From (3.1.56), because of the boundedness of $\nabla_x V$, we obtain the inequality

$$V(t, x(t)) \leq V(t, f(t)) + M \int_0^t |k(t, s, x(s))| \, ds \qquad (3.1.57)$$

Let us now assume that k satisfies an estimate of the form

$$|k(t, s, x)| \leq K(t, s, |x|) \qquad (3.1.58)$$

in the domain $\Delta_a \times R^n$, with $K(t, s, r)$ as in Theorem 3.1.4. Then (3.1.57) and (3.1.58) yield

$$V(t, x(t)) \leq V(t, f(t)) + M \int_0^t K(t, s, |x(s)|) \, ds. \qquad (3.1.59)$$

Of course, (3.1.59) is valid on the whole interval of existence of the solution $x(t)$ – assumed to be saturated.

In order to obtain an adequate comparison equation, we shall further assume that $V(t, x)$ satisfies a lower estimate of the form

$$\mu(|x|) \leq V(t, x), \quad (t, x) \in \Delta_a \times R^n, \qquad (3.1.60)$$

where $\mu: R_+ \to R_+$ is continuous, strictly increasing, $\mu(0) = 0$, and $\mu(r) \to \infty$ as $r \to \infty$.

Then (3.1.59) and (3.1.60) lead to the integral inequality

$$V(t, x(t)) \leq V_0(t) + M \int_0^t K(t, s, \mu^{-1}(V(s, x(s)))) \, ds, \qquad (3.1.61)$$

in which $V_0(t) = V(t, f(t))$ is known.

The conclusion is the same as in Theorem 3.1.4. Namely, any maximally defined solution of (3.1.1) does exist on the interval $[0, a')$, $a' \leq a$, on which the maximal solution of

$$v(t) = V_0(t) + M \int_0^t K(t, s, \mu^{-1}(v(s))) \, ds \qquad (3.1.62)$$

is defined.

In particular, the requirements formulated above for $V(t, x)$ are verified by $V(t, x) = |x|$, except for the existence of the gradient at $x = 0$. But this lack of smoothness for V at $x = 0$ does not really matter because a Lipschitz condition is obviously verified. Of course, this choice for $V(t, x)$ corresponds to Theorem 3.1.4.

If instead of the boundedness of $\nabla_x V$ we assume only the Lipshitz type condition for $V(t, x)$, with respect to the variable x, then (3.1.57) is still valid, and the comparison equation can be dealt with as above.

A more general situation occurs when we allow the gradient of V to be unbounded, but a certain upper estimate is imposed. In this case, the comparison equation will have the form

$$V(t, x(t)) \leq V(t, f(t)) + \int_0^t \tilde{k}(t, s, V(s, x(s))) \, ds, \qquad (3.1.63)$$

where $\tilde{k}(t, s, v)$ is such that

$$|\nabla_x \tilde{V} \cdot k(t, s, x(s))| \leq \tilde{k}(t, s, V(s, x(s))), \qquad (3.1.64)$$

with an estimate of the form (3.1.60) holding.

The last basic problem we want to discuss in this section is that of *continuous dependence* of solutions with respect to the data. More precisely, we are interested in estimating the change in the solution of equation (3.1.1), when $f(t)$ and $k(t, s, x)$ are allowed to change. As one might expect, under adequate assumptions the answer is positive.

First, we shall limit our considerations to the case in which $k(t, s, x)$ is locally Lipschitz in x. As we know, this assumption implies uniqueness.

Besides equation (3.1.1), we consider now the 'perturbed' equation

$$y(t) = f(t) + h(t) + \int_0^t [k(t, s, y(s)) + r(t, s, y(s))] \, ds, \qquad (3.1.65)$$

in which $h(t)$ and $r(t, s, x)$ are regarded as 'small' perturbations of $f(t)$ and $k(t, s, x)$, respectively. We assume that f and h are continuous maps from $[0, a)$ into R, while k and r are continuous from $\Delta_a \times R$ into R, with $k(t, s, x)$ satisfying a local Lipschitz condition in $\Delta_a \times R$ (i.e., in each compact part of $\Delta_a \times R$).

Let $x(t)$ be the (unique) solution of equation (3.1.1). We assume it is maximally defined, and its interval of existence is $[0, a')$, $a' \leq a$. Let $y(t)$ be any solution of the equation (3.1.65), also maximally defined. Then, on the common interval of existence of $x(t)$ and $y(t)$ we have

$$y(t) - x(t) = h(t) + \int_0^t [k(t, s, y(s)) - k(t, s, x(s))] \, ds + \int_0^t r(t, s, y(s)) \, ds,$$

which obviously implies

$$|y(t) - x(t)| \le |h(t)| + L \int_0^t |y(s) - x(s)| \, ds + \int_0^t |r(t, s, y(s))| \, ds. \qquad (3.1.66)$$

The way the Lipschitz constant L is chosen in (3.1.66) will be made precise immediately. Namely, let T be a fixed number $0 < T < a'$. It can be chosen as close as we want to a' (and if $a' = a = \infty$, then T is any positive number). For an aribtrary $\varepsilon > 0$, let $K \subset \varDelta_a$ be the compact set defined by $K = \{(t, x);$ $0 \le t \le T, |x - x(t)| \le \varepsilon\}$.

What we want to prove now is the fact that the (saturated) solution $y(t)$ of (3.1.65) has its graph in K, for $0 \le t \le T$, provided h and r are small enough. Since the graph of a saturated solution must leave any compact set in $\varDelta_a$, it will be shown that the graph of $y(t)$ leaves K through a boundary point $(T, y(T))$. This will imply that any maximally defined solution $y(t)$ of (3.1.65) is defined at least on the interval $[0, T]$, and remains close to $x(t)$ on that interval, as soon as h and r are small enough.

Let us now assume

$$|h(t)| \le \delta, \quad t \in [0, T], \qquad (3.1.67)$$

and

$$|r(t, s, x)| \le \delta, \quad (t, x) \in K, 0 \le s \le t. \qquad (3.1.68)$$

Then (3.1.66) yields the estimate

$$|y(t) - x(t)| \le \delta(1 + T) + L \int_0^t |y(s) - x(s)| \, ds, \qquad (3.1.69)$$

for those $t > 0$ for which the graph of y remains in K. Let us point out that for small enough δ, more precisely for $\delta < \varepsilon$, this graph starts in K because $|y(0) - x(0)| = |h(0)| \le \delta$, and therefore remains there for sufficiently small t.

But (3.1.69) is a Gronwall type inequality, and we obtain

$$|y(t) - x(t)| \le \delta(1 + T)e^{LT} \qquad (3.1.70)$$

for all $t > 0$ such that $(t, y(t)) \in K$. If δ is chosen small enough, such that

$$\delta < \varepsilon(1 + T)^{-1}e^{-LT} = \delta(\varepsilon, T), \qquad (3.1.71)$$

then (3.1.70) shows that $(t, y(t)) \in K$ for $t \in [0, T]$. In other words,

$$|y(t) - x(t)| < \varepsilon, \quad t \in [0, T], \qquad (3.1.72)$$

provided h and r satisfy (3.1.67) and (3.1.68), where $\delta = \delta(\varepsilon, T)$ is given by (3.1.71).

In summarizing the above discussion, we can state the following result on the continuous dependence of solutions of integral equations of the form (3.1.1), with respect to the functions f and k.

Proposition 3.1.5. *Consider the integral equation* (3.1.1), *with* $f \in C([0, a), R)$, $k \in C(\Delta_a \times R, R)$ *and such that k is locally Lipschitz with respect to x in $\Delta_a \times R$. Let $x(t)$ be the unique saturated solution of* (3.1.1), *defined on* $[0, a')$, $a' \leq a$, *and let T be a positive number with $T < a'$. Then for every $\varepsilon > 0$, there exists $\delta = \delta(\varepsilon, T) > 0$, such that any solution $y(t)$ of the 'perturbed' equation* (3.1.65) *satisfies the estimate* (3.1.72), *provided $h \in C([0, a), R)$, and $r \in C(\Delta_a \times R, R)$ satisfy* (3.1.67) *and* (3.1.68).

The above result shows that perturbing an equation with kernel of Lipschitz type changes its solution continuously with respect to perturbations, even though the uniqueness property might be lost for the perturbed equations.

Now we have to consider the more general case, when equation (3.1.1) itself does not possess uniqueness.

Besides equation (3.1.1), we will consider the auxiliary integral equations

$$x_n(t) = f_n(t) + \int_0^t k_n(t, s, x_n(s))\, ds, \qquad (3.1.73)$$

where the following property holds:

$$f_n(t) \to f(t), \quad k_n(t, s, x) \to k(t, s, x) \text{ as } n \to \infty, \qquad (3.1.74)$$

the type of convergence being that of uniform convergence, or another kind of convergence which should be precisely defined. The sequences $\{f_n(t)\}$ and $\{k_n(t, s, x)\}$ appearing in (3.1.74) are just any sequences of continuous functions satisfying that condition. If $x_n(t)$ denotes any solution of (3.1.73), one can prove that the sequence $\{x_n(t)\}$, on some interval $[0, \delta]$, $\delta > 0$, is compact (i.e., the sequence is uniformly bounded and equicontinuous on $[0, \delta]$). Hence, from $\{x_n(t)\}$ one can extract a subsequence which converges uniformly on $[0, \delta]$ to some continuous function $x(t)$. It is easily seen from (3.1.73) that $x(t)$ is a solution of (3.1.1). The case when $k(t, s, x)$ is Lipschitz continuous in x was clarified in Proposition 3.1.5. But, when $k(t, s, x)$ is merely continuous in $\bar{D}$, all we can say is that $x(t)$ is *one* of the solutions of (3.1.1). This kind of result, very likely the best under the solely continuity assumption on $k(t, s, x)$, does not appear to be very profound. It can be shown that the process described above leads to *any* solution of (3.1.1), provided we choose conveniently the data (the following construction is due to O. Staffans). For instance, if $x(t)$ is a solution of (3.1.1), then we consider an arbitrary sequence $\{x_n(t)\}$ which converges uniformly to $x(t)$ on some interval $[0, \delta]$. Then let $k_n(t, s, x)$ be such that the second condition (3.1.74) holds, uniformly on a sufficiently large compact set containing the graph of the solution $x(t)$ on $[0, \delta]$. Of course, we can assume each $k_n(t, s, x)$ is Lipschitz in x. Finally, let us define $f_n(t)$ by means of (3.1.73), in terms of $x_n(t)$ and $k_n(t, s, x)$. It is now obvious that the sequence of equations (3.1.73) leads to equation (3.1.1) by letting $n \to \infty$, and precisely to the preassigned solution $x(t)$ of this equation. From the discussion conducted above we can see that the continuous dependence of the solution, with respect to k and f, does not have the same useful meaning as in the case of uniqueness for (3.1.1).

A complimentary discussion to that conducted above for continuous data could be envisaged when the convergence

$$x_n(t) \to x(t) \text{ as } n \to \infty, \tag{3.1.75}$$

is not meant in the uniform sense.

We shall now consider a rather special situation, when equation (3.1.1) has the form

$$x(t) = x_0 + \int_0^t g(x(s)) \, ds, \tag{3.1.76}$$

with $g(x) > 0$ a continuous function on some interval $[x_0, X] \subset R$. Obviously, (3.1.76) is equivalent to $x' = g(x)$, $x(0) = x_0$; in other words, we are dealing with the case of an autonomous differential equation.

Since $g(x) > 0$, any solution of (3.1.76) is strictly increasing on some interval $[0, \delta]$, $\delta > 0$. One can easily establish a connection between X and δ, but this is not important at the moment.

If we denote $M = \sup g(x)$, $x \in [x_0, X]$, then from the uniform continuity of g on $[x_0, X]$ we derive the existence of a $\delta_n > 0$, such that

$$|g(x) - g(y)| < \frac{1}{n} \quad \text{for} \quad |x - y| \le M\delta_n, \tag{3.1.77}$$

on the interval $[0, \delta]$.

For fixed $n \ge 1$, we consider a partition of the interval $[0, \delta]$, say

$$0 = t_0^n < t_1^n < \cdots < t_{j_n}^n = \delta, \tag{3.1.78}$$

such that $\max(t_{i+1}^n - t_i^n) \le \delta_n$, and $t_i^n = i\delta_n$, $i = 1, 2, \ldots, j_n - 1$, where j_n is uniquely determined by the inequalities $(j_n - 1)\delta_n < \delta \le j_n \delta_n$. In case $\delta < j_n \delta_n$, one takes $t_{j_n}^n$ as shown in (3.1.78).

For the fixed solution $x(t)$ of (3.1.76) one denotes $x_i^n = x(t_i^n)$, $i = 0, 1, \ldots, j_n$. Let $\bar{t}_i^n$ be the first number in (t_i^n, t_{i+1}^n) such that

$$x(t_{i+1}^n) - x(t_i^n) = (t_{i+1}^n - t_i^n) x'(\bar{t}_i^n). \tag{3.1.79}$$

This is possible because the derivative $x'(t)$ is a continuous function.

We now introduce the numbers $\bar{x}_i^n$ by means of $\bar{x}_i^n = x(\bar{t}_i^n)$. Therefore, in view of (3.1.79) we can write

$$x_{i+1}^n - x_i^n = (t_{i+1}^n - t_i^n) g(\bar{x}_i^n), \tag{3.1.80}$$

for $i = 0, 1, 2, \ldots, j_n$. Notice that the monotonicity of $x(t)$ implies the inequalities (strict) $x_i^n < \bar{x}_i^n < x_{i+1}^n$, $i = 0, 1, \ldots, j_n$.

We can now construct both $x_n(t)$ and $g_n(x)$, as follows:

$$g_n(x) = g(\bar{x}_i^n), \quad x_i^n \le x < x_{i+1}^n, \tag{3.1.81}$$

and

$$x_n(t) = x_i^n + (t - t_i^n)g(\bar{x}_i^n), \quad t_i^n \le t < t_{i+1}^n. \tag{3.1.82}$$

First, notice that $x_n'(t) = g(\bar{x}_i^n) = g_n(x_n(t))$, because $x_i^n < x_n(t) < x_{i+1}^n$ for $t_i^n < t < t_{i+1}^n$. Therefore, apart from the points t_i^n, $x_n(t)$ is a solution of $x' = g_n(x)$. In any case, we have

$$x_n(t) = x^0 + \int_0^t g_n(x_n(s))\,ds, \quad 0 \le t \le \delta, \tag{3.1.83}$$

and $x_n(t)$ are obviously continuous solutions for (3.1.83).

It remains to be shown that (3.1.75) is true. The fact that $g_n(x) \to g(x)$, uniformly on $[x_0, X]$, is almost obvious. Indeed, let $x \in [x_i^n, x_{i+1}^n)$ for some i. Then $|g_n(x) - g(x)| = |g(\bar{x}_i^n) - g(x)| < n^{-1}$ because $|\bar{x}_i^n - x| < x_{i+1}^n - x_i^n$, and

$$|x_{i+1}^n - x_i^n| = |x(t_{i+1}^n) - x(t_i^n)| = \left| \int_{t_i^n}^{t_{i+1}^n} g(x(s))\,ds \right| \le M\delta_n.$$

On the other hand, we can write for $t \in (t_i^n, t_{i+1}^n]$:

$$x_n(t) - x(t) = x(t_i^n) + (t - t_i^n)g(x(\bar{t}_i^n)) - x(t_i^n)$$

$$- \int_{t_i^n}^t g(x(s))\,ds = (t - t_i^n)[g(x(\bar{t}_i^n)) - g(x(t^*))],$$

with $t^* \in [t_i^n, t]$, which obviously implies

$$|x_n(t) - x(t)| \le Mn^{-2}, \tag{3.1.84}$$

if we also keep in mind that $t_i^n, t^* \in (t_i^n, t_{i+1}^n)$ and $|g_n(x) - g(x)| < n^{-1}$.

Let us point out that the example constructed above (due to N. H. Pavel) shows that *any* solution of (3.1.76) can be obtained as a uniform limit of Euler polygonals. This result seems to be new even if we limit our considerations to ordinary differential equations. The procedure used is inspired by the problem of the existence of solutions on closed sets, and leads to an estimate for the error.

3.2 Abstract Volterra equations and some special cases

In this section, we shall discuss existence and related problems for equations involving Volterra operators of an abstract kind, as defined in Section 2.3. Both functional and functional-differential equations of the form

$$x(t) = (Vx)(t), \quad t \in [0, T), \tag{3.2.1}$$

respectively

$$\dot{x}(t) = (Vx)(t), x(0) = x^0, \quad t \in [0, T), \tag{3.2.2}$$

will be considered, where V denotes a Volterra operator acting on a convenient function space, $x^0 \in R^n$, and x is a map form $[0, T)$, $0 < T \leq \infty$, into R^n. Of course, the case when x is defined only on a subinterval $[0, T') \subset [0, T)$ makes sense, a feature which will allow us to deal with local solutions.

It is obvious that equation (3.2.2), together with the associated initial condition, can be reduced to an equation of the form (3.2.1). Namely, if we have in mind continuous solutions, and V in (3.2.2) is acting from the space of continuous functions on some interval $[0, T)$ into itself, then (3.2.2) can be integrated. The result is

$$x(t) = x^0 + \int_0^t (Vx)(s)\, ds. \tag{3.2.3}$$

If we denote

$$(V_1 x)(t) = x^0 + \int_0^t (Vx)(s)\, ds, \tag{3.2.4}$$

we obtain from (3.2.3)

$$x(t) = (V_1 x)(t), \tag{3.2.5}$$

and since V is by assumption a Volterra type operator, so is V_1. Of course, (3.2.5) leads (by differentiation) to (3.2.2), under the same continuity assumptions. If spaces of measurable functions are considered, the validity of equation (3.2.1) or (3.2.2) has to be understood only a.e. on $[0, T)$.

There are various kinds of hypotheses under which the existence problem for (3.2.1) or (3.2.2) can be discussed. We shall consider first the case when the Volterra operator is acting on the space of continuous functions, and then the case when it is acting on spaces of measurable (locally integrable) functions. It is also interesting to discuss the case when V takes a space of continuous functions into a space of measurable functions. This last case makes particular sense for the functional-differential equation (3.2.2).

Notice that the following problem is still open: under what conditions can we assert that a Volterra type operator V_1 is representable in the form (3.2.4), with V another Volterra operator?

In order to formulate the existence results for abstract Volterra type equations of the form (3.2.1), we need to make the necessary assumptions on the operator

V. We shall consider, first of all, the case when the operator *V* is acting on the space of continuous functions on a given interval $[0, a]$, $a > 0$.

The following hypotheses are basic with regard to the investigation of equation (3.2.1).

(1) $V: C([0, a], R^n) \to C([0, a], R^n)$ is a Volterra type operator;
(2) V is continuous in the topology of the space $C([0, a], R^n)$;
(3) V is compact in the topology of $C([0, a], R^n)$, which means that it takes any bounded set in $C([0, a], R^n)$ into a relatively compact set in the same space.

One may think that these conditions are sufficient for the existence of a solution to equation (3.2.1), especially if we have in mind the conditions required by the Schauder–Tychonoff fixed point theorem. But conditions (1), (2), and (3) do not suffice for the existence of a convex subset $S \subset C([a, b], R^n)$, such that $VS \subset S$. Something has to be added to provide for this first condition in order to secure the existence for (3.2.1).

Actually, if we consider the Volterra integral operator $(Vx)(t) = f(t) + \int_0^t k(t, s, x(s)) \, ds$, as we did in Section 3.1, we notice that $(Vx)(0) = f(0) = \text{const.}$, for every $x \in C([0, a], R^n)$. This fact suggests the following condition, in addition to conditions (1), (2), and (3):

(4) $(Vx)(0) = (Vy)(0)$, for any $x, y \in C([0, a], R^n)$.

The following result is true for equation (3.2.1).

Theorem 3.2.1. *Let V be an operator satisfying the conditions* (1)–(4) *stated above. Then, there exists $\delta > 0$, $\delta \leq a$, such that equation (3.2.1) has at least one solution defined on $[0, \delta]$.*

Proof. We shall use the Schauder–Tychonoff fixed point theorem (see Section 2.4). It can be stated, for our purpose, as follows: 'If V is a continuous operator on a closed convex set $S \in C([0, \delta], R^n)$, such that VS is relatively compact, and $VS \subset S$, then V has at least one fixed point in S.'

Under our assumptions, a convenient δ has to be found, and then a set S has to be constructed in such a manner that the conditions of Schauder–Tychonoff are satisfied.

Since $\delta \leq a$, and instead of the original space $C([0, a], R^n)$ we might have to use $C([0, \delta], R^n)$, we will preserve the same notation V for the operator induced by the initial operator V on the space $C([0, \delta], R^n)$. This is possible because V satisfies condition (1), i.e., for $x(s) = y(s)$, $0 \leq s \leq t$, $(Vx)(t) = (Vy)(t)$. Consequently, we can consider the 'restriction' of the operator V to any space $C([0, \delta], R^n)$, with $\delta < a$.

Let us denote by x^0, $x^0 \in R^n$, the common value of $(Vx)(t)$ at $t = 0$, $x \in C([0, a], R^n)$. Let S be the ball of radius $r > 0$ in $C([0, a], R^n)$, centered at x^0. Since S is a bounded set in $C([0, a], R^n)$, the compactness of the operator V (see condition (3) above) allows us to determine $\delta > 0$, such that $\|(Vx)(t) - x^0\| = \|(Vx)(t) - (Vx)(0)\| < \frac{1}{2}r$ for all $x \in S$, provided $0 \le t \le \delta = \delta(r)$. Therefore, restricting our considerations to the interval $[0, \delta]$, or to the space $C([0, \delta], R^n)$, we obtain the following fact: the operator V takes the ball $S \subset C([0, \delta], R^n)$ into a relatively compact subset of it: $VS \subset S$. Since S is obviously closed and convex in $C([0, \delta], R^n)$, while the operator V is continuous and compact, the Schauder–Tychonoff theorem yields the existence of (at least) one fixed point in S.

Remark 1. Instead of assuming V is defined on the whole space $C([0, a], R^n)$, one can consider the case when V is given only on an open subset of that space, or on the closure of an open set (for instance, on a ball like S). Then Theorem 3.2.1 yields existence results similar to Theorem 3.1.1, in which case

$$(Vx)(t) = f(t) + \int_0^t k(t, s, x(s)) \, ds. \qquad (3.2.6)$$

Remark 2. More general results than Theorem 3.1.1 can be obtained from Theorem 3.2.1. For instance, the operator V given by (3.2.6) could be chosen in such a way that it takes $C([0, a], R^n)$ into itself. However, $k(t, s, x)$ need not be continuous, as assumed above. In [1], Miller makes the following assumptions on $k(t, s, x)$:

(a) $f \in C([0, a], R^n)$;
(b) $k: \Delta_a \times R^n \to R^n$ is measurable in (t, s, x), and is continuous in x for each fixed $(t, s) \in \Delta_a$;
(c) for each bonded set $B \subset R^n$, there exists a nonnegative measurable function $m(t, s)$, such that

$$|k(t, s, x)| \le m(t, s), \quad (t, s) \in \Delta_a, \, x \in B,$$

and

$$\sup \int_0^t m(t, s) \, ds < \infty, \quad t \in [0, a];$$

(d) for every $t_0 \in [0, a]$, and $x \in C([0, a], R^n)$, $|x(t)| \le M$ ($M > 0$ arbitrary), one has

$$\sup \int_0^a |k(t, s, x(s)) - k(t_0, s, x(s))| \, ds \to 0 \text{ as } t \to t_0. \qquad (3.2.7)$$

It can be easily checked that under conditions (a), (b), (c), (d) the operator V given by (3.2.6) takes $C([0, a], R^n)$ into itself, and verifies other conditions required in Theorem 3.2.1. In particular, condition (3.2.7) implies the compactness property (3).

Conditions (a)–(d) on the operator V given by (3.2.6) are what one usually calls Carathéodory type conditions.

Remark 3. Instead of condition (4) on the operator V, one can assume only the existence of an $\bar{x}(t)$ such that $\bar{x}(0) = (V\bar{x})(0)$. See Neustadt [1].

We shall now consider the abstract Volterra equation (3.2.1), and assume that the operator V is acting between function spaces whose elements are not necessarily continuous maps. More specifically, we shall deal with the case when the operator V takes the space $L^2_{\text{loc}}([0, T), R^n)$ into itself. It is then obvious that equation (3.2.1) will be understood as holding almost everywhere. Accordingly, in the definition of Volterra operator we have to make a change. Namely, V will be called of Volterra type (or causal) if, from $x(t) = y(t)$ almost everywhere on $[0, \tau]$, $\tau < T$, it follows that $(Vx)(t) = (Vy)(t)$ almost everywhere on $[0, \tau]$.

It is more or less obvious that making the kind of hypotheses encountered in Theorem 3.2.1, we can expect the existence of a solution to equation (3.2.1) in the space $L^2_{\text{loc}}([0, T), R^n)$. While formulating hypotheses similar to the hypotheses (1)–(3) in Theorem 3.2.1 does not pose any problem, we see that the fourth hypothesis in the above theorem does not make sense for measurable functions (more precisely, for the classes of equivalent functions that constitute the elements of the space). Instead of the hypothesis (4) in Theorem 3.2.1, we have to substitute another condition that will assume, for instance, the inclusion $VS \subset S$, with $S \subset L^2_{\text{loc}}([0, T), R^n)$ conveniently constructed.

Of course, this is just one possibility in dealing with this matter, but it turns out to be quite useful. Moreover, it will allow us to obtain a global existence theorem for equation (3.2.1), as well as a local one.

Theorem 3.2.2. *Consider the equation (3.2.1), and assume V is an operator from $L^2_{\text{loc}}([0, T), R^n)$ into itself, verifying the following conditions:*

(1) *V is an operator of Volterra type;*
(2) *V is continuous in the topology of the space $L^2_{\text{loc}}([0, T), R^n)$;*
(3) *V is compact on $L^2_{\text{loc}}([0, T), R^n)$;*
(4) *there exist two functions A, B: $[0, T) \to R_+$, with A continuous and positive, and B locally integrable, such that $x \in L_{\text{loc}}([0, T), R^n)$ and*

$$\int_0^t |x(s)|^2 \, ds \leq A(t), \quad t \in [0, T) \tag{3.2.8}$$

   imply

$$|(Vx)(t)|^2 \leq B(t) \quad a.e. \ on \ [0, T), \tag{3.2.9}$$

and furthermore

$$\int_0^t B(s)\,ds \le A(t), \quad t \in [0, T). \tag{3.2.10}$$

Then there exists a solution $x \in L^2_{\mathrm{loc}}([0, T), R^n)$ of the equation (3.2.1), such that (3.2.8) holds.

Proof. The underlying space is the locally convex space $L^2_{\mathrm{loc}}([0, T), R^n)$, and the closed convex set $S \subset L^2_{\mathrm{loc}}([0, T), R^n)$ is defined by

$$S = \left\{ x \in L^2_{\mathrm{loc}}([0, T), R^n); \int_0^t |x(s)|^2\,ds \le A(t), t \in [0, T) \right\} \tag{3.2.11}$$

From (3.2.8)–(3.2.10) one obtains

$$VS \subset S. \tag{3.2.12}$$

Since V is continuous and compact on $L^2_{\mathrm{loc}}([0, T), R^n)$, the fixed point theorem of Tychonoff (in locally convex spaces!) applies to the operator V and set S. Therefore, we conclude that there exists $x \in S$, such that (3.2.1) is verified.

As far as the local result is concerned, we should notice that the existence of the pair A, B, with the properties specified in the statement of Theorem 3.2.2, except for (3.2.10), will always imply the validity of (3.2.10) in some right neighborhood $[0, \delta]$, $\delta > 0$.

This ends the proof of Theorem 3.2.2. In order to check how efficient this result is, let us consider the classical Volterra linear equation

$$x(t) = f(t) + \int_0^t k(t, s)x(s)\,ds, \quad t \in [0, T), \tag{3.2.13}$$

with $f \in L^2_{\mathrm{loc}}([0, T], R^n)$, and $k(t, s)$ a matrix-valued kernel, of type n by n, whose entries are in $L^2_{\mathrm{loc}}(D, R)$, where $D = \{(t, s)|0 \le s \le t < T\}$.

Hence, for every $t_1 < T$, one has

$$\int_0^{t_1} \int_0^t |k(t, s)|^2\,ds\,dt \le M(t_1) < \infty. \tag{3.2.14}$$

It is easily seen that the operator V given by

$$(Vx)(t) = f(t) + \int_0^t k(t, s)x(s)\,ds, \quad t \in [0, T), \tag{3.2.15}$$

is a Volterra operator on $L^2_{\mathrm{loc}}([0, T), R^n)$.

In order to prove its continuity on the same space, we notice that for any $t_1 < T$ one has

$$\int_0^{t_1} |(Vx)(t) - (Vy)(t)|^2 \, dt$$

$$\leq \left(\int_0^{t_1} \int_0^t |k(t,s)|^2 \, ds \, dt \right) \left(\int_0^{t_1} |x(t) - y(t)|^2 \, dt \right). \tag{3.2.16}$$

The compactness of the operator V on $L^2_{loc}([0,T), R^n)$ follows from the estimates

$$\int_0^{t_1} |(Vx)(t)|^2 \, dt \leq 2 \int_0^{t_1} |f(t)|^2 \, dt$$

$$+ 2 \left(\int_0^{t_1} \int_0^t |k(t,s)|^2 \, ds \, dt \right) \left(\int_0^{t_1} |x(t)|^2 \, dt \right), \tag{3.2.17}$$

and, for $h > 0$,

$$\int_0^{t_1} |(Vx)(t+h) - (Vx)(t)|^2 \, dt \leq 3 \int_0^{t_1} |f(t+h) - f(t)|^2 \, dt$$

$$+ 3A(t_1 + h) \int_0^{t_1} \int_t^{t+h} |k(t+h,s)|^2 \, ds \, dt$$

$$+ 3 \left(\int_0^{t_1} |x(s)|^2 \, ds \right) \left(\int_0^{t_1} \int_0^t |k(t+h,s) - k(t,s)|^2 \, ds \, dt \right). \tag{3.2.18}$$

Indeed, (3.2.17) shows that for any $x \in L^2_{loc}([0,T), R^n)$, with $\int_0^{t_1} |x(s)|^2 \, ds \leq M$, the set $\{Vx\}$ remains bounded in L^2-norm. From (3.2.18) one obtains

$$\lim_{h \to 0} \int_0^{t_1} |(Vx)(t+h) - (Vx)(t)|^2 \, dt = 0, \tag{3.2.19}$$

uniformly with x, such that $\int_0^{t_1} |x(s)|^2 \, ds \leq M$.

While the first and third terms in the right hand side of (3.2.18) obviously tend to zero as $h \to 0$, we notice that in the second term the integral is taken on a set of measure that tends to zero with h.

It remains to construct the functions $A(t)$ and $B(t)$, such that condition (4) of Theorem 3.2.2 is satisfied.

If we start with an arbitrary $A(t)$, then it is easily seen that $B(t)$ can be chosen as

$$B(t) = 2|f(t)|^2 + 2A(t) \int_0^t |k(t,s)|^2 \, ds. \tag{3.2.20}$$

Let us denote

$$\lambda(t) = \int_0^t |k(t,s)|^2 \, ds, \quad t \in [0,T). \tag{3.2.21}$$

Since we assumed $|k|$ to be in $L_{\text{loc}}^2(D, R)$, $x(t)$ is locally integrable on $[0, T)$. If we now take into account the condition (3.2.10), we obtain for $A(t)$ the following integral inequality:

$$A(t) \geq 2 \int_0^t A(s)\lambda(s)\,ds + 2 \int_0^t |f(s)|^2\,ds, \quad t \in [0, T). \tag{3.2.22}$$

For (3.2.22) we need to construct a continuous solution $A(t) > 0, t \in [0, T)$.

Notice that, for any $\varepsilon > 0$, the inequality

$$A(t) \geq 2 \int_0^t A(s)[\lambda(s) + \varepsilon]\,ds + 2 \int_0^t |f(s)|^2\,ds + \varepsilon \tag{3.2.23}$$

is stronger than (3.2.22). A solution $A(t)$ for (3.2.23) is also a solution for (3.2.22). We shall now reduce (3.2.23) to a differential inequality, by letting

$$y(t) = \int_0^t A(s)[\lambda(s) + \varepsilon]\,ds, \tag{3.2.24}$$

which implies $y(0) = 0$, and

$$y'(t) \geq 2[\lambda(t) + \varepsilon]y(t) + \left[2 \int_0^t |f(s)|^2\,ds + \varepsilon\right][\lambda(t) + \varepsilon] \tag{3.2.25}$$

for $t \in [0, T)$. More precisely, (3.2.25) has to be considered only a.e. on $[0, T)$. Obviously, a solution $y(t)$ for (3.2.25) can be obtained if we solve the linear differential equation corresponding to the inequality, with the initial condition $y(0) = 0$. This choice gives

$$A(t) = y'(t)[\lambda(t) + \varepsilon]^{-1} = 2y(t) + 2 \int_0^t |f(s)|^2\,ds + \varepsilon, \tag{3.2.26}$$

which shows that $A(t) > 0$ is continuous.

Summarizing the above discussion in relation to equation (3.2.13), with $f \in L_{\text{loc}}^2([0, T), R^n)$, and $|k| \in L_{\text{loc}}^2(D, R)$, we can state the existence of a solution $x \in L_{\text{loc}}^2([0, T), R^n)$ which verifies the equation almost everywhere on $[0, T)$.

We leave to the reader the task of showing that the solution is unique in $L_{\text{loc}}^2([0, T), R^n)$.

An L^2-existence theory for (3.2.13), using the method of successive approximations, can be found in Tricomi's book [1]. The resolvent kernel is also constructed under various assumptions in Neustadt [1], Schwabik et al. [1], Miller [1].

Finally, we point out the fact that Theorem 3.2.2 can be applied to the nonlinear case

$$x(t) = f(t) + \int_0^t k(t, s, x(s))\,ds, \tag{3.2.27}$$

assuming among other things the validity of an estimate of the form

$$|k(t, s, x)| \leq k_0(t, s)|x|,\qquad (3.2.28)$$

where $k_0 \in L^2_{loc}(D, R)$. The so-called Carathéodory's conditions must be imposed on $k(t, s, x)$, in order to be able to construct the pair of functions $A(t)$ and $B(t)$. The reader is encouraged to try to get through the details.

We shall discuss now the existence problem for functional differential equations of Volterra type. Such an equation is usually given in the form (3.2.2), with the associated initial condition:

$$\dot{x}(t) = (Vx)(t), \ t \in [0, T), \quad x(0) = x^0 \in R^n.\qquad (3.2.2)$$

As we mentioned earlier in this section, it is adequate to assume that V is acting on a space of continuous functions on $[0, T)$, or on a subinterval $[0, \tau) \subset [0, T)$, while the range of V is in a space of measurable (locally integrable) functions defined on $[0, T)$, or on a convenient subinterval. This kind of assumption is easily understandable if we take into account (3.2.2), which shows that $x(t)$ must be absolutely continuous.

Since (3.2.2) is equivalent to the Volterra equation $x(t) = (V_1 x)(t)$, with V_1 given by (3.2.4), i.e. $(V_1 x)(t) = x^0 + \int_0^t (Vx)(s) \, ds$, we realize that the existence problem is basically solved for (3.2.2). Indeed, equation (3.2.4) is a Volterra equation of the form (3.2.1), and therefore we can apply both Theorems 3.2.1 and 3.2.2 to produce existence results for the problem (3.2.2). It is interesting to point out that V_1 is acting from the space of absolutely continuous functions into itself, even though V does not possess this property (under our assumption). Therefore, applying Theorem 3.2.1 to the equation (3.2.4) makes sense.

We shall give a theorem that is closer in formulation to Theorem 3.2.2, and which can be also proved by means of the Schauder–Tychonoff fixed point method.

The following assumptions will be made about the operator V in equation (3.2.2).

A_1. V is a continuous Volterra operator from the space $C([0, T), R^n)$ into the Banach (or Frechet) space $B \subset L_{loc}([0, T), R^n)$, the topology of B being stronger than the topology of $L_{loc}([0, T), R^n)$.

A_2. There exists a pair of maps from $[0, T)$ into R, say $A(t)$ and $B(t)$, such that $A(t)$ is continuous and positive, while $B(t)$ is locally integrable and nonnegative, with the property that $x \in C([0, T), R^n)$ and

$$|x(t)| \leq A(t), \quad t \in [0, T)\qquad (3.2.29)$$

implies

$$|(Vx)(t)| \leq B(t) \quad \text{a.e. on } [0, T),\qquad (3.2.30)$$

where

$$A(t) - A(0) \geq \int_0^t B(s)\,ds, \quad t \in [0, T). \qquad (3.2.31)$$

The following result can now be formulated regarding equation (3.2.2).

Theorem 3.2.3. *Assume that conditions A_1 and A_2 hold with respect to the operator V. Then the problem (3.2.2) has a solution $x \in AC_{\mathrm{loc}}([0, T), R^n) \subset C([0, T), R^n)$, such that (3.2.29) holds, provided $|x^0| < A(0)$.*

Proof. We shall deal with the equivalent equation (3.2.4), involving the Volterra operator V_1.

It is obvious that inequality (3.2.29) defines in $C([0, T), R^n)$ a convex closed set, say S. The operator V_1 takes, under our assumptions, the set S into itself: $V_1 S \subset S$. We have only to show that V_1 is continuous on S, and $V_1 S$ is relatively compact in $C([0, T), R^n)$.

The continuity of V_1 follows easily from the continuity of V. Indeed, if $x^m \to x$ in $C([0, T), R^n)$, then $(Vx^m)(t) \to (Vx)(t)$ in B, as $m \to \infty$, But convergence in B implies convergence in $L_{\mathrm{loc}}([0, T), R^n)$, i.e., L^1-convergence on each $[0, t_1] \subset [0, T)$. This means that $(V_1 x^m)(t) \to (V_1 x)(t)$ in $C([0, T), R^n)$, because $(V_1 x)(0) = x^0$ for any $x \in C([0, T), R^n)$.

In order to show that $V_1 S$ is relatively compact in $C([0, T), R^n)$, we need to show that this set is uniformly bounded and equicontinuous on each $[0, t_1] \subset [0, T)$.

Since $x \in S$ implies $\|(Vx)(t)\| \leq B(t)$ a.e. on $[0, T)$, one easily obtains from A_2

$$|(V_1 x)(t)| \leq A(t), \quad t \in [0, T),$$

and

$$|(V_1 x)(t + h) - (V_1 x)(t)| \leq \left| \int_t^{t+h} B(s)\,ds \right|.$$

The local integrability of B suffices to assure the equicontinuity of $V_1 S$.

The proof of Theorem 3.2.3 is thereby completed.

As an application of Theorem 3.2.3, let us consider the existence problem for one of the most often encountered integrodifferential equation, namely

$$\dot{x}(t) = f\left(t, x(t), \int_0^t k(t, s)x(s)\,ds\right), \quad x(0) = x^0, \qquad (3.2.32)$$

in which, $x, f \in R^n$, $t \in [0, T)$, and $k(t, s)$ is a matrix-valued kernel defined on $D = \{(t, s) | 0 \leq s \leq t < T\}$. We assume that $f(t, x, y)$ is measurable in t for fixed (x, y), and continuous in (x, y) for fixed t, a.e. on $[0, T)$. Moreover, we assume that

$$\|f(t, x, y)\| \leq \alpha(t)|x| + \beta(t)|y| + \gamma(t), \qquad (3.2.33)$$

where α, β, γ are nonnegative locally integrable functions on $[0, T)$. The kernel $k(t, s)$ is assumed continuous on D, even though measurability conditions could be allowed.

The operator V is defined by

$$(Vx)(t) = f\left(t, x(t), \int_0^t k(t, s)x(s)\,ds\right), \qquad (3.2.34)$$

and it is obviously a Volterra operator. The continuity of V, from $C([0, T), R^n)$ into $L_{loc}([0, T), R^n)$, is easily obtained by using the estimate (3.2.33).

It remains to construct the pair of functions $A(t)$ and $B(t)$, with the properties indicated in the hypothesis A_2. In other words, assuming $|x(t)| \leq A(t)$ on $[0, T)$, with $x \in C([0, T), R^n)$, we must satisfy the inequality

$$A(t) - A(0) \geq \int_0^t \left[\alpha(s)A(s) + \beta(s) \int_0^s |k(s, u)|A(u)\,du + \gamma(s)\right] ds. \qquad (3.2.35)$$

In order to handle this inequality, we will assume that $A(t)$ is nondecreasing on $[0, T)$. This allows us to deal with the stronger inequality

$$A(t) - A(0) \geq \int_0^t \left[\alpha(s) + \beta(s) \int_0^s |k(s, u)|\,du\right] A(s)\,ds + \int_0^t \gamma(s)\,ds. \qquad (3.2.36)$$

The inequality (3.2.36) has the same form as inequality (3.2.22). Therefore, an absolutely continuous, positive and nondecreasing function $A(t)$ can be constructed as shown above. The function $B(t)$ can obviously be chosen as

$$B(t) = \alpha(t)A(t) + \beta(t)A(t) \int_0^t |k(t, s)|\,ds + \gamma(t). \qquad (3.2.37)$$

Remark. The result of Theorem 3.2.3 was given, in a slightly different form, in Corduneanu [12]. The linear estimate on f helps in obtaining a global result. Without such an estimate, only a local result can be obtained.

Let us consider now functional-differential equations of Volterra type depending upon a functional parameter:

$$\dot{x}(t) = V(x, \phi)(t), \quad t \in [0, T), \qquad (3.2.38)$$

with the usual initial condition

$$x(0) = x^0 \in R^n. \qquad (3.2.39)$$

The functional parameter ϕ is supposed to belong to a certain function space S, endowed with convenient structures in order to make possible the study of dependence of the solution with respect to ϕ. More precise assumptions will be made in the sequel.

Our immediate goal is to show that differential equations with delay can be treated, at least with regard to the existence of solutions, as functional-differential equations depending upon a parameter. As an illustration of this, consider the infinite-delay equation

$$\dot{x}(t) = \int_{-\infty}^{t} k(t,s)x(s)\,ds + f(t), \quad t \in R_+, \tag{3.2.40}$$

with the initial condition (3.2.39), and the functional initial data

$$x(t) = \phi(t), \quad t \in R_-. \tag{3.2.41}$$

Formally, this problem is equivalent to the following Volterra functional-differential equation

$$\dot{x}(t) = \int_0^t k(t,s)x(s)\,ds + \int_{-\infty}^0 k(t,s)\phi(s)\,ds + f(t)$$

on $t > 0$, with (numerical) initial data (3.2.39). In other words, $V(x,\phi)$ is given by

$$V(x,\phi)(t) = \int_0^t k(t,s)x(s)\,ds + \int_{-\infty}^0 k(t,s)\phi(s)\,ds + f(t),$$

a formula which illustrates the following feature: $(x,\phi) \to V(x,\phi)$ is a map from the product space $X \times S$, with X a function space whose elements are given on R_+ and take values in R^n into another function space Y. If we take into account the discussion carried out above in relation to equation (3.2.2), it is a reasonable assumption that Y has its elements defined on R_+, with values in R^n. In contrast to X, whose elements have to be 'smoother', the elements of Y will be assumed to be only locally integrable.

Going back to equation (3.2.38), under initial condition (3.2.39), we will see that the existence of solutions can be obtained without difficulty by applying Theorem 3.2.3. Indeed, if we assume that $V(x,\phi)$ in (3.2.38) satisfies for every fixed $\phi \in S$ the conditions A_1 and A_2, then the existence of a solution to (3.2.38), with the initial condition (3.2.39) is assured by Theorem 3.2.3. Such a result does not really provide valuable information, and extra hypotheses are needed in order to make it meaningful.

Without intending to cover the greatest generality, let us assume the following conditions are satisfied by the operator V in (3.2.38).

(1) V is a Volterra operator (with regard to the first argument) from $C([0,T],R^n) \times S$ into $L_{loc}([0,T],R^n)$, where S stands for a Banach or Frechet function space whose norm or linearly invariant metric is denoted by $|\cdot|_S$.

(2) There exists a nonnegative locally integrable function $\lambda(t)$ on $[0,T]$, such that

$$|V(x,\phi)(t) - V(y,\psi)(t)| \le \lambda(t)\left[\sup_{u \in [0,t]} |x(u) - y(u)| + |\phi - \psi|_S\right] \quad (3.2.42)$$

a.e. on $[0, T)$, and for any $x, y \in C([0, T), R^n)$, and any $\phi, \psi, \in S$.

Let us point out that (3.2.42) implies that $V(x, \phi)$ is Volterra in x.

Hence, the method of successive approximations is applicable to the integral equation equivalent to (3.2.38), (3.2.39), namely

$$x(t) = x^0 + \int_0^t V(x, \phi)(s)\,ds, \quad t \in [0, T), \quad (3.2.43)$$

and its convergence in $C([0, T), R^n)$ is an elementary result, which is left to the reader.

We shall establish an inequality for $|x(t) - y(t)|$, where x denotes the solution of (3.2.38) corresponding to $\phi \in S$ and $x^0 \in R^n$, while y is the solution of the same equation (3.2.38), corresponding to $\psi \in S$, and $y^0 \in R^n$ as initial value.

From (3.2.43) and the similar relationship corresponding to y, one obtains in view of (3.2.42):

$$|x(t) - y(t)| \le |x^0 - y^0| + \int_0^t \lambda(s)\left[\sup_{u \in [0,S]} |x(u) - y(u)| + |\phi - \psi|_S\right]ds$$

which easily yields for $t \in [0, T)$

$$\sup_{s \in [0,t]} |x(s) - y(s)| \le |x^0 - y^0| + |\phi - \psi|_S \int_0^t \lambda(s)\,ds$$

$$+ \int_0^t \lambda(s) \sup_{u \in [0,s]} |x(u) - y(u)|\,ds. \quad (3.2.44)$$

The inequality (3.2.44) is an integral inequality of Gronwall's type. As seen in Section 1.3, it implies

$$\sup_{s \in [0,t]} |x(s) - y(s)| \le K(t)(|x^0 - y^0| + |\phi - \psi|_S) \quad (3.2.45)$$

for some nonnegative continuous function $K(t)$.

Consequently, *under conditions (1) and (2) for the operator $V(x, \phi)$ the (unique) solution of the problem (3.2.38), (3.2.39) is Lipschitz continuous (on any finite interval!) with respect to the initial data, and the functional parameter.*

Let us show now how existence results for delay equations can be obtained from the existence of a solution to equation (3.2.38), with the initial condition (3.2.39).

We shall choose the case of equations with unbounded (infinite) delay. The case of finite delays can be dealt with similarly.

Let us consider the equation

$$\dot{x}(t) = f(t, x_t), \quad t \in [0, T), \quad (3.2.46)$$

under initial data

$$x_0 = \phi \in S, \quad x(0) = x^0 \in R^n, \tag{3.2.47}$$

with $x \in R^n$, $S = S(R_-, R^n)$ being a function space to be made precise below, $x_t(u) = x(t + u)$ for any $u \in R_-$, and $f: [0, T) \times S \rightarrow R^n$ a map with the property $f(t, x_t) \in L_{loc}([0, T), R^n)$, for any $x \in C([0, T), R^n)$.

Of course, this is the most common description of functional-differential equations with unbounded (infinite) delay (see, for instance, Corduneanu and Lakshmikantham [1]).

In order to reduce the delay equation to an equation of Volterra type depending upon a parameter, we have to define the operator $V(x, \phi)$ in terms of f and ϕ from (3.2.46) and (3.2.47), in such a way that any solution of (3.2.46), (3.2.47) is a solution of (3.2.38), (3.2.39), and vice versa.

Let us now proceed formally and, for $x \in C([0, T), R^n)$ and $\phi \in S$, let

$$V(x, \phi)(t) = f(t, x_t), \quad t \in [0, T), \tag{3.2.48}$$

where according to (4.2.47)

$$x_0 = \phi \in S, \tag{3.2.49}$$

and for every $t \in [0, T)$

$$x_t(u) = \begin{cases} \phi(t + u), & u < -t, \\ x(t + u), & -t \le u \le 0. \end{cases} \tag{3.2.50}$$

From this notation it is obvious that the infinite delay equation (3.2.46) becomes (3.2.38), while the first initial condition (3.2.47) is now absorbed in $V(x, \phi)$. The second condition (3.2.47) is identical to (3.2.39), and therefore the reduction is achieved.

It remains to clarify a few details that will make the above reduction meaningful, in other words, we have to provide adequate conditions on (3.2.46), (3.2.47), such that the resulting $V(x, \phi)$ satisfies conditions that will allow us to ascertain the existence of solutions for (3.2.38), (3.2.39).

Of course, the above problem does not have a unique answer. It depends on what kind of conditions we expect for $V(x, \phi)$. Since our aim is only to illustrate how the above scheme works, we will choose conditions that do not lead to significant technical difficulties. We shall soon realize that in order to obtain a convenient $V(x, \phi)$, besides the conditions to be imposed on $f(t, \phi)$ in (3.2.46), certain assumptions have to be made on the space S of initial functions.

Let us assume first that f is a continuous map from $[0, T) \times S$ into R^n, satisfying a Lipschitz type condition

$$|f(t, \phi) - f(t, \psi)| \le \lambda(t)|\phi - \psi|_S, \tag{3.2.51}$$

with $\lambda(t)$ a nonnegative continuous function on $[0, T)$.

Now let $x, y \in C([0, T], R^n)$, and $\phi, \psi \in S$ be arbitrary, and define $x_0 = \phi$, $y_0 = \psi$, in agreement with (3.2.49). Then, from (3.2.48), (3.2.51), we obtain for $t \in [0, T)$

$$|V(x, \phi)(t) - V(y, \psi)(t)| = |f(t, x_t) - f(t, y_t)| \le \lambda(t)|x_t - y_t|_S, \quad (3.2.52)$$

provided $x_t, y_t \in S$.

Of course, we are interested in deriving (3.2.42) from (3.2.52), which appears to be immediate when an inequality of the form

$$|x_t - y_t|_S \le k(t) \left\{ \sup_{u \in [0,t]} |x(u) - y(u)| + |\phi - \psi|_S \right\} \quad (3.2.53)$$

is valid for any $t \in [0, T)$, with $k(t)$ a continuous nonnegative function, or, more generally, a nonnegative locally integrable function on $[0, T)$.

We shall now impose some restrictions on the function space S, such that (3.2.53) is satisfied for any $x, y \in C([0, T], R^n)$, and $\phi, \psi \in S$.

Notice first that

$$x_t(u) = x_t(u)\chi_{[-t, 0]}(u) + \phi(t + u)\chi_{(-\infty, -t)}(u),$$

where $\chi_{(a, b)}(u)$ is the characteristic function of the interval (a, b). Hence

$$|x_t - y_t|_S \le |[x_t(u) - y_t(u)]\chi_{[-t, 0]}(u)|_S$$
$$+ |[\phi(t + u) - \psi(t + u)]\chi_{(-\infty, -t)}(u)|_S, \quad (3.2.54)$$

and (3.2.53) will follow from (3.2.54) if we assume

$$|x_t(u)\chi_{[-t, 0]}(u)| \le m(t) \sup_{u \in [0,t]} |x(u)| \quad (3.2.55)$$

for any $x \in C([0, T], R^n)$ and a fixed locally (continuous) integrable function $m(t)$ on $[0, T)$, and also

$$|\phi(t + u)\chi_{(-\infty, -t)}(u)|_S \le n(t)|\phi|_S, \quad (3.2.56)$$

for any $\phi \in S$, and a function $n(t)$ with the same properties as $m(t)$ in (3.2.55).

Combining (3.2.54)–(3.2.56), one derives (3.2.53), with $k(t) = \max\{m(t), n(t)\}$. In turn, (3.2.52) and (3.2.53) imply (3.2.42), with $\lambda(t)k(t)$ instead of $\lambda(t)$.

Summing up the above discussion, we can state the following existence and uniqueness result for equations with infinite delay of the form (3.2.46), under initial data (3.2.49).

Proposition 3.2.4. *Consider the infinite delay equation (3.2.46), under initial data (3.2.47), and assume the following conditions are satisfied:*

(a) *$f: [0, T) \times S \to R^n$ is a continuous map, such that (3.2.51) holds for some nonnegative locally integrable function λ;*

(b) *the 'phase space'* $S = S(R_-, R^n)$ *is either a Banach space or a Frechet space,
in which the norm or the linearly invariant metric is denoted by* $|\cdot|_S$, *such
that the following properties are valid:*

  (1) *for every* $\phi \in S$ *and* $x \in C([0, T), R^n)$, x_t *defined by* (3.2.50) *belongs to* S,
      and the map $t \to x_t$ *is continuous;*

  (2) *there exists* $m: [0, T) \to R_+$, *continuous, such that* (3.2.55) *holds for any*
      $x \in C([0, T), R^n)$;

  (3) *there exists* $n: [0, T) \to R_+$, *continuous, such that* (3.2.56) *holds for any*
      $\phi \in S$.

Then, there exists a unique solution $x = x(t, x^0, \phi)$ *of the problem* (3.2.46), (3.2.47)
defined on $[0, T)$. *This solution is continuously differentiable with respect to t and
uniformly Lipschitz continuous in* (x^0, ϕ) *on any compact interval of* $[0, T)$.

Remark 1. Condition (3.2.55) makes sense because of assumption (1) in (b).
Indeed, $x_t(u)\chi_{[-t,0]}(u) \in S$, because it coincides with zero on $(-\infty, -t)$, and with
$x(t + u)$ on $[-t, 0]$. It is, in fact, the construction of x_t defined by (3.2.50),
corresponding to $\phi = 0 \in S$. Similar comment is valid for (3.2.56), in which case
ϕ is arbitrary in S, while $x(t) \equiv 0$.

Remark 2. The reader familiar with the theory of equations with infinite delay,
will easily realize that in (b) we have retrieved the axioms of the phase space (see,
for instance, Hale and Kato [1]) starting from the idea of treating infinite delay
equations as functional-differential equations depending upon a parameter.

Remark 3. It is obvious that the conditions of Proposition 3.2.4 cover the linear
case. In other words, if f is linear in ϕ, and therefore (3.2.46) becomes

$$\dot{x}(t) = L(t, x_t) + f(t), \quad t \in [0, T), \tag{3.2.57}$$

in which

$$|L(t, \phi)| \leq \lambda(t)|\phi|_S, \quad \phi \in S, \tag{3.2.58}$$

for some nonnegative *continuous* $\lambda(t)$ on $[0, T)$, the existence and uniqueness of
the solution of (3.2.57) verifying (3.2.47) are guaranteed. Also, the continuous
dependence on (x^0, ϕ) is assured.

Remark 4. It is useful to provide some examples of function spaces that satisfy
the conditions (b) of Proposition 3.2.4.
  A first choice is $S = L^1(R_-, R^n)$, with $1 \leq p < \infty$. Indeed, for $\phi \in L^1(R_-, R^n)$
and $x \in C([0, T), R^n)$, $x_t \in L^1(R_-, R^n)$ for each $t \in [0, T)$, and it can be represented
as

$$x_t(u) = x(t + u)\chi_{[-t,0]}(u) + \phi(t + u)\chi_{(-\infty, -t)}(u).$$

If we keep in mind the well known property $\lim_{h \to 0} |x(t) - x(t + h)|_{L^p} = 0$ for $x \in L^p$, $1 \le p < \infty$, then it is obvious that $t \to x_t$ is continuous as a map from $[0, T)$ into S.

We need only to check the validity of conditions (2) and (3) in (b). This is particularly simple, and the reader will easily find out that one can choose $m(t) = t^{1/p}$, and $n(t) = 1$.

Another possible choice is when S consists of integrable functions on R_-, with values in R^n, such that their restrictions to some fixed interval $[-r, 0]$, $r > 0$, are continuous. S is endowed in such case with a mixed norm, namely

$$|\phi|_S = \int_{-\infty}^{-r} |\phi(u)| \, du + \sup_{u \in [-r, 0]} |\phi(u)|. \qquad (3.2.59)$$

In this case, $\phi(0)$ is defined for each $\phi \in S$, and the initial condition $x(0) = x^0 \in R^n$ is usually replaced by the more natural condition $x(0) = \phi(0)$. In this way, $x_t \in S$ for any $t \in [0, T)$, and $t \to x_t$ is continuous. To check the last assertion we refer again to the same property as in the case $S = L^p$, plus the uniform continuity of continuous functions on $[-r, 0]$.

With regard to property (2) in (b), one can easily see that for

$$m(t) = \begin{cases} 1, & 0 \le t \le r, \\ 1 + t - r, & r < t < T, \end{cases}$$

(3.2.55) is satisfied, while $n(t) = 1$ is convenient for condition (3.2.56).

The mixed norms like (3.2.59), and their analogues for $1 < p < \infty$, have been extensively used in regard to infinite delay equations.

3.3 Linear Volterra equations: the resolvent and some applications

We shall consider the linear Volterra equation

$$x(t) = (Lx)(t) + f(t), \quad t \in [0, T), \qquad (3.3.1)$$

where L is a linear Volterra operator on the space $C([0, T), R^n)$, and $f \in C([0, T), R^n)$, satisfying further conditions to be specified below.

Our aim is to show that equation (3.3.1) has a unique solution $x(t) \in C([0, T), R^n)$, for every $f(t) \in C([0, T), R^n)$, and this solution can be expressed by means of the formula

$$x(t) = f(t) + (Rf)(t), \quad t \in [0, T), \qquad (3.3.2)$$

where R is also a linear Volterra type operator on $C([0, T), R^n)$, called the resolvent of L.

In order to achieve this, we shall proceed as follows: *first*, we will obtain the existence of a local solution by applying Theorem 3.2.1; *second*, we will prove the uniqueness of the solution for linear equations; *third*, we will show that any solution of (3.3.1) can be continued until we obtain a solution of the same equation, defined on the maximal interval $[0, T)$; *fourth*, we will prove the existence of the resolvent operator R, and the validity of formula (3.3.2).

Then several illustrations and applications of the formula (3.3.2) will be considered: either constructing R when L has a particular form, or using the representation to derive a variation of the constants formula for linear functional-differential equations of the form

$$\dot{x}(t) = (Lx)(t) + f(t), \quad x(0) = x^0 \in R^n. \tag{3.3.3}$$

The *first step* of our discussion requires the proof of local existence for (3.3.1), which does not pose any problem if we accept the conditions (1)–(4) of Theorem 3.3.1. More precisely, we assume that $L: C([0, T), R^n) \to C([0, T), R^n)$ is causal, compact, and of fixed initial value: $(Lx)(0) = (Ly)(0)$ for any $x, y \in C([0, T), R^n)$. Since the compactness of linear operators implies continuity, while $(Lx)(0) = (L\theta)(0) = \theta$ (the null vector), conditions (1)–(4) of Theorem 3.2.1 are satisfied by $Vx = Lx + f$, and therefore (3.3.1) has a solution on some interval $[0, \delta]$, $0 < \delta < T$.

The *second step* consists in proving that the homogeneous equation $x(t) = (Lx)(t)$, associated with (3.3.1), has only the zero solution on $[0, T)$. This result can be obtained if we preserve the hypotheses on L, stated in the first step of the discussion.

Indeed, if we assume that the equation $x(t) = (Lx)(t)$ has a continuous solution $x_1(t)$ on some interval $[0, t_1] \subset [0, T)$, such that $x_1(t) \not\equiv 0$ on this interval, then we can reach a contradiction.

As noticed above, one must have $x_1(0) = (Lx_1)(0) = \theta$. On the other hand, there must be a τ, $0 \leq \tau < t_1$, such that $x_1(t) = \theta$ for any t with $0 \leq t \leq \tau$, while in any right neighborhood of τ there are points such that $x_1(t) \neq \theta$.

Because of the compactness of L, there exists $\delta_1 > 0$ with the following property: for each $x(t) \in C([0, t_1], R^n)$ with $|x(t)| \leq 1$ on $[0, t_1]$, one has

$$|(Lx)(t) - (Lx)(s)| < 1, \tag{3.3.4}$$

for any $t, s \in [0, t_1]$, such that $|t - s| < \delta_1$.

Going back now to the solution $x_1(t)$, we choose a positive number $\delta < \min\{\delta_1, t_1 - \tau\}$, and notice that we can assume without loss of generality

$$\sup|x_1(t)| = 1, \quad t \in [\tau, \tau + \delta]. \tag{3.3.5}$$

Indeed, since $x(t) = (Lx)(t)$ is linear and homogeneous, $cx_1(t)$ is also a solution for any real c. Denote by $s, s \in [\tau, \tau + \delta]$, a number such that $|x_1(s)| = 1$. Then

$$|x_1(s) - x_1(\tau)| = 1, \quad |s - \tau| \leq \delta < \delta_1. \tag{3.3.6}$$

Now define $x_2(t) = x_1(t), 0 \leq t \leq \tau + \delta, x_2(t) = x_1(\tau + \delta), \tau + \delta < t \leq t_1$. One has $x_2(t) \in C([0, t_1], R^n), |x_2(t)| \leq 1, t \in [0, t_1]$. Therefore (3.3.4) should be satisfied by $x_2(t)$. But (3.3.6) shows that (3.3.4) cannot be true for $x_2(t)$, which coincides with $x_1(t)$ on $[0, \tau + \delta]$, and moreover $x_1(t) = (Lx_1)(t)$ on that interval.

It remains only to point out that $t_1 < T$ is arbitrary, in order to conclude that equation (3.3.1) possesses the property of uniqueness in $C([0, T), R^n)$.

The *third step* of our discussion is aimed at showing that any solution of (3.3.1) can be assumed, without loss of generality, to be maximally defined on the whole interval $[0, T)$. This is a well-known feature for all kinds of linear equations that we are familiar with (ordinary differential equations, integral equations, integro-differential equations etc.).

First of all, notice that in the first step of our discussion we saw that equation (3.3.1) has a local solution. In other words, for some $t_1, 0 < t_1 < T$, there exists a function $x(t) \in C([0, t_1], R^n)$, such that $x(t) = (Lx)(t) + f(t)$, for all $t \in [0, t_1]$.

Let us now denote by $\bar{t}, 0 < \bar{t} \le T$, the least upper bound of those $t_1, 0 < t_1 < T$, such that equation (3.3.1) has a solution $x(t)$, defined on $[0, t_1]$. We can easily show that equation (3.3.1) has a unique solution $\bar{x}(t)$ defined on $[0, \bar{t})$. Obviously, the interval $[0, \bar{t})$ is the maximal interval of existence for the solution of equation (3.3.1). Then it will be shown that $\bar{t} = T$ (because of the linearity!).

The definition of $\bar{x}(t)$ is as follows: let $t' < \bar{t}$ be an arbitrary number, as close as we wish with respect to $\bar{t}$, or as large as we wish if $\bar{t} = \infty$. Then equation (3.3.1) must have a solution $x(t)$ defined on $[0, t_1]$, with $t' < t_1$. We set $\bar{x}(t') = x(t')$, and note that $\bar{x}(t')$ is uniquely determined, regardless of the (possible) choice of $x(t)$. Indeed, the uniqueness result, discussed in the second step, guarantees that. It is obvious that $\bar{x}(t)$ is a solution of (3.3.1) on $[0, \bar{t})$.

In order to conclude the third step of our discussion, it remains to show that, necessarily, we have $\bar{t} = T$. This result can be achieved if we rely again on the compactness of the operator L in (3.3.1).

Notice the fact that

$$\limsup |\bar{x}(t)| = \infty \text{ as } t \uparrow \bar{t}. \tag{3.3.7}$$

If (3.3.7) does not hold, then the family of functions defined by $x_S(t) = \bar{x}(t)$, $0 \le t \le s < \bar{t}$, $x_S(t) = \bar{x}(s)$, $s \le t \le \bar{t} + \varepsilon_0$ (for some $\varepsilon_0 > 0$ fixed) is uniformly bounded on $[0, \bar{t} + \varepsilon_0]$. From the compactness of L we easily obtain the existence of $\lim \bar{x}(t) = \bar{x}_0 \in R^n$, as $t \uparrow \bar{t}$, as a result of Cauchy's criterion.

Hence, $\bar{x}(t)$ could be assumed continuous on $[0, \bar{t}]$, letting $\bar{x}(\bar{t}) = \bar{x}_0$. Then $\bar{x}(t)$ becomes a continuous solution on $[0, \bar{t}]$ for (3.3.1), $(V\bar{x})(\bar{t}) = \bar{x}(\bar{t})$, and therefore it could be continued beyond $\bar{t}$ (see Remark 3 to Theorem 3.2.1). This contradicts the fact that $[0, \bar{t})$ is the maximal interval of existence for $\bar{x}(t)$, which means that (3.3.7) must hold.

Let us now assume that $\bar{t} < T$, and choose t_1 such that $\bar{t} < t_1 < T$. From the compactness of L we have

$$|(Lx)(t) - (Lx)(s)| < \tfrac{1}{4} \tag{3.3.8}$$

for all $x \in C([0,t_1], R^n)$, with $|x(t)| \le 1$ on $[0,t_1]$, and t, $s \in [0,t_1]$ such that $|t - s| < \delta$, where $\delta > 0$ is a small number ($\delta < 2\bar{t}$ at least). Denote $\beta = \sup\{|\bar{x}(t)|; 0 \le t \le \bar{t} - \delta/2\}$, and $\gamma = 1 + 2\beta + 8 \sup\{|f(t)|; 0 \le t \le \bar{t} - \delta/2\}$. By $\tilde{t}$ we denote the smallest t such that $|\bar{x}(t)| = \gamma$. According to (3.3.7), there are such values of t in any left neighborhood of $\bar{t}$, and obviously $\bar{t} > \tilde{t} > \bar{t} - \delta/2$.

We now define $x_1(t) \in C([0,t_1], R^n)$ by letting $x_1(t) = \gamma^{-1}\bar{x}(t)$ on $[0,\tilde{t}]$, and $x_1(t) = \gamma^{-1}\bar{x}(\tilde{t})$ on $[\tilde{t}, t_1]$. We have immediately

$$|x_1(t)| \le 1 \text{ on } [0,t_1], \quad \left|x_1\left(\bar{t} - \frac{\delta}{2}\right)\right| < \frac{1}{2}, \quad |x_1(\tilde{t})| = 1.$$

Consequently,

$$\left|x_1(\tilde{t}) - x_1\left(\bar{t} - \frac{\delta}{2}\right)\right| > \frac{1}{2}, \tag{3.3.9}$$

while (3.3.8) yields

$$\left|(Lx_1)(\tilde{t}) - (Lx_1)\left(\bar{t} - \frac{\delta}{2}\right)\right| < \frac{1}{4}. \tag{3.3.10}$$

Notice now that (3.3.1) leads to

$$x_1(t) = (Lx_1)(t) + \gamma^{-1}f(t), \quad t \in [0,\tilde{t}], \tag{3.3.11}$$

and consequently

$$\left|x_1(\tilde{t}) - x_1\left(\bar{t} - \frac{\delta}{2}\right)\right| \le \left|(Lx_1)(\tilde{t}) - (Lx_1)\left(\bar{t} - \frac{\delta}{2}\right)\right|$$

$$+ \gamma^{-1}\left|f(\tilde{t}) - f\left(\bar{t} - \frac{\delta}{2}\right)\right| < \frac{1}{4}$$

$$+ \gamma^{-1}2\sup\left\{|f(t)|; 0 \le t \le \bar{t} - \frac{\delta}{2}\right\}$$

$$\le \frac{1}{4} + \frac{1}{4} = \frac{1}{2},$$

a contradiction to (3.3.9).

This ends the third step of our discussion.

Finally, the *fourth step* can now be carried out. In other words, we have to prove the existence of the (Volterra) operator R on $C([0, T), R^n)$, such that the unique solution of (3.3.1) is given by (3.3.2).

We shall adopt the operator notation as specified above, and rely to some extent on operator algebra.

Equation (3.3.1) can be written in the form

$$(I - L)x = f, \quad I = \text{Identity operator.} \tag{3.3.12}$$

Since $I - L$ is a continuous operator from $C([0, T], R^n)$ onto itself – because equation (3.3.1) has a unique solution in $C([0, T], R^n)$ for any f belonging to this space – it follows that (see, for instance Yosida [1]) there exists a continuous inverse $(I - L)^{-1}$. Hence, (3.3.12) leads to $x = (I - L)^{-1}f$, and if we let

$$R = (I - L)^{-1} - I, \tag{3.3.13}$$

we can write the solution of (3.3.1) in the form $x = f + Rf$, which is exactly (3.3.2).

It remains to show that R is of Volterra type, and compact. This follows from the identity

$$R = L(I - L)^{-1}, \tag{3.3.14}$$

which is an immediate consequence of (3.3.13). The compactness of R is a consequence of the well-known property that the product of a continuous operator and a compact one is compact. The fact that R is a Volterra operator also follows from (3.3.14), in which both factors are Volterra operators (check that $(I - L)^{-1}$ is a Volterra operator if L is).

The conclusion of the discussion conducted above in relation to the linear equation (3.3.1) can be formulated in the following theorem.

Theorem 3.3.1. *Consider equation (3.3.1) in $C([0, T], R^n)$, where L is a compact linear Volterra operator. Then (3.3.1) has a unique solution in $C([0, T], R^n)$ for every $f \in C([0, T], R^n)$. This solution is given by (3.3.2), where the resolvent operator $R = L(I - L)^{-1}$ is also compact and of Volterra type.*

Remark 1. In the algebra of linear bounded operators on $C([0, T], R^n)$, the relationship between L and R can be also written as $(I - L)(I + R) = I$, which shows that the resolvent operator corresponding to $-R$ is $-L$. The relationship can be also written as $L - R + LR = 0$, which reminds us of the integral equation of the resolvent kernel (see Section 1.3).

Remark 2. In the case of the classical Volterra operator

$$(Lx)(t) = \int_0^t k(t, s)x(s)\,ds, \quad t \in [0, T), \tag{3.3.15}$$

with $k(t, s)$ a matrix whose entries are continuous on $\Delta = \{(t, s); 0 \le s \le t < T\}$, we have constructed the resolvent, and found out that it can be expressed in the same form as L:

$$(Rf)(t) = \int_0^t \tilde{k}(t, s)f(s)\,ds, \quad t \in [0, T), \tag{3.3.16}$$

where $\tilde{k}$ is the resolvent kernel of k (see again Section 1.3).

Remark 3.   If instead of the space $C([0, T), R^n)$ we deal with $L_{loc}^{\infty}([0, T), R^n)$, and the operator L is still given by (3.3.15), with $k(t, s)$ having its entries in $L_{loc}^{\infty}(\Delta, R)$, then R is given by (3.3.16), with $\tilde{k}(t, s)$ defined in the same way as in the continuous case, the uniform convergence being substituted with convergence in $L_{loc}^{\infty}(\Delta, \mathscr{L}(R^n, R^n))$. See Neustadt [1] for details.

Remark 4.   The resolvent (kernel) operator has been constructed, under various hypotheses, in Schwabik *et al.* [1], the underlying space being the space of functions with bounded variation, and in Hönig [1], when the underlying space consists of regulated functions (i.e., having only discontinuities of the first kind). In both cases the operator is represented by

$$(Lx)(t) = \int_0^t d_s k(t, s) x(s), \tag{3.3.17}$$

where the index s shows the variable with respect to which integration is performed.

For instance, in Neustadt [1] it is shown that the resolvent is determined by the kernel $R(t, s)$ which is the unique solution of

$$R(t, s) = I + \int_s^t d_u k(t, u) R(u, s), \tag{3.3.18}$$

with kR standing for the matrix product in the case of finite dimensional spaces, or the operator product in the case of Banach spaces of infinite dimension.

The precise conditions under which such scheme works are given in the above indicated references (Schwabik *et al.*, and Hönig).

Let us consider now the functional differential equation (3.3.3), in which L stands for a linear Volterra operator from $L_{loc}^2([0, T), R^n)$ into itself, and $f \in L_{loc}^2([0, T), R^n)$. This equation together with the initial condition $x(0) = x^0 \in R^n$, can be reduced by integration to a Volterra linear equation of the form (3.3.1):

$$x(t) = \int_0^t (Lx)(s)\, ds + x^0 + \int_0^t f(s)\, ds, \quad t \in [0, T). \tag{3.3.19}$$

We need only notice that

$$(L_1 x)(t) = \int_0^t (Lx)(s)\, ds, \quad t \in [0, T), \tag{3.3.20}$$

is a linear operator of Volterra type on $L_{loc}^2([0, T), R^n)$. More precisely, its range consists of locally absolutely continuous functions from $[0, T)$ into R^n, whose derivatives (a.e.) are in $L_{loc}^2([0, T), R^n)$.

If we let $g(t) = x^0 + \int_0^t f(s)\, ds$, $t \in [0, T)$, then equation (3.3.19) becomes

$$x(t) = \int_0^t (Lx)(s)\,ds + g(t), \quad t \in [0, T). \tag{3.3.21}$$

For equation (3.3.21), the classic scheme of successive approximations is applicable, and uniform convergence is assured on each compact interval $[0, t_1] \subset [0, T)$.

Indeed, we can start with $x^0(t) = g(t)$, and for $m \geq 1$ let

$$x^m(t) = \int_0^t (Lx^{m-1})(s)\,ds + g(t), \quad t \in [0, T). \tag{3.3.22}$$

The sequence $\{x^m(t)\}$, $m \geq 0$, consists of locally absolutely continuous functions from $[0, T)$ into R^n, which means $\{x^m(t)\} \subset C([0, T), R^n)$. Convergence will be established in this space, which is a subspace of $L^2_{loc}([0, T), R^n)$ and has a stronger topology.

Before we prove the above assertion, let us notice the following fact about the operator L: there exists a positive nondecreasing function $L(t)$, $t \in [0, T)$, such that

$$\int_0^t |(Lx)(s)|^2\,ds \leq L(t_1) \int_0^t |x(s)|^2\,ds, \quad 0 \leq t \leq t_1, \tag{3.3.23}$$

for any $x \in L^2_{loc}([0, T), R^n)$.

Indeed, let us write the continuity condition for L on $L^2_{loc}([0, T), R^n)$:

$$\int_0^t |(Lx)(s)|^2\,ds \leq L(t) \int_0^t |x(s)|^2\,ds, \quad t \in [0, T), \tag{3.3.24}$$

where $L(t) > 0$ is understood to be the smallest possible in (3.3.24). If we now write the condition (3.3.24) at $t + h$, $h > 0$, we have

$$\int_0^{t+h} |(Lx)(s)|^2\,ds \leq L(t + h) \int_0^{t+h} |x(s)|^2\,ds. \tag{3.3.25}$$

But if $x \in L^2_{loc}([0, T), R^n)$, then for any $0 < t < T$ the function $x_t(s) = x(s)$, $0 \leq s \leq t$, $x_t(s) = 0$ for $s > t$, is also in $L^2_{loc}([0, T), R^n)$. Therefore, (3.3.25) yields

$$\int_0^{t+h} |(Lx_t)(s)|^2\,ds = \int_0^t |(Lx)(s)|^2\,ds + \int_t^{t+h} |(Lx_t)(s)|^2\,ds$$

$$\leq L(t + h) \int_0^t |x(s)|^2\,ds,$$

for any $x \in L^2_{loc}([0, T), R^n)$. Comparing the inequality just obtained with the inequality (3.3.24) we conclude $L(t) \leq L(t + h)$.

The inequality (3.3.24) leads immediately to the following inequality for the successive approximations defined above:

$$|x^{m+1}(t) - x^m(t)|^2 \le L(t_1) \int_0^t |x^m(s) - x^{m-1}(s)|^2 \, ds, \quad m \ge 1, 0 \le t \le t_1. \quad (3.3.26)$$

Since we have for some $M > 0$

$$|x^1(t) - x^0(t)|^2 \le M, \quad t \in [0, t_1], \quad (3.3.27)$$

the inequality (3.3.26) easily leads to

$$|x^{m+1}(t) - x^m(t)|^2 \le M \frac{(Lt)^m}{m!}, \quad t \in [0, t_1], \quad (3.3.28)$$

for $m \ge 1$.

The estimate (3.3.28) shows not only the convergence of the series $\sum |x^{m+1}(t) - x^m(t)|^2$, but also the (absolute and uniform) convergence of the series $\sum [x^{m+1}(t) - x^m(t)]$. Indeed, we have to notice that the series $\sum (Lt_1)^{m/2}(m!)^{-1/2}$ is convergent.

If we let

$$x(t) = \lim_{m \to \infty} x^m(t), \quad t \in [0, T), \quad (3.3.29)$$

then $x(t)$ is a solution of (3.3.19) or, equivalently, a locally absolutely continuous solution of (3.3.1), with $x(0) = x^0$.

The uniqueness of the solution also follows from the method of successive approximations. We omit the details which are standard.

The above discussion related to equation (3.3.3) can be summarized as follows.

Theorem 3.3.2. *Consider the Volterra functional differential equation* (3.3.3) *under the assumptions: L is a linear continuous operator from $L^2_{\text{loc}}([0, T), R^n)$ into itself, and $f \in L^2_{\text{loc}}([0, T), R^n)$.*

Then, there exists a unique solution $x(t) \in C([0, T), R^n)$, which is locally absolutely continuous on $[0, T)$, and such that $x(0) = x^0 \in R^n$.

Remark 1. Since (3.3.3) is equivalent to the 'integral' equation (3.3.19), one can think of applying the result in Theorem 3.3.1 to the last equation. This approach can be used, but requires slightly different assumptions from those involved in Theorem 3.3.2.

Remark 2. The integral equation (3.3.21) can be written in the classical form, namely

$$x(t) = \int_0^t k(t, s)x(s) \, ds + g(t), \quad t \in [0, T), \quad (3.3.30)$$

where the kernel k is determined by the operator L. The representation

$$\int_0^t (Lx)(s)\, ds = \int_0^t k(t,s)x(s)\, ds, \quad t \in [0, T), \tag{3.3.31}$$

for any $x \in L^2_{loc}([0, T), R^n)$ was obtained in Section 2.3.

Therefore, if we denote by $\tilde{k}$ the resolvent kernel associated with k, the solution of (3.3.30) can be represented in the form

$$x(t) = g(t) + \int_0^t \tilde{k}(t,s)g(s)\, ds. \tag{3.3.32}$$

Taking into account that $g(t) = x^0 + \int_0^t f(s)\, ds$, the formula (3.3.32) can be rewritten as

$$x(t) = x^0 + \int_0^t f(u)\, du + \int_0^t \tilde{k}(t,s)\left[x^0 + \int_0^s f(u)\, du \right] ds,$$

which leads immediately to

$$x(t) = X(t,0)x^0 + \int_0^t X(t,s)f(s)\, ds, \tag{3.3.33}$$

where

$$X(t,s) = I + \int_s^t \tilde{k}(t,u)\, du \quad \text{for } 0 \le s \le t < T. \tag{3.3.34}$$

The formula (3.3.33) is the *variation of constants formula* for equation (3.3.3). The (matrix-valued) function $X(t,s)$ defined by (3.3.34), which in the case of ordinary differential equations is given by $X(t,s) = X(t)X^{-1}(s)$, with $X(t)$ a fundamental matrix for $\dot{x} = A(t)x$, does verify certain functional relations, such as $\partial X/\partial s = -\tilde{k}(t,s)$, a.e. on $[0, t]$, for all $t \in (0, T)$. If one takes into account the integral equations of the resolvent kernel, then further relationships can be obtained for $X(t,s)$.

Remark 3. If instead of equation (3.3.3) one considers a similar equation involving a functional parameter ϕ, with ϕ belonging to the space S, say

$$\dot{x}(t) = (Lx)(t) + (B\phi)(t) + f(t), \tag{3.3.35}$$

where $B: S \to L^2_{loc}([0, T), R^n)$ is a linear operator, then the solution of (3.3.35) can be represented as

$$x(t) = X(t,0)x^0 + \int_0^t X(t,s)f(s)\, ds + \int_0^t X(t,s)(B\phi)(s)\, ds, \tag{3.3.36}$$

with $X(t,s)$ defined by (3.3.34).

In concluding this section, we shall apply the above results – particularly the formula (3.3.36) – to the case of functional differential equations with infinite delay. Such equations were considered briefly in the preceding section in relation to the existence of solutions, and they are usually written in the form (3.2.46), with initial data (3.2.47).

Since only linear equations are now of interest to us, we shall write such equations in the form (3.2.57), namely

$$\dot{x}(t) = L(t, x_t) + f(t), \quad t \in [0, T),$$ (3.3.37)

with initial conditions

$$x_0 = \phi \in S, \quad x(0+) = x^0 \in R^n.$$ (3.3.38)

We shall assume that $L(t, \phi)$ is linear in the second argument, continuous in $(t, \phi) \in [0, T) \times S$, and satisfying an inequality of the form

$$|L(t, \phi)| \leq \lambda(t)|\phi|_S,$$ (3.3.39)

where λ is nonnegative and locally square integrable on $[0, T)$. The function f in (3.3.37) will be assumed continuous on $[0, T)$, even though more general assumptions could work in this respect. Finally, the function space S will be chosen as described in Proposition 3.2.4, condition (b), and it is in fact the space of initial functions.

Since the representation

$$x_t(u) = x_t(u)\chi_{[-t,0]}(u) + \phi(t + u)\chi_{(-\infty, -t)}(u)$$

holds for any $\phi \in S$ and $x \in C([0, T), R^n)$, we can rewrite equation (3.2.57) in the form (3.3.35), where

$$(Lx)(t) = L(t, x_t\chi_{[-t, 0]}), \quad t \in [0, T)$$ (3.3.40)

and

$$(B\phi)(t) = L(t, \phi(t + \cdot)\chi_{(-\infty, -t)}(\cdot)).$$ (3.3.41)

Therefore, taking into account the results established above in this section, as well as the general existence result given in Proposition 3.2.4, we can say that equation (3.3.37) has a unique solution (continuously differentiable in t) on $[0, T)$ which is representable in the form

$$x(t) = X(t, 0)x^0 + \int_0^t X(t, s)f(s)\,\mathrm{d}s + \int_0^t X(t, s)L(s, \phi(s + \cdot)\chi_{(-\infty, -s)}(\cdot))\,\mathrm{d}s.$$

(3.3.42)

Obviously, the procedure to obtain (3.3.42) can be adapted to different hypotheses from those used above.

3.4 A singular perturbation approach to abstract Volterra equations

We shall now illustrate the method of singular perturbation in providing some existence results for the abstract Volterra equations

$$x(t) = (Vx)(t), \quad t \in [0, a], \tag{3.4.1}$$

and its perturbed versions, where V satisfies the conditions listed at the beginning of Section 3.2, but not necessarily with respect to the space $C([0, a], R^n)$. The spaces $L^p([0, a], R^n)$, $1 \leq p < \infty$, will be also dealt with.

The singular perturbation method should be regarded, in this case, as a regularization/approximation procedure for the solutions of equation (3.4.1), even though what we have in mind is to obtain an alternative proof of the basic existence result in the continuous case, and new results in L^p-spaces.

Roughly speaking, we are trying to construct approximate solutions to (3.4.1) by means of the solutions of the functional-differential equation of Volterra type

$$\varepsilon \dot{x}_\varepsilon(t) = -x_\varepsilon(t) + (Vx_\varepsilon)(t), \tag{3.4.2}$$

where $\varepsilon > 0$ is a small parameter. As we shall find out subsequently, this procedure can be relatively simply validated.

Concerning the initial condition to be associated with (3.4.2), we notice the following feature in the case when V is acting on the space of continuous maps $C([0, a], R^n)$: since V has the property of being of constant initial value, i.e., there exists $x^0 \in R^n$ such that $(Vx)(0) = x^0$ for any $x \in C([0, a], R^n)$, the solution of (3.4.1), if any, will satisfy the initial condition $x(0) = x^0$. Hence, it is a reasonable assumption to impose on $x_\varepsilon(t)$ the initial condition

$$x_\varepsilon(0) = x^0 = \text{the fixed initial value of } V. \tag{3.4.3}$$

If V is given on spaces of measurable functions such as $L^p([0, a], R^n)$, $1 \leq p < \infty$, we shall see that it is immaterial which solution we choose among the solutions of (3.4.2), in order to approximate the solution of (3.4.1).

Notice first that the functional-differential equation (3.4.2) can be put in integral form, namely

$$x_\varepsilon(t) = Ce^{-t/\varepsilon} + \frac{1}{\varepsilon} \int_0^t e^{-(t-s)/\varepsilon} (Vx_\varepsilon)(s) \, ds, \tag{3.4.4}$$

where C is an arbitrary constant vector in R^n.

If we introduce the notation

$$\phi(t, \varepsilon) = \begin{cases} 0, & t < 0 \\ \dfrac{1}{\varepsilon} e^{-t/\varepsilon}, & t \in R_+, \end{cases} \tag{3.4.5}$$

then (3.4.4), with $C = x^0$, becomes

$$x_\varepsilon(t) = x^0 e^{-t/\varepsilon} + \int_0^t \phi(t - s, \varepsilon)(Vx_\varepsilon)(s) \, ds. \tag{3.4.6}$$

It is useful to point out the following properties of the (scalar) function $\phi(t, \varepsilon)$:

(a) $\int_0^t \phi(s, \varepsilon)\,ds = 1 - e^{-t/\varepsilon}, \quad t \in R_+;$

(b) $\lim_{\varepsilon \to 0} \int_0^T \phi(s, \varepsilon)\,ds = 1, \quad T > 0;$

(c) $\lim_{\varepsilon \to 0} \int_\delta^T \phi(s, \varepsilon)\,ds = 0, \quad \delta > 0, T > 0.$

Because of the properties (a), (b), (c), the family of functions $\{\phi(t, \varepsilon); \varepsilon > 0\}$ provides an *approximate identity* for the convolution, on any interval $[0, T]$. Therefore, as shown in Edwards [2], for any $\{\varepsilon_m\}$, such that $\varepsilon_m \downarrow 0$, we have

$$\lim_{m \to \infty} \int_0^t \phi(t - s, \varepsilon_m)f(s)\,ds = f(t), \quad f \in C([0, T], R^n), f(0) = 0, \quad (3.4.7)$$

uniformly on $[0, T]$, and

$$\lim_{m \to \infty} \int_0^t \phi(t - s, \varepsilon_m)f(s)\,ds = f(t), \quad f \in L^p([0, T], R^n), \quad (3.4.8)$$

in the convergence of $L^p([0, T], R^n)$, $1 \le p < \infty$. (3.4.8) holds for almost all $t \in [0, T]$, as shown for instance in Sadosky [1].

Let us go back now to equation (3.4.6), and assume that V is a Volterra type operator, continuous and compact on $C([0, a], R^n)$, with fixed initial value x^0. If we take into account (a), then (3.4.6) can be rewritten as

$$x_\varepsilon(t) - x^0 = \int_0^t \phi(t - s, \varepsilon)[(Vx_\varepsilon)(s) - x^0]\,ds. \quad (3.4.9)$$

The form (3.4.9) of the equation (3.4.6) is very convenient for the application of the Schauder fixed point theorem. More precisely, we shall obtain a local existence theorem for (3.4.6), or, equivalently, for the functional-differential equation (3.4.2), under initial condition (3.4.3).

First, notice that, for any $\varepsilon > 0$, the operator appearing in the right hand side of (3.4.9) is a continuous compact operator on $C([0, a], R^n)$. Indeed, if we let

$$(T_\varepsilon u)(t) = x^0 + \int_0^t \phi(t - s, \varepsilon)[(Vu)(s) - x^0]\,ds, \quad (3.4.10)$$

for any $u \in C([0, a], R^n)$, then T_ε is the product of two continuous operators on $C([0, a], R^n)$, one of which is also compact [in our case, $(Vu)(t) - x^0$]. Hence, T_ε is a continuous compact operator on $C([0, a], R^n)$. It remains to show only that T_ε takes into itself a certain convex closed set of $C([0, a], R^n)$.

We shall consider the ball of radius $r > 0$, centered at x^0:

$$S_r = \{x | x \in C([0, a], R^n), |x(t) - x^0| \le r\}.$$

The number r can be chosen arbitrarily, or we may assume that V is given only on the ball S_r.

Since the operator V (and therefore $V - x^0$) is compact on S_r, there results the existence of a positive number $\eta = \eta(r) > 0$, such that $|(Vu)(t) - (Vu)(s)| < r$ for each $u \in S_r$, as soon as $|t - s| \leq \eta(r)$, $t, s \in [0, a]$. But V is of fixed initial value x^0, and consequently $|(Vu)(s) - x^0| < r$ for $0 \leq s \leq \eta$.

Therefore, if we restrict V and T_ε to the interval $[0, \eta]$, and we take into account property (a) of the approximate identity, we obtain from (3.4.10)

$$|(T_\varepsilon u)(t) - x^0| < r(1 - e^{-t/\varepsilon}) < r, \quad 0 \leq t \leq \eta,$$

which obviously shows that $T_\varepsilon S_r \subset S_r$, where S_r consists now of functions restricted to $[0, \eta]$.

On the other hand, S_r is closed and convex in $C([0, \eta], R^n)$, which implies, as a result of Schauder's fixed point theorem, the existence of at least one solution $x_\varepsilon(t)$ of equation (3.4.9), such that $x_\varepsilon \in S_r$.

Now let $\{\varepsilon_m\}$ be a sequence of positive numbers such that $\varepsilon_m \downarrow 0$. For each m, let $x_{\varepsilon_m}(t)$ be a solution of (3.4.9) defined on $[0, \eta]$, and such that $x_{\varepsilon_m}(t) \in S_r$, $t \in [0, \eta]$,

$$x_{\varepsilon_m}(t) - x^0 = \int_0^t \phi(t - s, \varepsilon_m)[(Vx_{\varepsilon_m})(s) - x^0]\,ds. \qquad (3.4.11)$$

We can take advantage of the compactness of the operator V, and assume – without loss of generality – that $\{Vx_{\varepsilon_m}\}$ is convergent in $C([0, \eta], R^n)$. Then, because of property (3.4.7) for the approximate identity of the convolution, the existence of the limit in the right hand side of (3.4.11) implies the convergence of $\{x_{\varepsilon_m}(t)\}$ to some $x(t) \in S_r$, and therefore, letting $m \to \infty$ in (3.4.11), we obtain $x(t) - x^0 = (Vx)(t) - x^0$. This is the same as (3.4.1).

The discussion conducted above clarifies completely the case of continuous solutions to (3.4.1). The result is the same as in Theorem 3.2.1. The extra information provided by the above procedure consists mainly in the possibility of constructing approximate (or regularized) solutions for the abstract equation (3.4.1).

We can now discuss the existence problem for equation (3.4.1) in the case of L^p-spaces, $1 \leq p < \infty$.

First, we shall notice that $|e^{-t/\varepsilon}|_p \to 0$ as $\varepsilon \to 0$, $1 \leq p < \infty$. This means that instead of the integral equation (3.4.6), we can limit our considerations to the simpler equation

$$x_\varepsilon(t) = \int_0^t \phi(t - s, \varepsilon)(Vx_\varepsilon)(s)\,ds, \quad t \in [0, a]. \qquad (3.4.12)$$

In other words, for the functional-differential equation (3.4.2) we choose the zero initial condition instead of (3.4.3). We should point out that this last condition does not make sense when we are looking for measurable solutions (not continuous!).

Equation (3.4.12) can be shown to possess a solution $x_\varepsilon(t)$ in $L^p([0,a],R^n)$, provided adequate assumptions are made on the operator V. Also, local existence is what we should expect, unless really restrictive conditions are accepted on the operator V.

We will assume that V is a continuous compact operator of Volterra type on $L^p([0,a],R^n)$, $1 \le p < \infty$, with a certain fixed p, and $\varepsilon > 0$ is fixed. Then we consider the operator

$$(T_\varepsilon u)(t) = \int_0^t \phi(t - s, \varepsilon)(Vu)(s)\,ds \qquad (3.4.13)$$

on a certain ball, say

$$B_r = \{x \,|\, x \in L^p([0,a],R^n), |x|_p \le r\}, \qquad (3.4.14)$$

and prove that T_ε, or rather its restriction to some $L^p([0,\delta],R^n)$ with $\delta \le a$, has a fixed point.

Indeed, if we use a well-known property of convolutions, namely, for $f \in L^1$ and $g \in L^p$, $1 \le p < \infty$, one has $f * g \in L^p$, and $|f * g|_p \le |f|_1 |g|_p$, then we see that T_ε is continuous and compact on $L^p([0,a],R^n)$. On the basis of the above inequality one can easily show that B_r, possibly with $[0,a]$ replaced by a smaller interval, is taken into itself by the operator T_ε. We obtain from (3.4.13) on $[0,t]$:

$$\int_0^t |(T_\varepsilon u)(s)|^p\,ds \le (1 - e^{-t/\varepsilon}) \int_0^t |(Vu)(s)|^p\,ds \qquad (3.4.15)$$

for any $t \in [0,a]$, and $u \in B_r$. It is now obvious that the inclusion $T_\varepsilon B_r \subset B_r$ will follow from

$$\int_0^t |(Vu)(s)|^p\,ds \le r^p, \quad u \in B_r, \qquad (3.4.16)$$

if $t \in [0,\delta]$, for some $\delta \le a$.

The validity of the inequality (3.4.16) on some interval $[0,\delta]$ can be established based on the compactness of the set of images $\{Vu | u \in B_r\} \subset L^p$. Indeed, since the set of images is relatively compact in L^p, it admits a finite α-net for every $\alpha > 0$. On the other hand, if $\{u_k\}$, $k = 1, 2, \ldots, m$, is the finite α-net, then (3.4.16) holds for every $u = u_k$, $k = 1, 2, \ldots, m$, for some $\delta > 0$, $\delta \le a$. For an aribtrary $u \in B_r$ we then have

$$\int_0^t |(Vu)(s)|^p\,ds < 2\alpha, \quad 0 \le t \le \delta. \qquad (3.4.17)$$

It suffices to choose $2\alpha \le r^p$, in order to establish the validity of (3.4.16).

The Schauder fixed point theorem can now be applied to the operator T_ε and the ball B_r, on the interval $[0, \delta]$, with δ constructed as shown above. Hence, the existence in $L^p([0, \delta], R^n)$ for (3.4.12) is established for every $\varepsilon > 0$. It should be pointed out that δ does not depend on ε, as seen from the construction.

The next, and last, step in proving the existence in L^p, $1 \leq p < \infty$, for (3.4.1), proceeds in a completely analogous way to the continuous case examined above. The only differences consist in dealing with L^p-compactness instead of compactness with respect to uniform convergence, and the application of the property (3.4.8) of the approximate identity for convolution, instead of (3.4.7).

Summarizing the discussion of equation (3.4.1), conducted in this section, we can state the following basic result.

Theorem 3.4.1. *Let the operator V in equation (3.4.1) be of Volterra type, continuous and compact from E into itself, where E stands for $C([0, a], R^n)$, or $L^p([0, a], R^n)$, $1 \leq p < \infty$. In the case $E = C$ one also assumes that V has fixed initial value. Then, there exists at least one solution of (3.4.1) that belongs to the space E (possibly restricted to some interval $[0, \delta]$, $\delta \leq a$).*

Moreover, the singularly perturbed equation (3.4.2) is solvable in E for $\varepsilon > 0$, and its solutions can be regarded as regularized solutions for (3.4.11).

Remark 1.   As noticed above in this section, in relation to the type of convergence involved in the basic property of the approximate identity for convolution, if we deal with the L^p-spaces we can also assert the convergence almost everywhere for the sequence of approximate solutions to the exact solution of (3.4.1).

Remark 2.   The continuous case requested one extra condition (the fixed initial value property) for the operator V. This can, of course, be explained if we keep in mind that the L^p-spaces are much richer than the space C, and therefore fewer restrictions are necessary to secure the existence of a solution.

Since the continuous case for equation (3.4.1) has been discussed separately in Section 3.2, and the classical type of Volterra integral equations has been dealt with in detail in Section 3.1, we shall consider here only one example illustrating the case of L^p-spaces. This example is, in fact, an extension of the result stated in Section 3.2 regarding the nonlinear equation

$$x(t) = f(t) + \int_0^t k(t, s) g(s, x(s)) \, ds, \ t \in [0, a], \qquad (3.4.18)$$

known as the Volterra–Hammerstein equation.

The following assumptions will be made with regard to equation (3.4.18):

(a) $f \in L^p([0, a], R^n)$, $1 < p < \infty$;
(b) the $n \times n$ matrix kernel $k(t, s)$ is measurable on $0 \leq s \leq t \leq a$, and such that

146 Basic theory of Volterra equations: integral and abstract

$$\left(\int_s^a |k(t,s)|^p \, dt \right)^{1/p} \in L^q([0,a], R), \qquad (3.4.19)$$

where $p^{-1} + q^{-1} = 1$;

(c) g is a map from $[0,a] \times R^n$ into R^n, satisfying the Carathéodory condition, and such that

$$|g(t,x)| \leq c(t) + b|x|^{p/q}, \qquad (3.4.20)$$

where $c \in L^q([0,a], R)$, and $b > 0$.

Under conditions (a), (b), and (c) one can easily see (on the basis of Section 2.3) that the operator

$$(Tx)(t) = f(t) + \int_0^t k(t,s)g(s,x(s)) \, ds$$

is a continuous compact Volterra operator on $L^p([0,a], R^n)$. Therefore, Theorem 3.4.1 applies directly and yields the existence of a local solution x (on some $[0,\delta]$, $\delta \leq a$) for equation (3.4.18), such that $x \in L^p([0,\delta], R^n)$.

In concluding this section, we shall deal with the Volterra linear equation of the first kind

$$\int_0^t A(t,s)x(s) \, ds = f(t), \quad t \in [0,T], \qquad (3.4.21)$$

where $x, f \in R^n$, and $A \in \mathscr{L}(R^n, R^n)$ for every $(t,s) \in \Delta = \{(u,v); 0 \leq u \leq v \leq T\}$. We shall apply the technique of singular perturbation in order to approximate the solution of (3.4.21).

The following conditions will be assumed in relation to (3.4.21):

(1) $f \in C^{(1)}([0,T], R^n)$, $f(0) = 0$, with f' defined only as a right derivative at $t = 0$, and as the left derivative at $t = T$;
(2) $A \in C(\Delta, \mathscr{L}(R^n, R^n))$, and $\partial A/\partial t \in C(\Delta, \mathscr{L}(R^n, R^n))$, with similar precisions as above for the derivative $\partial A/\partial t$.
(3) $A(t,t) = I$ (the unit matrix), $t \in [0,T]$.

Condition (3) is not a real restriction when $\det A(t,t) \neq 0$ on $[0,T]$. Indeed, in order to assure (3), one can multiply both sides of (3.4.21) by the inverse of the matrix $A(t,t)$.

It is well known that, under conditions (1), (2), (3), equation (3.4.21) is equivalent to the Volterra linear equation of the second kind

$$x(t) + \int_0^t \frac{\partial}{\partial t} A(t,s)x(s) \, ds = f'(t), \qquad (3.4.22)$$

which has a unique solution in $C([0,T], R^n)$.

Consider now the Volterra equation of the second kind

$$\varepsilon x_\varepsilon(t) + \int_0^t A(t,s)x_\varepsilon(s)\,\mathrm{d}s = f(t), \quad t \in [0, T], \tag{3.4.23}$$

where $\varepsilon > 0$ is a 'small' parameter. This equation has a unique solution $x_\varepsilon(t)$ for every $\varepsilon > 0$, and it is defined on the whole interval $[0, T]$. We are interested in estimating the function

$$y_\varepsilon(t) = x_\varepsilon(t) - x(t), \quad t \in [0, T],$$

where $x(t)$ is the unique solution of the equation (3.4.21).

From (3.4.21) and (3.4.23) we obtain the equation

$$y_\varepsilon(t) + \int_0^t \varepsilon^{-1}A(t,s)y_\varepsilon(s)\,\mathrm{d}s = -x(t). \tag{3.4.24}$$

Let us now convolve both sides of (3.4.24) with the function $\phi(t, \varepsilon)$ defined by (3.4.5), and then subtract the result from (3.4.24).

The following integral equation is obtained

$$y_\varepsilon(t) + \int_0^t A_\varepsilon(t,s)y_\varepsilon(s)\,\mathrm{d}s = -x(t) + \int_0^t \phi(t - s, \varepsilon)x(s)\,\mathrm{d}s, \tag{3.4.25}$$

in which the kernel $A_\varepsilon(t,s)$ is given by

$$A_\varepsilon(t,s) = \varepsilon^{-1}A(t,s) - \phi(t - s, \varepsilon)I - \varepsilon^{-1}\int_s^t \phi(t - \tau, \varepsilon)A(\tau, s)\,\mathrm{d}\tau. \tag{3.4.26}$$

Integrating by parts in the last integral on the right hand side of (3.4.26), and taking condition (3) into account, $A_\varepsilon(t, s)$ can be written in the form

$$A_\varepsilon(t,s) = \int_s^t \phi(t - \tau, \varepsilon)\frac{\partial}{\partial\tau}A(\tau, s)\,\mathrm{d}\tau. \tag{3.4.27}$$

Because of condition (2), the derivative $\partial A/\partial\tau$ is bounded in the norm on Δ, which allows us to obtain from (3.4.27) an inequality of the form

$$|A_\varepsilon(t,s)| \leq C[1 - \mathrm{e}^{(t-s)/\varepsilon}], \tag{3.4.28}$$

where $|\cdot|$ is a matrix norm (for instance, the Euclidean one).

Equation (3.4.25) and inequality (3.4.28) lead to the following Gronwall type inequality for $y_\varepsilon(t)$:

$$|y_\varepsilon(t)| \leq K(\varepsilon) + C\int_0^t |y_\varepsilon(s)|\,\mathrm{d}s, \quad t \in [0, T], \tag{3.4.29}$$

where

$$k(\varepsilon) = \sup_{t \in [0, T]} \left| -x(t) + \int_0^t \phi(t - s, \varepsilon)x(s)\,\mathrm{d}s \right|. \tag{3.4.30}$$

Inequality (3.4.29) implies

$$|y_\varepsilon(t)| \le K(\varepsilon)e^{CT}, \quad t \in [0, T], \tag{3.4.31}$$

which means

$$|x_\varepsilon(t) - x(t)| \le K(\varepsilon)e^{CT}, \quad t \in [0, T]. \tag{3.4.32}$$

Taking into account the properties of an approximate identity for convolution, from (3.4.32) we derive $\lim x_\varepsilon(t) = x(t)$ as $\varepsilon \to 0$, provided $x(0) = 0$ (see formula (3.4.7)).

Consequently, under the extra assumption $x(0) = 0$, $x_\varepsilon(t)$ converges uniformly on $[0, T]$ to the solution $x(t)$ of the first kind of Volterra equation (3.4.21). If $x(0) \ne 0$, then uniform convergence can be shown to hold on any interval $[t_0, T]$, with $t_0 > 0$.

A more general problem related to the singular perturbation method is the following:

Given the first kind of Volterra equation in abstract form

$$(Lx)(t) = f(t), \quad t \in [0, T], \tag{3.4.33}$$

and considering the perturbed equation

$$\varepsilon x_\varepsilon(t) + (Lx_\varepsilon)(t) = f(t), \quad t \in [0, T], \tag{3.4.34}$$

find conditions under which $x_\varepsilon(t) \to x(t)$ as $\varepsilon \to 0$ in convenient norms. One could also choose as perturbed equation the following one:

$$\varepsilon\dot{x}_\varepsilon(t) + (Lx_\varepsilon)(t) = f(t), \quad t \in [0, T]. \tag{3.4.35}$$

More information about the singular perturbation (regularization) method applied to Volterra equations can be found in the book by Lavrentiev, Romanov, and Shishatski [1].

Bibliographical notes

The topics discussed in Section 3.1 pertain to the classical heritage in the theory of Volterra integral equations. Most books on integral equations do include some results on Volterra classical equations, particularly the existence (and sometimes) uniqueness theorems. In Tricomi's book [1], the existence theorem is proven in the case of measurable solutions (square integrable). The book by Miller [1] provides more recent contributions, and also investigates the problem of the continuation of solutions, as well as the behavior of saturated solutions at the endpoint. Further investigations into these basic problems concerning classical Volterra equations can be found in many journal papers. The survey paper by Caljuk [2] contains other useful references related to the basic problems discussed in Section 3.1. Of the journal papers on the basic theory of Volterra integral equations, sometimes including results related to integral inequalities, I mention the following: Sato [1], Azbelev and Caljuk [1], Artstein [1], [2], Gollwitzer and Hager [1], Kiffe [1], [2]; the monographs by Kurpel' and Shuvar [1], Mamedov and Ashirov [1], Miller and Sell [1] are also useful references. The comparison method and the theory of continuous dependence of solutions with respect to the data seems to be as yet insufficiently developed. See Burton [1], [2] and Staffans [8].

In Section 3.2 a theory of abstract Volterra equations is sketched. This theory originates in Tonelli's paper [2], and the ideas contained in this paper have been continued by Cinquini [1] and Graffi [1], who considered existence, uniqueness and continuous dependence problems related to the abstract Volterra equation. Tychonoff [1] devoted a notable paper to this subject emphasizing the role that such equations are playing in the description of physical phenomena (which are causal). More recently, Neustadt [1] investigated some basic problems related to equations involving abstract Volterra operators, pointing our their significance for the development of a control theory for systems governed by them. During the 1970s and 1980s many authors contributed to the development of this theory (in Soviet literature such equations are sometimes called 'Tychonoff equations'), especially in journal papers: Caljuk [1], Bachurskaja and Caljuk [1], Corduneanu [2], [12], Karakostas [1], Szufla [1], [2], Turinici [1], [2], [3], Visintin [1].

Section 3.2 also contains a discussion of the problem of how delay equations can be treated as Volterra functional-differential equations. The case of infinite delay equations is treated in some detail, because their theory is significantly more complicated than the theory of equations with finite delay (when the choice of the phase space, or of the initial space, is not as difficult as it is for infinite delays). It turns out that this approach is efficient enough and also allows a higher degree of generality in formulating the problem (it is almost trivial to show that any delay equation can be regarded as a Volterra type equation), a better understanding of the theory and the solution of some problems.

In Section 3.3 the linear case of equations with Volterra operators is discussed, and the resolvent operator is constructed (again, in the abstract setting). The proof is that given by Neustadt [1]. The case of linear functional-differential equations with Volterra operators is also considered in this section, a global existence result being obtained. One of the most interesting results of this section is related to the construction of the variation of constants formula for the abstract Volterra linear differential equation. This formula is then applied to the special case of equations with infinite delays, with an initial space given axiomatically. The literature pertaining to this problem is very rich. I will mention here the basic paper by Hale and Kato [1], the paper by Naito [1], as well as those of Sawano [1] and Tyshkevich [1]. The survey paper by Corduneanu and Lakshmikantham [1] contains most of the significant contributions to this topic until 1980. A more recent survey paper is due to Azbelev [1]; see also Azbelev and Maksimov [1]. For a different approach to the problem of infinite delays see Coffman and Schäffer [1], and Schäffer [1]. The paper by Marcus and Mizel [1] also deals with problems related to infinite delays.

In Section 3.4, a singular perturbation approach to abstract Volterra equations is illustrated. These results appeared for the first time in C. Corduneanu [14]. The procedure was inspired by reading the paper of Reich [1]. The significance of the method in the abstract setting follows from the fact that the solution of the abstract Volterra equation, even in spaces of measurable functions, can always be approximated by means of absolutely continuous functions. Of course, the method has been applied earlier, but only the classical Volterra operator has been considered. Recent contributions to this subject are due to Angell and Olmstead [1], and Imanaliev [1]. The book by Groetsch [2] is a good source for the regularization procedure for integral equations, the case of

Fredholm type equations being considered. The application of the regularization method to the Volterra linear equation of the first kind given in this section is due to Imomnazarov [1]. With regard to Volterra integral equations of the first kind, see the recent contributions of Magnickii [1] and Myshkis [1].

4

Some special classes of integral and integrodifferential equations

In this chapter we shall discuss various kinds of asymptotic behavior of the solutions for several classes of nonlinear integral equations. Usually, such equations will be considered on a half-axis, or on the whole real axis (as it follows from the nature of the problem when, for instance, we deal with existence of almost periodic solutions).

Each kind of asymptotic behavior is usually described by the fact that the solution belongs to a certain function space (among those investigated in Chapter 2). This means that we shall look for the existence of solutions in such function spaces.

Most of the integral equations to be considered will belong to the so-called Hammerstein type, which in abstract form could be written as $x = f + LGx$, where L is a linear operator, and G is – generally – nonlinear.

We shall use fixed point methods, as well as other tools available from functional analysis (for instance, monotone operators).

Attention will be paid to some classes of convolution equations, for which the frequency domain technique can be successfully applied.

Boundary value problems for ordinary differential equations and Volterra functional differential equations will be also investigated within this chapter.

4.1 Hammerstein equations and admissibility technique

We shall begin with some classes of nonlinear integral equations that can be written in the form

$$x(t) = h(t) + \int_0^\infty k(t, s)f(s; x)\,ds, \tag{4.1.1}$$

where x, h take values in R^n, k is an n by n matrix whose entries are real-valued functions defined on $R_+ \times R_+$, and f stands for an operator acting between convenient function spaces with elements defined on R_+.

The case $k(t,s) = 0$ (the zero matrix) for $t > s$ leads to the Hammerstein–Volterra equation

$$x(t) = h(t) + \int_0^t k(t,s)f(s;x)\,ds. \qquad (4.1.2)$$

In both equations (4.1.1) and (4.1.2), we have a superposition of operators, one of which is a linear integral operator. This is the reason for relying on admissibility results in relation to such operators.

Let us also point out that the procedure we shall follow in this section is applicable to equations of the form

$$x(t) = h(t;x) + \int_0^t k(t,s)x(s)\,ds, \qquad (4.1.3)$$

to mention only the Volterra case. If we denote by $\tilde{k}(t,s)$ the resolvent kernel corresponding to $k(t,s)$, then an equivalent form of the equation (4.1.3) is

$$x(t) = h(t;x) + \int_0^t \tilde{k}(t,s)h(s;x)\,ds \qquad (4.1.4)$$

Before we consider the abstract form of Hammerstein equations, let us provide some relatively simple results relating to the equations mentioned above. These results are obtainable by means of the contraction mapping principle, taking into account the admissibility results given in Section 2.3.

First, let us deal with the case of equations of the form (4.1.1). In this case, we shall rely on the admissibility result given in Theorem 2.3.2.

Theorem 4.1.1. *Consider the integral equation* (4.1.1), *under the following assumptions*:

(1) *the triplet* (E_g, C_G, K) *satisfies the conditions* (a), (b), (c) *of Theorem* 2.3.2, *which assure the strong stability of the linear system* (E_g, C_G, K);
(2) *the map* $f: C_G \to E_g$ *is Lipschitz continuous, i.e.*

$$|f(\cdot;x) - f(\cdot;y)|_{E_g} \le \lambda|x - y|_{C_G}, \qquad (4.1.5)$$

   with λ *sufficiently small*;
(3) $h \in C_G$,

Then equation (4.1.1) *has a unique solution* $x \in C_G$.

Proof. On the space $C_G(R_+, R^n)$, we consider the operator

$$(Tx)(t) = h(t) + \int_0^\infty k(t,s)f(s;x)\,ds, \qquad (4.1.6)$$

which is obviously defined on the whole space. It is easy to show that $T: C_G \rightarrow C_G$ is a contraction, provided λ is sufficiently small. Indeed, (4.1.6) yields for $x, y \in C_G$:

$$(Tx)(t) - (Ty)(t) = \int_0^\infty k(t, s)[f(s; x) - f(s; y)] \, ds. \qquad (4.1.7)$$

As seen in Theorem 2.3.2, (4.1.7) implies the inequality:

$$|Tx - Ty|_{C_G} \le \lambda |G_{-1}(t)k(t; \cdot)g(\cdot)|_{E'} |x - y|_{C_G}. \qquad (4.1.8)$$

Therefore, it suffices to assume

$$\lambda < (|G_{-1}(t)k(t; \cdot)g(\cdot)|_{E'})^{-1}, \qquad (4.1.9)$$

in order to assure that T is a contraction on C_G.

This ends the proof of Theorem 4.1.1. We shall now consider equation (4.1.2), which is of Volterra type, provided f is a Volterra operator.

Theorem 4.1.2. *Assume that the following conditions are satisfied with respect to equation (4.1.2):*

(1) *the triplet (C_g, C_G, K) satisfies the conditions required by Theorem 2.3.1, which assure the (strong) admissibility of the linear system (C_g, C_G, K);*
(2) *the map $f: C_G \rightarrow C_g$ is Lipschitz continuous, i.e.*

$$|f(\cdot; x) - f(\cdot; y)|_{C_G} \le \lambda |x - y|_{C_G}, \qquad (4.1.10)$$

for any $x, y \in C_G$, with λ sufficiently small;
(3) *$h \in C_G$.*

Then equation (4.1.2) has a unique solution $x \in C_G$.

The proof can be immediately reduced to the contraction mapping principle for the operator

$$(Tx)(t) = h(t) + \int_0^t k(t, s)f(s; x) \, ds.$$

The details are left to the reader, including finding an upper bound for λ in (4.1.10).

It is to be noted that C_g is not an E_g, in which case Theorem 4.1.2 would be a special case of Theorem 4.1.1.

The degree of generality reached in Theorems 4.1.1 and 4.1.2 is considerable. Most of the results based on admissibility technique in C. Corduneanu [4] are special cases of these two theorems. In particular, the weight functions g, G can be chosen of scalar type (i.e., $g = g_0 I$ with g_0 scalar and I the unit matrix).

Let us notice now that equation (4.1.3) is a perturbed version of the linear equation

154 Some special classes of integral and integrodifferential equations

$$x(t) = h(t) + \int_0^t k(t,s)x(s)\,ds, \qquad (4.1.11)$$

in which the term $h(t)$ is replaced by $h(t; x)$ – in general, a nonlinear operator acting on the space in which the solution is sought.

Besides the admissibility concept we defined in Chapter 2, in relation to an operator and a pair of spaces, we shall now consider the *admissibility* with respect to the integral equation (4.1.11). Namely, we shall say that the pair of function spaces (B, D) is *admissible* with respect to (4.1.11), if for every $h \in B$, the (unique) solution x of (4.1.11) belongs to D.

Notice that this concept of admissibility is somewhat more general than the *stability* concept for integral equations considered, for instance, in Vinokurov [1], [2].

The following result provides a necessary and sufficient condition for the admissibility of the pair (C_g, C_G), with respect to equation (4.1.11):

Theorem 4.1.3. *The pair of spaces (C_g, C_G) is admissible with respect to equation (4.1.11) with continuous kernel $k(t,s)$ in Δ, if and only if the following conditions are verified:*

(1) $P_G(t)g(t) = g(t)$, $\forall t \in R_+$;

(2) $P_G(t)\tilde{k}(t,s)g(s) = \tilde{k}(t,s)g(s)$, $\forall(t,s) \in \Delta$;

(3) *there exists a positive constant A, such that*

$$|G_{-1}(t)g(t)| + \int_0^t |G_{-1}(t)\tilde{k}(t,s)g(s)|\,ds \leq A, \quad \forall t \in R_+. \qquad (4.1.12)$$

The proof of Theorem 4.1.3 can be carried out following the same lines as in the case when g and G are scalar weight functions (see C. Corduneanu [4]).

We notice the fact that the conditions of Theorem 4.1.3 involve the resolvent kernel $\tilde{k}(t,s)$, associated with $k(t,s)$, which is not a desirable feature. Indeed, given $k(t,s)$, $\tilde{k}(t,s)$ is uniquely determined (see, for instance, C. Corduneanu [4], [11], Tricomi [1]) but finding it and checking conditions (2) and (3) of the theorem might be quite difficult.

Therefore, it remains as an open problem to find conditions for admissibility involving only the assigned data.

Based on the above discussion, it is possible to conduct an investigation of the nonlinear equation (4.1.3), obtained by perturbing the free term $h(t)$ in the linear equation (4.1.11). Using again the contraction mapping principle, one can easily obtain an existence (and uniqueness) result for (4.1.3). We leave this task to the reader.

We shall now define a more general concept of admissibility with respect to an equation, following Cushing [1], [2]. We will use an abstract formulation of the problem, applying the result to the case of integral equations.

Let us consider the equation

$$Lx = h, \tag{4.1.13}$$

where $x, h \in F$, with F a Fréchet space, and L a closed linear operator that maps one-to-one F onto F. This means (closed graph theorem) that both L and L^{-1} are continuous from F onto F.

For instance, if we let

$$(Lx)(t) = \dot{x}(t) - \int_0^t k(t,s)x(s)\,ds, \tag{4.1.14}$$

then equation (4.1.13) reduces to equation (4.1.11). Assuming k to be continuous on Δ, and taking $F = C(R_+, R^n)$, one finds out easily that L in (4.1.14) satisfies the conditions imposed on L in (4.1.13). The results given in Section 3.4 provide the basis for reaching the above conclusion, even in the case of an abstract Volterra operator (instead of $\int_0^t k(t,s)x(s)\,ds$).

The new concept of admissibility will be formulated in relation to equation (4.1.13). Let us denote by B and D some normed subspaces (not necessarily closed) of F, and assume that B admits a decomposition $B = B_1 \oplus B_2$.

The following assumption is the basic support of the new concept of *admissibility*:

H: For any $h_2 \in B_2$, there exists $h_1 \in B_1$ such that $h_1 + h_2 \in L(D)$.

In other words, for each element $h_2 \in B_2$, there exists at least one element $h_1 \in B_1$, with the property that for $h = h_1 + h_2 \in B$, equation (4.1.13) has at least one solution $x \in D$.

Let us notice the fact that, in a sense to be specified below, this new concept of admissibility with respect to an equation can be reduced to the admissibility concept defined before we stated Theorem 4.1.3. Indeed, if we look for solutions in D to equation (4.1.13), when $h \in B \cap L(D) = B_0$, then the pair (B_0, D) is obviously admissible with respect to L (or equation (4.1.13)) and, regardless of which concept of stability we refer to, we ultimately end up with the same set of solutions for (4.1.13).

Nevertheless, starting with a given decomposition $B = B_1 \oplus B_2$ of the space B gives us more options in conducting the investigation of equation (4.1.13). Simultaneously, one obtains a generalization of the admissibility concept in ordinary differential equations as defined and used by Massera and Schäffer [1]. Classical concepts such as conditional stability, partial boundedness (i.e., only part of the components of the solution are bounded) are fully covered by using Cushing's concept of admissibility with respect to an equation.

Before we define and can take advantage of the new concept of admissibility, let us make an important remark.

Let C be any subspace of B_1, complementary to the subspace $B_1 \cap L(D)$.

Then, for any $h_2 \in B_2$, there exists a unique $h_1 \in C$ for which $h_1 + h_2 \in L(D)$. If we let $h_1 = Ah_2$, the map A is linear from B_2 into C.

Indeed, for any $h_2 \in B_2$ there exists according to H an element $h_1^* \in B_1$, and $u^* \in D$, such that $Lu^* = h_1^* + h_2$. Let us represent $h_1^* = f + g$, with $f \in B_1 \cap L(D)$ and $g \in C$. Denote by $\bar{u}$ the element of D such that $L\bar{u} = f$, and let $u = u^* - \bar{u} \in D$. Since $Lu = g + h_2$, one has $g + h_2 \in L(D)$. To see that g is unique in C, assume for $g' \in C$ we have also $Lu' = g' + L_2$, for some $u' \in D$. Then $L(u - u') = g - g' \in B_1 \cap L(D)$, and since $g - g' \in C$ one must have $g = g'$. The linearity of A is a consequence of the linearity of L.

Definition. We shall say that the pair (B, D), with $B = B_1 \oplus B_2$, is *admissible* with respect to equation (4.1.13), if: both operators L and L^{-1} are continuous from F to F; for any $h_2 \in B_2$ there exists $h_1 \in B_1$ such that $h_1 + h_2 \in L(D)$; and the operator A, from B_2 into C, is continuous.

Remark. The continuity of L and L^{-1} is a simple consequence of the closed graph theorem, if we assume only the closedness of L and the fact it maps one-to-one F onto F.

It is significant to notice the fact that the continuity of the operator A is a consequence of certain (rather mild) restrictions imposed on the subspaces involved in the above definition of admissibility.

Proposition 4.1.4. *If B and D are Banach spaces and B_2, C are closed (as subspaces of B), then A is a continuous operator from B_2 into C.*

Proof. Let us consider the space $E = \{x; x \in D, Lx \in C \oplus B_2\} \subset D$ with the norm $|x|_E = |x|_D + |Lx|_B$. We prove first that E is a Banach space. The properties of the norm being obviously verified, we need only notice that the completeness of D and $C \oplus B_2$ implies the completeness of E.

If $p: C \oplus B_2 \to B_2$ is the projection operator (hence bounded), let us consider the linear operator $\tilde{L} = pL: E \to B_2$. It is bounded, being the product of two bounded operators. We notice that $\tilde{L}$ is one-to-one and onto B_2. Indeed, if $pLx = pLy$, then $p(Lx - Ly) = \theta$, or $L(x - y) \in C$. Since $L(x - y) \in L(D) \cap B_1$ and C is complementary to $L(D) \cap B_1$, we have $Lx = Ly$. But L is invertible and therefore $x = y$. That $\tilde{L}$ is onto B_2 can be seen immediately from its definition and the admissibility assumption. Hence, $\tilde{L}$ is invertible and $\tilde{L}^{-1}$ is a continuous operator from B_2 into E (Banach's theorem on the inverse).

If $x \in E$ implies $Lx = h_1 + h_2$, with $h_1 \in C$ and $h_2 \in B_2$, then $h_1 = Ah_2$ according to the definition of A. Since $\tilde{L}x = pLx = h_2$, it means $x = \tilde{L}h_2$, and therefore

$$|Ah_2 + h_2|_{B_2} = |Lx|_{B_2} \leq |\tilde{L}^{-1}h_2|_{B_1} + |Lx|_{B_2}$$

$$= |\tilde{L}^{-1}h_2|_E \leq |\tilde{L}^{-1}||h|_{B_2}.$$

This means

$$|Ah_2|_{B_2} \leq (|\tilde{L}^{-1}| + 1)|h|_{B_2},$$

which shows that A is bounded.

This ends the proof of Proposition 4.1.4.

We shall now state the main result related to the new concept of stability defined above in this section, and discussed in Proposition 4.1.4.

Theorem 4.1.5. *Consider the equation*

$$Lx = h + Nx, \qquad (4.1.15)$$

where $N: D \to B_2$ *is such that*

$$|Nx - Ny|_B \leq \lambda |x - y|_D, \quad \forall x, y \in S(r), \qquad (4.1.16)$$

with $S(r) = \{x; x \in D, |x|_D \leq r\}$, *and* $\lambda > 0$.

Assume further that L *and* L^{-1} *are continuous on* F, *while the pair of spaces* (B, D), *with* $B = B_1 \oplus B_2$, *is admissible with respect to* L, *and both* D *and* B_2 *are Banach spaces.*

Then the following conclusions hold: (i) *there exists* $k > 0$, *such that if* $\lambda < k$ *and* $|N\theta|_B < k$, *to each* $h_2 \in LS(kr)$ *there corresponds* $h_1 \in C$ *with the property that* (4.1.15) *has a unique solution in* $S(r)$ *for* $h = h_1 + h_2$; (ii) *there exist constants* $\tilde{r}$ *and* $\tilde{k}$ *satisfying* $0 < \tilde{r} < r$, $0 < \tilde{k} < k$, *such that* $\lambda < k$, $|N\theta|_B < \tilde{k}$ *imply the existence of a homeomorphism* Q *between the set of those* $y \in S(kr)$ *for which* $Ly \in B_2$, *and the set of those* $x \in S(\tilde{r})$ *for which* $Lx - Nx \in B_2$. *Moreover,* $x = Qy$ *satisfies* $P(Lx - Nx) = Ly$, *where* P *is the projector from* B_1 *onto* $B_1 \cap L(D)$, *with respect to* C.

Proof. (i) The linear operator $\tilde{L} = L^{-1}(A + I)$, from B_2 into D, is continuous. Indeed, one can easily see that $\tilde{L}$ is closed. Assuming $h_n \to \tilde{h} \in B_2$ and $\tilde{L}h_n \to h_0 \in D$, then $Ah_n + h_n \to (A + I)\tilde{h}$ in B_2, and therefore in F. Since L^{-1} is continuous on F, one has $\tilde{L}h_n \to \tilde{L}\tilde{h}$, which implies $h_0 = \tilde{L}\tilde{h}$. Hence $\tilde{L}$ is closed, and by the closed graph theorem one obtains its continuity.

We shall now define an operator T from $S(r)$ into itself, which will result in a contraction under our assumptions. Given $h_2 \in LS(kr)$, let $y = L^{-1}h_2$. Then define T by

$$Tx = y + \tilde{L}Nx, \quad x \in S(r).$$

Let us now choose k in such a way that

$$k < \min\{|\tilde{L}|^{-1}, r(r + |\tilde{L}|(r + 1))^{-1}\}. \qquad (4.1.17)$$

Taking into account the estimate

$$|\tilde{L}Nx - \tilde{L}Nz|_D \leq k|\tilde{L}||x - z|_D, \qquad (4.1.18)$$

which holds for $x, z \in S(r)$, we obtain from (4.1.16) and our assumptions:

$$|Tx|_D \leq |y|_D + |\tilde{L}|(|Nx - N\theta|_B + |N\theta|_B)$$

$$\leq kr + |\tilde{L}|(k|x|_D + k) < r$$

for any $x \in S(r)$, or $TS(r) \subset S(r)$. But (4.1.18) shows that T is a contraction, if (4.1.17) is assumed.

The completeness of D shows that there exists a unique $x \in S(r)$ with $Tx = x$. Then $LTx = Lx$, which is the same as (4.1.15) for $h_1 = ANx$, because $LTx = Ly + L(\tilde{L}Nx) = h_2 + (A + I)Nx = h_1 + h_2 + Nx = h + Nx$.

(ii) For each $y \in S(kr)$, with $Ly \in B_2$, we have according to (i) a unique fixed point x of T in $S(r)$. Let $x = Qy$. Since x is satisfying $x = y + \tilde{L}Nx$, we can see easily that $Qy = Qz$ implies $y = z$. For y, z in the domain of Q we have when $u = Qz$:

$$Qy - Qz = x - u = y - z + \tilde{L}(NQy - NQz),$$

which implies according to our assumptions (see (4.1.16)).

$$|Qy - Qz|_D \leq (1 - k|\tilde{L}|)^{-1}|y - z|_D, \tag{4.1.19}$$

where k satisfies (4.1.17). One obtains also by similar manipulations

$$|Q^{-1}x - Q^{-1}u|_D \leq (1 + k|\tilde{L}|)|x - u|_D,$$

which together with (4.1.19) yield the continuity of both Q and its inverse Q^{-1}. Since Q is one-to-one, Q is homeomorphic from the set of those $y \in S(kr)$ for which $Ly \in B_2$ onto the image set.

We now choose $\tilde{k} < \min\{k, |\tilde{L}|/2kr\}$ and $\tilde{r} < \min\{r, 1/2kr(1 + k|\tilde{L}|)\}$. We need to show that the set of those $y \in S(kr)$ for which $Ly \in B_2$ is taken by Q onto the set of those $x \in S(\tilde{r})$, with $Lx - Nx \in B_2$. Choose any x satisfying both conditions specified above. We will show that $y = x - \tilde{L}Nx \in D$, and $Qy = x$, provided $|N\theta|_B \leq \tilde{k}$. We then have $Ly = Lx - Nx - ANx$, taking into account $\tilde{L} = L^{-1}(A + I)$. This implies $Ly \in B_2$ because $Lx - Nx \in B_2$ and $ANx \in B_2$. We have further

$$|y|_D \leq |x|_D + |\tilde{L}|(k|x|_D + |N\theta|_B) \leq kr,$$

because of the choice we made for $\tilde{r}$ and $\tilde{k}$. Hence, $y \in S(kr)$ and $Ly \in B_2$, which means Qy is defined and $Qy = x^* \in S(\tilde{r})$. We need to show $x^* = x = $ the element used above to construct y. By the definition of Q and T we have $x^* = Tx^* = y + \tilde{L}Nx^*$, which leads to $x^* = x - \tilde{L}Nx + \tilde{L}Nx^*$. We now obtain $|x^* - x|_D \leq k|\tilde{L}||x^* - x|_D$, which implies $x^* = x$, because of our assumption $k|\tilde{L}| < 1$. Therefore, Q is onto the set of those x for which $x \in S(\tilde{r})$, and $Lx - Nx \in B_2$, which is exactly what we wanted to prove.

The relationship $P(Lx - Nx) = Ly$ for $x = Qy$ follows immediately from the definition of Q: indeed, from $x = y + \tilde{L}Nx$ we derive $Ly = Lx - L(\tilde{L}Nx) =$

$Lx - Nx - ANx$, and taking into account that C is a complementary subspace of B_2 with respect to $B_2 \cap L(D)$, one obtains $PLy = Ly = P(Lx - Nx) - P(ANx)$. Since $A: B_2 \to C$, one has $P(ANx) = \theta$, and the relationship is established.

Remark. Part (ii) of Theorem 4.1.5 states, in fact, the existence of a local homeomorphism between the set of those elements of B yielding solutions in D for the linear equation $Lx = h$, and the set of those elements in B yielding solutions in D for the nonlinear equation (4.1.15).

Theorem 4.1.5 is a very general result in the sense that the operator L and the spaces B, D can be chosen in a variety of ways.

Let us consider first the case when the operator L is given by (4.1.14), and assume the matrix kernel $k(t, s)$ is continuous on Δ. As noticed in this section, if one takes $F = C(R_+, R^n)$, then L and L^{-1} are both continuous from F to F. This assertion is easily verifiable if one takes into account the fact that the (unique) solution of the equation $Lx = f$ can be represented by means of the resolvent kernel $\tilde{k}(t, s)$ in the form

$$x(t) = f(t) + \int_0^t \tilde{k}(t, s) f(s) \, ds. \tag{4.1.20}$$

This shows that when $f_m \to f$ in $C(R_+, R^n)$, i.e., it converges uniformly on any compact interval of R_+, one has $x_m \to x$ in the same topology. By $x_m(t)$ we denote the solution of $Lx_m = f_m$, $m \geq 1$.

We shall now choose the spaces B and D, letting $B = D = C_\ell(R_+, R^n)$. Since $C_\ell = R^n \oplus C_0$, we can let $B_1 = R^n$, $B_2 = C_0$, obviously, C_0 is closed in C_ℓ.

Therefore, we want solutions in $C_\ell(R_+, R^n)$ for the equation (4.1.15) or its linear counterpart when the term $h \in C_\ell(R_+, R^n)$. But this property is not expected to hold for *any* $h \in C_\ell(R_+, R^n)$, in which case full admissibility would take place. According to our admissibility assumption, for every $h_2 \in C_0(R_+, R^n)$, there must be an element $x^0 \in R^n$, such that the linear equation $Lx = x^0 + h_2$ has its solution in $C_\ell(R_+, R^n)$. For given $h_2 \in C_0$, there might be several $x^0 \in R^n$ with the stated property. If we want uniqueness for x^0, then we have to restrict it to a subspace $C \subset R^n$ which complements the subspace $R^n \cap L(C_0)$. Since both B_2 and C are closed (C is finite dimensional), the conditions required in Theorem 4.1.5 are satisfied.

It remains only to choose the operator N occurring in (4.1.5) in such a way that Theorem 4.1.5 can be applied. Since $N: D \to B_2$ and must satisfy (4.1.16), we can take, for instance,

$$(Nx)(t) = \int_t^\infty k_0(t, s) x(s) \, ds, \quad t \in R_+,$$

where the kernel $k_0(t, s)$ is measurable on $R_+ \times R_+$, satisfies an estimate of the form $|k_0(t, s)| \leq K(s)$, with $K \in L^1(0, \infty)$, and

$$\int_t^\infty |k_0(t + h, s) - k_0(t, s)| \, ds \to 0 \quad \text{as } h \to 0.$$

The last condition guarantees that N takes functions from C_ℓ into continuous functions on R_+. As a result of these conditions on N, it is clear that $NC_\ell \subset C_0$. A condition of the form (4.1.16) is also secured, but one has to assume that

$$\int_0^\infty K(s) \, ds = \lambda$$

is small enough.

With the choices shown above for various elements occurring in the statement of Theorem 4.1.5, equation (4.1.15) now becomes

$$x(t) = \int_0^t k(t, s) x(s) \, ds + \int_t^\infty k_0(t, s) x(s) \, ds + h(t), \qquad (4.1.21)$$

and the main conclusion of the theorem asserts: for each $h_2 \in C_0(R_+, R^n)$, there exists $x^0 \in C \subset R^n$ uniquely determined ($x^0 = Ah_2$), such that equation (4.1.21) has a unique solution $x \in C_\ell(R_+, R^n)$ corresponding to $h(t) = x^0 + h_2(t)$; moreover, the set of all those functions $h(t)$, for which (4.1.21) has solutions in C_ℓ, is topologically equivalent (locally) with the corresponding set for the linear equation obtained from (4.1.21) when $k_0(t, s)$ is the zero matrix.

The admissibility theory for ordinary differential equations, as developed by Massera and Schäffer [1], is a special case of the above concept of admissibility.

Indeed, the differential system $\dot{x} = A(t)x + f(t)$ generates the family of integral equations

$$x(t) = x^0 + \int_0^t f(s) \, ds + \int_0^t A(s) x(s) \, ds, \qquad (4.1.22)$$

with $x^0 \in R^n$ arbitrary (or in a Banach space, if x takes values in such a space). If we let $h(t) = x^0 + \int_0^t f(s) \, ds$, with $f \in B_0$, a Banach space of measurable functions from R_+ into R^n, and $B = R^n \oplus B_2$, where $B_2 = \{\int_0^t f(s) \, ds \,|\, f \in B_0\}$ with an adequate topology (for instance, that of $C(R_+, R^n)$), then letting $B_1 = R^n$ we obtain the desired decomposition for B. By D we denote a Banach space of (measurable) functions from R_+ into R^n. It is then clear that $B_1 \cap L(D)$ consists of $\dot{x} = A(t)x$.

This special framework of admissibility leads to deeper results about the solutions of the differential system, and allows us to establish more meaningful relationships between the admissibility concept and the data involved in the problem (than in the case of integral equations). The reader is refered to Massera and Schäffer [1], [2].

4.2   Some problems related to resolvent kernels and applications

We know (Section 1.3) that the solution of the linear Volterra equation

$$x(t) = h(t) + \int_0^t k(t, s)x(s)\,ds, \quad t \in [0, T) \qquad (4.2.1)$$

can be represented by the formula

$$x(t) = h(t) + \int_0^t \tilde{k}(t, s)h(s)\,ds, \quad t \in [0, T), \qquad (4.2.2)$$

where $\tilde{k}(t, s)$ is the resolvent kernel associated with $k(t, s)$.

The existence (construction) of the resolvent kernel has been shown in Section 1.3, for continuous kernels, as well as for kernels which are locally L^2. It is, of course, possible to establish the existence of the resolvent kernel under different assumptions than those specified above. But more important, perhaps, than the existence of the resolvent kernel is some information about the properties that might help in finding adequate results on the behavior of solutions.

Let us illustrate the idea sketched above by examining the formula (4.2.2). For instance, if we know that $\tilde{k}(t, s)$ is such that

$$\int_0^t |\tilde{k}(t, s)|\,ds \le M < \infty, \quad t \in [0, T), \qquad (4.2.3)$$

then from (4.2.2) we easily derive that equation (4.2.1) has only bounded solutions on $[0, T)$ when $h(t)$ is bounded on the same interval. If instead of (4.2.3) we assume

$$\int_s^T |\tilde{k}(t, s)|\,dt \le N, \quad s \in [0, T)\lambda, \qquad (4.2.4)$$

then one obtains from (4.2.2) that the solution of (4.2.1), which corresponds to $h \in L^1(0, T)$, belongs to the same space. Indeed, one has only to prove, according to (4.2.2), that

$$\int_0^T dt \int_0^t |\tilde{k}(t, s)h(s)|\,ds < +\infty, \qquad (4.2.5)$$

for any $h \in L^1(0, T)$. If one takes (4.2.4) into account, then (4.2.5) follows immediately.

In this section we would like to discuss more general problems about the properties of the resolvent kernel than those outlined above.

With regard to the equation (4.2.1), we want to answer some questions of the following nature: what conditions should we impose on the kernel $k(t, s)$, such that the pair of spaces (L_g^1, L_G^1), or (L_g^∞, L_G^∞), where g and G are scalar weight functions, are admissible with respect to equation (4.2.1)? The admissibility here is understood in the simplest sense, i.e., for each $h \in L_g^1$ (or L_g^∞), the unique solution of (4.2.1) belongs to L_G^1 (or L_G^∞).

In the book C. Corduneanu [4], I dealt with such problems, and typical results are similar to Theorem 4.1.3 in the preceding section. Since the conditions involve the resolvent kernel $\tilde{k}(t, s)$, which does exist under rather general assumptions but might be hard to calculate, it is obviously very desirable to provide such results in a form that involves only the assigned data.

First, we shall prove a simple result which is the starting point of a series of auxiliary results to be established.

Lemma 4.2.1. *Assume the kernel $k(t, s)$ is measurable from $\Delta = \{(t, s); 0 \leq s \leq t < T\}$ into $\mathscr{L}(R^n, R^n)$, and such that*

$$\operatorname*{ess\,sup}_{0 \leq s < T} \int_0^t |k(t, s)| \, ds \leq \lambda < 1, \qquad (4.2.6)$$

Then, there exists a resolvent kernel $\tilde{k}(t, s)$ satisfying

$$\operatorname*{ess\,sup}_{0 \leq t < T} \int_0^t |\tilde{k}(t, s)| \, ds < +\infty. \qquad (4.2.7)$$

If the kernel $k(t, s)$ verifies

$$\operatorname*{ess\,sup}_{0 \leq s < T} \int_s^T |k(t, s)| \, dt \leq \lambda < 1, \qquad (4.2.8)$$

then there exists a resolvent kernel $\tilde{k}(t, s)$ that satisfies

$$\operatorname*{ess\,sup}_{0 \leq s < T} \int_s^T |\tilde{k}(t, s)| \, dt < +\infty. \qquad (4.2.9)$$

In both cases, $\tilde{k}(t, s)$ and $k(t, s)$ are related by

$$\tilde{k}(t, s) = k(t, s) + \int_s^t k(t, u)\tilde{k}(u, s) \, du, \qquad (4.2.10)$$

and

$$\tilde{k}(t, s) = k(t, s) + \int_s^t \tilde{k}(t, u)k(u, s) \, ds, \qquad (4.2.11)$$

which hold for almost all (t, s).

Proof. Let us consider the iterated kernels (see Section 1.3)

$$k_1(t,s) = k(t,s), \tag{4.2.12}$$

$$k_{m+1}(t,s) = \int_s^t k(t,u)k_m(u,s)\,ds, \quad n \geq 1, \tag{4.2.13}$$

which are defined a.e. on $\varDelta$. Taking into account (4.2.6), from (4.2.12), (4.2.13) one obtains by means of Fubini's theorem that $k_n(t,s)$, $n \geq 1$, are all measurable (of course, by induction with respect to n). Moreover, the following estimate can be obtained by induction:

$$\operatorname*{ess\,sup}_{0 \leq t < T} \int_0^t |k_m(t,s)|\,ds \leq \lambda^n. \tag{4.2.14}$$

If we set

$$\tilde{k}(t,s) = \sum_{m=1}^\infty k_m(t,s), \quad (t,s) \in \varDelta, \tag{4.2.15}$$

then from (4.2.14) we see that for almost all $t \in [0,\,T)$ the series (4.2.15) converges in $L^1(0,\,T)$ with respect to s. It is an elementary exercise to prove that (4.2.9) and (4.2.10) hold.

Notice now that the iterated kernels also verify the recurrent equation

$$k_{m+1}(t,s) = \int_s^t k_m(t,u)k(u,s)\,du, \quad m \geq 1. \tag{4.2.16}$$

To prove (4.2.16) one has to proceed by induction, and apply Fubini's theorem to derive the measurability of $k_m(t,s)$, $m \geq 1$.

By using (4.2.16), we easily obtain (4.2.11) if we take into account (4.2.14).

The next result is also an auxiliary type statement, which we formulate as

Lemma 4.2.2. *Assume that $k(t,s)$ verifies*

$$\operatorname*{ess\,sup}_{0 \leq t < T} \int_0^t |k(t,s)|\,ds < +\infty. \tag{4.2.17}$$

Let $\bar{k}(t,s)$ be a solution a.e. of (4.2.10) and (4.2.11) for $0 \leq s \leq t \leq \bar{T} < T$, such that

$$\operatorname*{ess\,sup}_{0 \leq t \leq \bar{T}} \int_0^t |\bar{k}(t,s)|\,ds < +\infty. \tag{4.2.18}$$

Further, let $\bar{\bar{k}}(t,s)$ be a solution a.e. of (4.2.10) and (4.2.11) for $\bar{T} \leq s \leq t < T$, satisfying

$$\operatorname*{ess\,sup}_{\bar{T} \leq t < T} \int_{\bar{T}}^t |\bar{\bar{k}}(t,s)|\,ds < +\infty. \tag{4.2.19}$$

Define $\tilde{k}(t,s) = \overline{k}(t,s)$ for $0 \leq s \leq t < \overline{T}$, $\tilde{k}(t,s) = \overline{\overline{k}}(t,s)$ for $\overline{T} \leq s \leq t < T$, and

$$\tilde{k}(t,s) = k(t,s) + \int_s^{\overline{T}} k(t,u)\overline{k}(u,s)\,du + \int_{\overline{T}}^t \overline{\overline{k}}(t,u)k(u,s)\,du$$

$$+ \int_{\overline{T}}^t \overline{\overline{k}}(t,u)\,du \int_s^{\overline{T}} k(u,v)\overline{k}(v,s)\,dv \qquad (4.2.20)$$

for $0 \leq s < \overline{T} < t < T$. Then $\tilde{k}(t,s)$ satisfies (4.2.7), and for almost all (t,s) it also satisfies (4.2.10) and (4.2.11).

A completely analogous statement holds when instead of (4.2.17) $k(t,s)$ satisfies

$$\operatorname*{ess\,sup}_{0 \leq s < T} \int_s^T |k(t,s)|\,dt < +\infty. \qquad (4.2.21)$$

Proof. First of all, let us explain the meaning of the formula (4.2.20) which helps us to construct the resolvent kernel for $0 \leq s < \overline{T} < t < T$. If we assume $\tilde{k}(t,s)$ is defined for $0 \leq s \leq t < T$, then equation (4.2.10) can be rewritten as

$$\tilde{k}(t,s) = k(t,s) + \int_s^{\overline{T}} k(t,u)\tilde{k}(u,s)\,du + \int_{\overline{T}}^t k(t,u)\tilde{k}(u,s)\,du.$$

Now, we can express $\tilde{k}(t,s)$ by means of resolvent formula if we fix s, and notice that $k(t,u)$ is the kernel of the equation for $\overline{T} \leq s \leq t < T$, while

$$k(t,s) + \int_s^{\overline{T}} k(t,u)\tilde{k}(u,s)\,du$$

stands for the free term of the equation (like h in equation (4.2.1)). One obtains

$$\tilde{k}(t,s) = k(t,s) + \int_s^{\overline{T}} k(t,u)\tilde{k}(u,s)\,du + \int_{\overline{T}}^t \tilde{k}(t,u)k(u,s)\,du$$

$$+ \int_{\overline{T}}^t \tilde{k}(t,u)\,du \int_s^{\overline{T}} k(u,v)\tilde{k}(v,s)\,dv,$$

which reduces to (4.2.20) if we substitute $\overline{k}(u,s)$ for $\tilde{k}(u,s)$ in the first integral in the right hand side, $\overline{\overline{k}}(t,u)$ for $\tilde{k}(t,u)$ in the second and third integrals, and again $\overline{k}(v,s)$ for $\tilde{k}(v,s)$ in the third integral.

Therefore, if a resolvent kernel $\tilde{k}(t,s)$ does exist for $0 \leq s \leq t < T$, such that it coincides with $\overline{k}(t,s)$ or $\overline{\overline{k}}(t,s)$ in the corresponding domains, then it must be constructed by (4.2.20) for $0 \leq s < \overline{T} < t < T$.

Now let $\tilde{k}(t,s)$ be constructed as shown in the statement of Lemma 4.2.2, and let us prove that it verifies the equations of the resolvent kernel (4.2.10) and (4.2.11). Of course, these must hold a.e. in (t,s), and it suffices to check only (4.2.10) because of the analogy between those equations. Moreover, (4.2.10) is verified by hypothesis for $0 \leq s \leq t \leq \overline{T}$, and for $\overline{T} \leq s \leq t < T$. Therefore it remains only to check the validity of the equation (4.2.10) for $0 \leq s < \bar{T} < t < T$, which means that we have to relate to the formula (4.2.20) only.

Formula (4.2.20) can be rewritten in the form

$$\tilde{k}(t,s) = k(t,s) + \int_s^{\bar{T}} k(t,u)\bar{k}(u,s)\,du + \int_{\bar{T}}^t \bar{\bar{k}}(t,u)\left[k(u,s) + \int_s^{\bar{T}} k(u,v)\bar{k}(v,s)\,dv \right]du$$

which means that it expresses the (unique) solution of the equation

$$k^*(t,s) = k(t,s) + \int_s^{\bar{T}} k(t,u)\bar{k}(u,s)\,du + \int_{\bar{T}}^t k(t,u)k^*(u,s)\,du.$$

Therefore we can write

$$\tilde{k}(t,s) = k(t,s) + \int_s^{\bar{T}} k(t,u)\bar{k}(u,s)\,du + \int_{\bar{T}}^t k(t,u)\tilde{k}(u,s)\,du \qquad (4.2.22)$$

for $\bar{T} \leq t < T$ and $0 \leq s \leq \bar{T}$. Of course this must hold a.e. with respect to (t,s), according to the assumptions we made on k and $\bar{k}$. But in the first integral in the right hand side of (4.2.22) $\bar{k}(u,s)$, $0 \leq s \leq u \leq \bar{T}$, can be substituted by $\tilde{k}(u,s)$ according to the definition of $\tilde{k}(t,s)$, $0 \leq s \leq t \leq \bar{T}$. Hence, (4.2.22) yields

$$\tilde{k}(t,s) = k(t,s) + \int_s^{\bar{T}} k(t,u)\tilde{k}(u,s) + \int_{\bar{T}}^t k(t,u)\tilde{k}(u,s)\,du$$

which obviously reduces to (4.2.10).

Since the proof of (4.2.11) can be accomplished in exactly the same way, we leave it to the reader.

Finally, we must show that $\tilde{k}(t,s)$, defined in Lemma 4.2.2, satisfies condition (4.2.7).

In order to avoid repetition, we omit the proof here. In the proof of Lemma 4.2.5, this part of the proof will be provided under more general conditions than assumed in Lemma 4.2.2.

We can now consider an application of Lemmas 4.2.1 and 4.2.2, which constitutes an intermediate result for what we want to prove in this section.

Proposition 4.2.3. *Assume that $k(t,s)$ satisfies (4.2.17) for $0 \leq s \leq t < T$, and suppose there exist two positive constants ε and γ with the property*

$$\operatorname*{ess\,sup}_{S < t < \bar{T}} \int_{\bar{T}}^t |k(t,s)|\,ds \leq \gamma < 1, \qquad (4.2.23)$$

when $0 \leq S < t < \bar{T} < T$, and $\bar{T} - S < \varepsilon$. Then there exists a unique measurable function $\tilde{k}(t,s)$ from $\Delta = \{(t,s): 0 \leq s \leq t < T\}$ into $\mathscr{L}(R^n, R^n)$, such that (4.2.7) holds and both (4.2.10) and (4.2.11) are satisfied a.e. in (t,s).

A similar statement holds when $k(t,s)$ verifies (4.2.8) instead of (4.2.17), and $\tilde{k}(t,s)$ will satisfy (4.2.9) instead of (4.2.7).

Proof. Let us divide the interval $[0, T)$ into a finite number of subintervals, the length of each subinterval being smaller than ε. Of course, this is possible only if $T < +\infty$. Then Lemma 4.2.1 guarantees the existence of a solution to both (4.2.10) and (4.2.11), such that a condition of the type (4.2.7) holds, on any subinterval of the subdivision. By applying Lemma 4.2.2 repeatedly, one obtains the existence of $\tilde{k}(t, s)$ in the whole Δ, such that (4.2.7), (4.2.10) and (4.2.11) are satisfied.

To prove the uniqueness of $\tilde{k}(t, s)$ as a solution of (4.2.10), (4.2.11), under condition (4.2.7), we proceed as follows: let $\tilde{k}_1(t, s)$ and $\tilde{k}_2(t, s)$ be two functions with the above mentioned properties; multiplying equation (4.2.11) (with $\tilde{k}$ replaced by $\tilde{k}_2$) by $\tilde{k}_1(t, s)$ on the right, and integrating from s to t, then, from Fubini's theorem and the fact that $\tilde{k}_1$ satisfies equation (4.2.10), with $\tilde{k}$ replaced by $\tilde{k}_1$, we obtain for almost all (t, s):

$$\int_s^t k(t, u)\tilde{k}_2(u, s)\,du = \int_0^t \tilde{k}_1(t, u)k(u, s)\,ds.$$

This equation, taken into account with (4.2.10) and (4.2.11) for $\tilde{k}_1$ and $\tilde{k}_2$, respectively, leads to $\tilde{k}_1(t, s) = \tilde{k}_2(t, s)$ a.e.

Remark. As we saw in Section 1.3, the existence of the resolvent kernel, i.e., of such a function that verifies equations (4.2.10) and (4.2.11), allows us to represent the (unique) solution of the equation

$$x(t) = h(t) + \int_0^t k(t, s)x(s)\,ds \qquad (4.2.1)$$

by the formula

$$x(t) = h(t) + \int_0^t \tilde{k}(t, s)h(s)\,ds, \qquad (4.2.2)$$

for every $h \in L(0, T)$, while the solution of the adjoint equation

$$y(s) = g(s) + \int_s^T y(t)k(t, s)\,dt,$$

with $g \in L(0, T)$ is given by the similar formula

$$y(s) = g(s) + \int_s^T g(t)\tilde{k}(t, s)\,dt.$$

The validity of these statements was obtained in Section 1.3 under continuity conditions, but the proof remains essentially the same under the measurability conditions assumed above.

In order to better emphasize the properties of the resolvent kernel $\tilde{k}(t, s)$ in terms of the properties of the kernel $k(t, s)$, we shall proceed now with the investigation of the case when weighted L^p-spaces with scalar weights are considered, instead of the usual L^p-spaces.

Lemma 4.2.4. *Let the assumptions in the first part of Proposition 4.2.3 hold and suppose further that for some $\delta > 0$*

$$\operatorname*{ess\,sup}_{0 \leq t < T} \int_0^t \left[\frac{G(s)}{G(t)} + \delta \frac{g(s)}{G(t)} \right] |k(t,s)|\, ds \leq 1. \tag{4.2.24}$$

Then the resolvent kernel $\tilde{k}(t,s)$ satisfies

$$\operatorname*{ess\,sup}_{0 \leq t < T} [G(t)]^{-1} \int_0^t |\tilde{k}(t,s)|\, g(s)\, ds \leq \delta^{-1}. \tag{4.2.25}$$

If the assumptions in the second part of Proposition 4.2.3 hold and for some $\delta > 0$

$$\operatorname*{ess\,sup}_{0 < s < T} \int_s^T \left[\delta \frac{g(s)}{G(t)} + \frac{g(s)}{g(t)} \right] |k(t,s)|\, dt \leq 1. \tag{4.2.26}$$

Then the resolvent kernel $\tilde{k}(t,s)$ satisfies

$$\operatorname*{ess\,sup}_{0 < s < T} g(s) \int_s^T [G(t)]^{-1} |\tilde{k}(t,s)|\, dt \leq \delta^{-1}. \tag{4.2.27}$$

Proof. As usual, we consider only the first part of the statement. Let us define $\varphi(s) = G(s) + \delta g(s)$. If we multiply (4.2.11) by $\varphi(s)$ and integrate, then on behalf of (4.2.24) one obtains

$$\operatorname*{ess\,sup}_{0 < t < T} [G(t)]^{-1} \int_0^t |\tilde{k}(t,s)|\, \varphi(s)\, ds$$

$$\leq \operatorname*{ess\,sup}_{0 < t < T} [G(t)]^{-1} \int_0^t |k(t,s)|\, \varphi(s)\, dt$$

$$+ \operatorname*{ess\,sup}_{0 < t < T} [G(t)]^{-1} \int_0^t |\tilde{k}(t,u)|\, du \int_0^u |k(u,s)|\, \varphi(s)\, ds$$

$$\leq 1 + \operatorname*{ess\,sup}_{0 < t < T} [G(t)]^{-1} \int_0^t |\tilde{k}(t,u)|\, G(u)\, du.$$

If the last term is finite, then shifting it into the first side of the inequality yields exactly (4.2.25). If the last term is infinite, then we restrict our considerations to an interval $[0, \bar{T}]$, $\bar{T} < T$, in which both G and G^{-1} are continuous and positive, and therefore we can apply Proposition 4.2.3. In this case we obtain

$$\operatorname*{ess\,sup}_{0 < t < \bar{T}} [G(t)]^{-1} \int_0^t |\tilde{k}(t,s)|\, g(s)\, ds \leq \delta^{-1},$$

and since $\bar{T}$ can be chosen arbitrarily close to T, (4.2.25) follows from the above inequality.

Lemma 4.2.5. *Assume that the conditions of the first part of Proposition 4.2.3 are satisfied, and that a $\bar{T} < T$ exists such that*

$$\operatorname*{ess\,sup}_{\bar{T} < t < T} \left[\frac{1}{g(t)} + \frac{1}{G(t)} \right] \int_0^{\bar{T}} |k(t,s)| \, [g(s) + G(s)] \, ds < \infty. \qquad (4.2.28)$$

Moreover, assume that $\tilde{k}(t,s)$ satisfies

$$\operatorname*{ess\,sup}_{0 \le t \le \bar{T}} [G(t)]^{-1} \int_0^t |\tilde{k}(t,s)| g(s) \, ds < \infty \qquad (4.2.29)$$

and

$$\operatorname*{ess\,sup}_{\bar{T} \le t < T} [G(t)]^{-1} \int_{\bar{T}}^t |\tilde{k}(t,s)| g(s) \, ds < +\infty. \qquad (4.2.30)$$

Then $\tilde{k}(t,s)$ satisfies

$$\operatorname*{ess\,sup}_{0 \le t < T} [G(t)]^{-1} \int_0^t |\tilde{k}(t,s)| g(s) \, ds < +\infty. \qquad (4.2.31)$$

A similar statement holds when the conditions of the second part of Proposition 4.2.3 are verified, and (4.2.28) is substituted by

$$\operatorname*{ess\,sup}_{0 < s < \bar{T}} [G(s) + g(s)] \int_{\bar{T}}^T \left[\frac{1}{G(t)} + \frac{1}{g(t)} \right] |k(t,s)| \, dt < \infty. \qquad (4.2.32)$$

Proof. Since $\tilde{k}(t,s)$ satisfies (4.2.29) and (4.2.30), all it remains to show is that

$$\operatorname*{ess\,sup}_{\bar{T} < t < T} [G(t)]^{-1} \int_0^T |\tilde{k}(t,s)| g(s) \, ds < \infty. \qquad (4.2.33)$$

Notice that for $0 \le s < \bar{T} < t < T$, $\tilde{k}(t,s)$ satisfies the equation

$$\tilde{k}(t,s) = k_0(t,s) + \int_s^{\bar{T}} k_0(t,u) \tilde{k}(u,s) \, du, \qquad (4.2.34)$$

where $k_0(t,s)$ is given a.e. by

$$k_0(t,s) = k(t,s) + \int_{\bar{T}}^t \tilde{k}(t,u) k(u,s) \, du. \qquad (4.2.35)$$

This follows easily from (4.2.11) by splitting the integral, and then applying the resolvent formula (see remark above) to the obtained equation. For almost all $t \in (\bar{T}, T)$ we have from (4.2.35)

$$[G(t)]^{-1} \int_0^{\bar{T}} |k_0(t,s)| [g(s) + G(s)] \, ds$$

$$\leq [G(t)]^{-1} \int_0^{\bar{T}} |k(t,s)| [g(s) + G(s)] \, ds$$

$$+ [G(t)]^{-1} \int_{\bar{T}}^t |\tilde{k}(t,u)| \, du \int_0^{\bar{T}} |k(u,s)| [g(s) + G(s)] \, ds,$$

and taking into account (4.2.30), (4.2.28) and (4.2.34) we derive

$$\operatorname*{ess\,sup}_{\bar{T} < t < T} [G(t)]^{-1} \int_0^{\bar{T}} |\tilde{k}(t,s)| g(s) \, ds$$

$$\leq \operatorname*{ess\,sup}_{\bar{T} < t < T} [G(t)]^{-1} \int_0^{\bar{T}} |k_0(t,s)| g(s) \, ds$$

$$+ \operatorname*{ess\,sup}_{\bar{T} < t < T} [G(t)]^{-1} \int_0^{\bar{T}} |k_0(t,u)| \, du \int_0^u |\tilde{k}(u,s)| g(s) \, ds < \infty. \quad (4.2.36)$$

This ends the proof of Lemma 4.2.5.

We can now state the main result of this section, which provides an answer to the question of finding admissibility conditions for the resolvent kernel $\tilde{k}(t,s)$, in terms of the given kernel $k(t,s)$.

Theorem 4.2.6. *Let $k(t,s)$ be a measurable map from $[0, T)$ into $\mathscr{L}(R^n, R^n)$, such that (4.2.17) is satisfied, and there exist two positive constants ε, γ for which (4.2.23) holds. Assume further that there exist $\delta > 0$ and $\bar{T} < T$ ($\bar{T} > 0$) such that*

$$\operatorname*{ess\,sup}_{\bar{T} < t < T} \int_{\bar{T}}^t \left[\frac{G(s)}{G(t)} + \delta \frac{g(s)}{G(t)} \right] |k(t,s)| \, ds \leq 1, \quad (4.2.37)$$

while (4.2.28) is true for any $\bar{T} < T$. Then the resolvent kernel $\tilde{k}(t,s)$ constructed in Proposition 4.2.3 satisfies the condition

$$\operatorname*{ess\,sup}_{0 \leq t \leq \bar{T}} [G(t)]^{-1} \int_0^t |\tilde{k}(t,s)| g(s) \, ds < +\infty. \quad (4.2.31)$$

A similar statement holds when $k(t,s)$ satisfies (4.2.21) instead of (4.2.17), as well as the analogous conditions to (4.2.37):

$$\operatorname*{ess\,sup}_{0 < s < T} \int_s^T \left[\delta \frac{g(s)}{G(t)} + \frac{G(s)}{G(t)} \right] |k(t,s)| \, dt \leq 1. \quad (4.2.38)$$

in which case the resolvent kernel $\tilde{k}(t,s)$ satisfies the condition

$$\operatorname*{ess\,sup}_{0 \leq s < T} g(s) \int_s^T [G(t)]^{-1} |\tilde{k}(t,s)| \, dt < \infty. \quad (4.2.39)$$

In both situations considered above the solution of (4.2.1), *expressed by* (4.2.2), *is unique for each* $h \in L_g^\infty([0, T), R^n)$ *and belongs to the space* $L_G^\infty([0, T), R^n)$, *respectively for each* $h \in L_G^1([0, T), R^n)$, *while a similar statement with respect to the adjoint equation to* (4.2.1), *i.e.*

$$y(s) = g(s) + \int_s^T k(t, s)y(t)\,\mathrm{d}t,$$

holds with $G(t)$ *replaced by* $[g(t)]^{-1}$, *if* $g(t)[G(t)]^{-1}$ *is bounded on* $[0, T)$.

Proof. Since our assumptions imply those of Proposition 4.2.3, the existence of $\tilde{k}(t, s)$ is guaranteed. Moreover, $\tilde{k}(t, s)$ satisfies (4.2.7).

Let us now consider some $\bar{T}$, $0 < \bar{T} < T$, and note that condition (4.2.7) implies condition (4.2.29). This follows from the fact that our weight functions g and G are continuous and positive on $[0, \bar{T}]$. From (4.2.37) and (4.2.28), by means of Lemma 4.2.4, one obtains condition (4.2.30) in Lemma 4.2.5. Now by applying Lemma 4.2.5 we find out that $\tilde{k}(t, s)$ satisfies condition (4.2.31).

It remains to prove the last statement of Theorem 4.2.6 concerning equation (4.2.1) and its adjoint. In other words, we need to show that (4.2.2) is a solution of (4.2.1) belonging to L_G^∞, for every $h \in L_G^\infty([0, T), R^n)$. If we notice that our hypotheses imply

$$\operatorname*{ess\,sup}_{0 < t < T} [G(t)]^{-1} \int_0^t |k(t, s)|\,[g(s) + G(s)]\,\mathrm{d}s < \infty,$$

then the proof continues on the same lines as in Section 1.3. More precisely, if we multiply (4.2.2) from the left by $k(u, t)$, integrate with respect to t from 0 to u, and use (4.2.10), then we obtain

$$\int_0^u k(u, t)x(t)\,\mathrm{d}t = \int_0^u \tilde{k}(u, t)h(t)\,\mathrm{d}t.$$

This equation, together with (4.2.2), leads immediately to (4.2.1). Also, the formula (4.2.2) can be obtained from (4.2.1) by means of (4.2.10), which shows the uniqueness.

4.3   Hammerstein equations on measure spaces

The results of existence we shall discuss in this section relate to integral equations of the form

$$u(x) + \int_\Omega k(x, y)f(y, u(y))\,\mathrm{d}y = h(x), \tag{4.3.1}$$

when $\mathrm{d}y$ stands for a σ-finite measure on Ω (see Yosida [1] or Edwards [1]). It would be more common notation if $\mathrm{d}y$ on the left hand side of (4.3.1) was replaced by $d\mu(y)$, where μ is the measure in the triplet (Ω, Σ, μ). Since no confusion can occur, we shall use the simpler notation in (4.3.1).

It will be assumed throughout this section that u and h are maps from Ω into R^n, and f is a map from $\Omega \times R^n$ into R^n, while k is a map from $\Omega \times \Omega$ into $\mathcal{L}(R^n, R^n)$, the space of $n \times n$ matrices with real entries.

As we have done several times in this book, let us introduce the operators

$$(Kv)(x) = \int_\Omega k(x, y)v(y)\,dy, \quad x \in \Omega, \tag{4.3.2}$$

and

$$(Fw)(x) = f(x, w(x)), \quad x \in \Omega, \tag{4.3.3}$$

which means that K is a linear integral operator (only formally), while F can be nonlinear. We shall refer in the sequel to F as the *Niemytzki operator* (see Section 2.3) generated by the map (function) f.

By using (4.3.2) and (4.3.3), the Hammerstein equation (4.3.1) can be rewritten in the operator form

$$(I + KF)u = h. \tag{4.3.4}$$

Of course, we need to make precise a series of concepts involved in the original integral equation (4.3.1), or in its abstract counterpart (4.3.4). For instance, what is the domain of the operator K, when is K compact or continuous, etc? All these details will be made precise in what follows, when we state various results related to (4.3.1).

Before we get into details concerning the integral equation (4.3.1), we shall provide a general existence result for the equation (4.3.4) due to Brézis and Browder [1–3]. Then, by choosing the data conveniently, we shall be able to obtain the desired results for (4.3.1).

Theorem 4.3.1. *Let X and Y be two real Banach spaces with a bilinear real pairing denoted by (y, x), such that $|(y, x)| \leq |y|_Y \cdot |x|_X$. Assume F is a continuous map of X into Y, taking bounded sets of X into bounded sets of Y. Further, assume K is a continuous map of Y into X, taking any bounded set of Y into a relatively compact set of X, and such that $(v, Kv) \geq 0$ for any $v \in Y$. If there exists a map $c: (0, \infty) \to [0, \infty)$, such that for each $k \geq 0$*

$$(Fu, u) \geq k|Fu|_Y - c(k), \quad u \in X, \tag{4.3.5}$$

then equation (4.3.4) has at least one solution $u \in X$, for each $h \in X$.

Proof. The proof will be carried out by means of the Leray–Schauder fixed point theorem (see Section 2.4), applied in the space X.

Notice first that the map $KF: X \to X$ is compact. Therefore, if we consider the equation with parameter

$$(I + \lambda KF)u = v, \tag{4.3.6}$$

with $\lambda \in [0, 1]$ and $v \in X$ fixed, it suffices to prove that a constant $R > 0$ can be found, such that for any solution u of (4.3.6) $|u|_X \leq R$.

If we assume u is a solution of (4.3.6), then the pairing of the spaces X, Y leads to

$$(Fu, u) + \lambda(Fu, KFu) = (Fu, v),$$

and since $(Fu, KFu) \geq 0$ we obtain

$$(Fu, u) \leq (Fu, v) \leq |Fu|_Y |v|_X.$$

If we choose now $k > |v|_X + 1$, then our hypothesis implies

$$(Fu, u) \geq k|Fu|_Y - c(k).$$

Comparing the last two inequalities we obtain

$$k|Fu|_Y \leq (Fu, u) + c(k) \leq |v|_X |Fu|_X + |v|_X,$$

which, because of the choice of k, reduces to

$$|Fu|_Y \leq c(k). \tag{4.3.7}$$

From (4.3.6) and $\lambda \in [0, 1]$ we derive

$$|u|_X \leq |KFu|_X + |v|_X. \tag{4.3.8}$$

Since K maps bounded sets into bounded sets, there exists a function $\mu: R_+ \to R_+$ such that

$$|Kw|_X \leq \mu(|w|_Y), \quad w \in Y. \tag{4.3.9}$$

From (4.3.7), (4.3.8) and (4.3.9) we easily obtain the estimate

$$|u|_X \leq \mu(c(k)) + |v|_X = R,$$

which ends the proof of Theorem 4.3.1.

In order to apply Theorem 4.3.1 to the integral equation (4.3.1), we need to make convenient hypotheses to obtain satisfactory properties for the operators K and F, given by (4.3.2) and (4.3.3). We will distinguish two situations: first, when the underlying space is the space $L^p(\Omega)$, $1 < p < \infty$; second, when the underlying space is $L^\infty(\Omega)$.

The hypotheses can be formulated as follows:

H^p: (a) K is a compact linear map from $L^{p'}(\Omega)$ to $L^p(\Omega)$, where p' is the
            conjugate index to p, $1 < p < \infty$;
        (b) F maps $L^p(\Omega)$ into $L^{p'}(\Omega)$.

H^∞: (a) K is a compact linear map from $L^1(\Omega)$ into $L^\infty(\Omega)$;
  (b) for each $R > 0$, there exists a function $g_R \in L^1(\Omega)$, such that

$$|f(y,u)| \le g_R(y), \quad y \in \Omega, \quad |u| \le R.$$

Remark 1. In terms of the kernel $k(t,s)$, generating the integral operator K, sufficient conditions for property (a) to be verified have been indicated in Section 2.3.

Remark 2. For condition (b) of hypothesis H^p, we refer the reader to Krasnoselskii's book [1].

As pointed out in Section 2.3, condition (b) implies the existence of a constant $\alpha > 0$, and a function $g \in L^{p'}(\Omega)$, such that

$$|f(y,u)| \le \alpha |u|^{p-1} + g(y), \tag{4.3.10}$$

holds for any $y \in \Omega$ and all real u.

Moreover, the Niemyzki's operator F turns out to be continuous from L^p to $L^{p'}$, and takes (obviously!) bounded sets into bounded sets.

Before we can apply Theorem 4.3.1 to equation (4.3.1), we need to derive two auxiliary results which are related to condition (4.3.5) of the theorem.

Lemma 4.3.2. *Assume f satisfies condition (b) of the hypothesis H^∞, and for some $c > 0$ the inequality $(f(y,u) \cdot u) \ge c|f(y,u)| \cdot |u|$ holds for $|u| \ge R_0 > 0$. Then, for each $k > 0$, there exists a constant $c(k)$ such that*

$$(Fu,u) \ge k|Fu|_{L^1} - c(k), \quad u \in L^\infty(\Omega).$$

Proof. We shall consider only the case when k is sufficiently large. Let $R \ge R_0$ be given. Then

$$(Fu,u) = \int_{|u| \le R} f(x,u(x)) \cdot u(x)\,dx + \int_{|u| \ge R} f(x,u(x)) \cdot u(x)\,dx$$

$$= I_1 + I_2.$$

For I_1 we have the following estimate:

$$|I_1| \le \int_{|u| \le R} g_R(x)|u(x)|\,dx \le R|g_R|_{L^1(\Omega)}.$$

For I_2, we can write, as a result of the inequality $f(x,u(x)) \cdot u(x) \ge c|f(x,u(x))| \cdot |u(x)| \ge cR|f(x,u(x))|$,

$$|I_2| \ge cR|Fu|_{L^1(\Omega)} - cR\int_{|u| \le R} |f(x,u(x))|\,dx$$

$$\ge cR|Fu|_{L^1(\Omega)} - cR|g_R|_{L^1(\Omega)}.$$

Combining the above inequalities we obtain

$$(Fu, u) \geq cR |Fu|_{L^1(\Omega)} - \gamma(R),$$

where

$$\gamma(R) = cR |g_R|_{L^1(\Omega)} + R |g_R|_{L^1(\Omega)}.$$

Now we take $k = cR$, and Lemma 4.3.2 is proven.

In the case of hypothesis H^p, a similar result to Lemma 4.3.2 can be obtained.

Lemma 4.3.3. *Assume f satisfies condition* (b) *of hypothesis H^p, and for some* $c > 0$ *the inequality* $(f(y, u) \cdot u) \geq c |f(y, u)| \cdot |u|$ *holds for* $|u| \geq R_0$. *Then if* $\text{meas}(\Omega) < \infty$, *there exist two constants* $c_1 > 0, c_2 \geq 0$ *such that*

$$(Fu, u) \geq c_1 |Fu|_{L^{p'}} - c_2, \quad u \in L^p(\Omega), \tag{4.3.11}$$

and, therefore, for each $k > 0$ *there exists* $c(k)$ *such that*

$$(Fu, u) \geq k |Fu|_{L^{p'}} - c(k), \quad u \in L^p(\Omega). \tag{4.3.12}$$

Proof. We write

$$(Fu, u) = \int_{|u| < R_0} f(x, u(x)) \cdot u(x)\, dx + \int_{|u| \geq R_0} f(x, u(x)) \cdot u(x)\, dx$$

$$= I_1 + I_2.$$

Since f satisfies (4.3.10), we can estimate I_1 as follows:

$$|I_1| \leq \alpha \int_{|u| < R_0} |u(x)|^p\, dx + \int_{|u| < R_0} |u(x)| g(x)\, dx$$

$$\leq \alpha R_0^p \,\text{meas}(\Omega) + |g|_{L^{p'}} \left(\int_{|u| < R_0} |u(x)|^p\, dx \right)^{1/p}$$

$$\leq \alpha R_0^p \,\text{meas}(\Omega) + \{\text{meas}(\Omega)\}^{1/p} R_0 |g|_{L^{p'}}.$$

For I_2 we have $f(x, u(x)) \cdot u(x) \geq c |f(x, u(x))| \cdot |u(x)|$, and taking (4.3.10) into account we obtain

$$|u(x)| \geq c_2 |f(x, u(x))|^{1/(p-1)} - c_3 g(x)^{1/(p-1)} \tag{4.3.13}$$

for suitable constants c_2, c_3. Indeed, if a, b, c are positive numbers, then $a \leq b + c$ implies $a^p \leq (b + c)^p \leq 2^p(b^p + c^p)$, for any $p > 0$. Applying this inequality to (4.3.10), with $(p - 1)^{-1}$ instead of p, we obtain (4.3.13).

If we take (4.3.13) into account, we obtain the inequality

$$f(x, u(x)) \cdot u(x) \geq c_4 |(Fu)(x)|^{1 + 1/(p-1)} - c_5 |(Fu)(x)| g(x)^{1/(p-1)}$$

which by integration leads to

$$I_2 \geq c_4 |Fu|^{p'}_{L^{p'}} - c_4 \int_{|u| \leq R_0} |(Fu)(x)|^{p'} \, dx - c_5 |Fu|_{L^{p'}} \cdot |g|_{L^{p'}}$$

$$\geq c_4 |Fu|^{p'}_{L^{p'}} - c_6 |Fu|_{L^{p'}} - c_7.$$

It remains only to notice that for each $\varepsilon > 0$, with $p' > 1$ ($p' = 1 + (p - 1)^{-1}$), the inequality $A \leq \varepsilon A^{p'} + c(\varepsilon)$ holds on $A \geq 0$, where $c(\varepsilon)$ is chosen conveniently. Then, from the last inequality for I_2, combined with that obtained above for I_1, we easily obtain (4.3.11). Finally, (4.3.11) implies (4.3.12).

This ends the proof of Lemma 4.3.3.

By means of the lemmas proven above, Theorem 4.3.1 can be applied to obtain the following results about the integral equation (4.3.1).

Theorem 4.3.4. *Consider equation* (4.3.1), *and assume the following conditions are satisfied*:

(1) $k \colon \Omega \times \Omega \to \mathscr{L}(R^n, R^n)$ *is a measurable map, such that the operator K defined by* (4.3.2) *is compact from $L^1(\Omega, R^n)$ into $L^\infty(\Omega, R^n)$, and*

$$(v, Kv) = \iint_{\Omega \times \Omega} k(x, y)v(y)v(x) \, dx \, dy \geq 0 \qquad (4.3.14)$$

for every $v \in L^1(\Omega, R^n)$;

(2) $f \colon \Omega \times R^n \to R^n$ *satisfies the Carathéodory condition, and for each $R > 0$ there exists $g_R \in L^1(\Omega, R)$, such that $|f(y, u)| \leq g_R(y)$ on Ω;*

(3) *there exist $\psi \in L^1(\Omega, R^n)$, and constants $R_0 > 0$, $c > 0$, such that $(f(y, u) - \psi(y)) \cdot u \geq c |f(y, u) - \psi(y)| \cdot |u|$ for $y \in \Omega$, $|u| \geq R_0$;*

(4) $h \in L^\infty(\Omega, R^n)$.

Then equation (4.3.1) *has a solution $u \in L^\infty(\Omega)$.*

Proof. This follows from Theorem 4.3.1, and Lemma 4.3.2.

Indeed, we choose $X = L^\infty(\Omega, R^n)$, $Y = L^1(\Omega, R^n)$. The pairing is given by

$$(x, y) = \int_\Omega x(t) \cdot y(t) \, dt,$$

and it obviously satisfies the requirement of Theorem 4.3.1.

It is now easy to check all the conditions in Theorem 4.3.1, by means of those listed in the statement of Theorem 4.3.4. The continuity of F, for instance, follows from applying Lebesque's dominated criterion. The operator K is the one given by (4.3.2). Finally, conditions (2) and (3) of Theorem 4.3.4 assure, by Lemma 4.3.2, the validity of the inequality (4.3.5) required by Theorem 4.3.1.

Since there is a certain difference between the way condition (3) is expressed and the similar condition in Lemma 4.3.2 (in which $\psi \equiv 0$), it is useful to notice that equation (4.3.1) is equivalent to

$$u(x) + \int_\Omega k(x,y)f_1(y,u(y))\,dy = h_1(x) \qquad (4.3.15)$$

where

$$f_1(y,u) = f(y,u) - \psi(y), \quad h_1(x) = h(x) - (K\psi)(y).$$

This remark ends the proof of Theorem 4.3.4.

Remark 1. It is, of course, interesting to have conditions under which $K: L^1(\Omega, R^n) \to L^\infty(\Omega, R^n)$ is a compact operator. While this question seems to be unanswered yet, we notice that the case $K: L^1(\Omega, R^n) \to C(\Omega, R^n)$ is clarified, at least under some extra assumptions. See, for instance, Krasnoselskii [1], C. Corduneanu [4].

Remark 2. This is concerned with condition (3) in the theorem. If we assume for f a monotonicity condition $(f(x,u) - f(x,v)) \cdot (u - v) \geq 0$, and notice that according to condition (2) one has $f(x,0) \in L^1(\Omega, R^n)$, then for $n = 1$ (scalar case), and $c = 1$, condition (3) is obviously verified: the monotonicity condition becomes $(f(x,u) - f(x,0)) \cdot u \geq 0$, and consequently $(f(x,u) - f(x,0)) \cdot u = |f(x,u) - f(x,0)| \cdot |u|$.

In the case $1 < p < \infty$, Theorem 4.3.1 and Lemma 4.3.3 yield the following result.

Theorem 4.3.5. *Consider equation* (4.3.1), *and assume the following conditions are satisfied*:

(1) $k: \Omega \times \Omega \to \mathscr{L}(R^n, R^n)$ *is measurable, and the operator K defined by* (4.3.2) *is a compact operator from $L^{p'}(\Omega, R^n)$ into $L^p(\Omega, R^n)$, such that* (4.3.14) *holds for any $v \in L^{p'}(\Omega, R^n)$*;

(2) $f: \Omega \times R^n \to R^n$ *satisfies the Carathéodory condition, and the Niemytzki operator F given by* (4.3.3) *takes $L^p(\Omega, R^n)$ into $L^{p'}(\Omega, R^n)$*;

(3) *there exist $\psi \in L^{p'}(\Omega, R^n)$, and constants $R_0 > 0$, $c > 0$, such that*

$$(f(x,u) - \psi(x)) \cdot u \geq c\,|f(x,u) - \psi(x)| \cdot |u|,$$

for $|u| \geq R_0$ and $x \in \Omega$;

(4) $h \in L^p(\Omega, R^n)$.

Then equation (4.3.1) *has a solution $u \in L^p(\Omega, R^n)$*.

Proof. In Theorem 4.3.1 we take $X = L^p(\Omega, R^n)$ and $Y = L^{p'}(\Omega, R^n)$, the pairing being given (as usual) by

$$(u,v) = \int_\Omega u(x) \cdot v(x)\,dx.$$

Based on Lemma 4.3.3, it is almost routine to check the conditions required by Theorem 4.3.1. In particular, the continuity of F, given by (4.3.3), is guaranteed by Krasnoselskii's result discussed in Section 2.3.

In both Theorems 4.3.4 and 4.3.5, the operator K is supposed to be compact. Certainly, this is quite a severe restriction and removing the compactness condition would lead to a substantial improvement of the existence results.

Of course, as long as we have to rely on Theorem 4.3.1, there is no hope of removing the compactness condition. Therefore, another tool from nonlinear analysis should be used. It turns out that the theory of *monotone operators* provides such a tool. The abstract result which now takes the place of Theorem 4.3.1 has already been formulated in Section 2.4.

While preserving some of the assumptions made in establishing Theorems 4.3.4 and 4.3.5, new concepts are necessary in order to be able to formulate the new result(s).

Definition. Let $A: X \to X^*$ a mapping, where X is a Banach space and X^* is its dual. A is called *angle-bounded* if there exists a positive constant σ, such that

$$(Au - Aw, w - v) \le \sigma(Au - Av, u - v), \tag{4.3.16}$$

for all $u, v, w \in X$.

Notice that (4.3.16) implies the monotonicity of the operator A. Indeed, if one takes $w = v$ in (4.3.16) one obtains $(Au - Av, u - v) \ge 0$ for any $u, v \in X$. Hence, angle-bounded operators constitute a subclass of the class of monotone operators.

Lemma 4.3.6. *Let $A: X \to X^*$ be angle-bounded and hemicontinuous. If $u, v \in X$ are such that $(Au - Av, u - v) = 0$, then $Au = Av$.*

Proof. If the condition in Lemma 4.3.6 is satisfied, then for all $w \in X$ we obtain from (4.3.16)

$$(Au - Aw, w - v) \le 0. \tag{4.3.17}$$

If we choose now $w = v + tz$, with $t > 0$, (4.3.17) yields

$$(Au - A(v + tz), z) \le 0,$$

and, letting $t \to 0$, we obtain because of the hemicontinuity of A the inequality $(Au - Av, z) \le 0$. Since the inequality holds for arbitrary $z \in X$, we must have $Au = Av$.

Lemma 4.3.6 will be used to prove uniqueness of the solution to equation (4.3.1).

Theorem 4.3.7. *Consider equation* (4.3.1), *and assume the following conditions are satisfied*:

(1) $k: \Omega \times \Omega \to \mathscr{L}(R^n, R^n)$ *is measurable, and the operator K given by* (4.3.2) *takes bounded sets from $L^{p'}(\Omega, R^n)$ into bounded sets of $L^p(\Omega, R^n)$, $1 < p < \infty$, also verifying the monotonicity condition* (4.3.14);

(2) $f: \Omega \times R^n \to R^n$ *satisfies the Carathéodory conditions, while the operator F defined by* (4.3.3) *is a σ-angle-bounded operator from $L^p(\Omega, R^n)$ into $L^{p'}(\Omega, R^n)$*;

(3) $h \in L^p(\Omega, R^n)$.

Then there exists a unique solution $u \in L^p(\Omega, R^n)$ of equation (4.3.1).

Proof. In order to apply Theorem 2.4.3 to obtain the existence of the solution, we shall consider the space $X = L^p(\Omega, R^n) \times L^{p'}(\Omega, R^n)$ as underlying space. It is obviously a reflexive Banach space.

In order to define a monotone operator A on X, let $Fu = v \in L^{p'}(\Omega, R^n)$ for $u \in L^p(\Omega, R^n)$, and notice that our equation $(I + KF)u = h$ is equivalent to the system

$$Fu - v = 0, \quad u + Kv = h. \tag{4.3.18}$$

We now define $A: X \to X$ by

$$A(u, v) = (Fu - v, u + Kv) \tag{4.3.19}$$

(Please note that parentheses in (4.3.19) have nothing to do with the pairing of the spaces L^p and $L^{p'}$.)

We need to prove that A is a monotone hemicontinuous operator from X into its dual $X^* = L^{p'}(\Omega, R^n) \times L^p(\Omega, R^n)$. The hemicontinuity is obvious, while the monotonicity means

$$(Fu - F\bar{u} - (v - \bar{v}), u - \bar{u}) + (u - \bar{u} + K(v - \bar{v}), v - \bar{v}) \geq 0$$

for $u, \bar{u} \in L^p$ and $v, \bar{v} \in L^{p'}$. The monotonicity of F and K implies the monotonicity of A.

It remains to prove the local boundedness of A^{-1} in order to obtain the existence of a solution to (4.3.18). In other words, if we consider the equation $A(u, v) = (f, g)$, we need to show that for (f, g) belonging to a bounded set in X^*, the solutions of $A(u, v) = (f, g)$ – if any – constitute a bounded set in X.

Since $A(u, v) = (f, g)$ means $Fu - v = f$ and $u + Kv = g$, from $u + K(Fu - f) = g$ we easily obtain

$$(Fu, u) + (Fu, K(Fu - f)) = (Fu, g),$$

and, since K is monotone, $(KFu, Fu) \geq 0$. Hence, we obtain

$$(Fu, u) \leq (Fu, Kf + g). \tag{4.3.20}$$

But F is σ-angle-bounded, and therefore we can write the inequality (4.3.16) for $v = 0$, i.e., $(Fu - Fw, w) \leq \sigma(Fu - F0, u)$, from which we obtain for any $\rho > 0$ and $u \in L^p(\Omega, R^n)$

$$\rho|Fu| = \sup_{|w|=\rho} (Fu, w) \leq \sigma(Fu - F0, u) + \rho \sup_{|w|=\rho} |Fw|, \qquad (4.3.21)$$

the norms involved in (4.3.21) being either $L^{p'}$-norms (for Fu and Fw), or L^p-norms (for w). As seen above, at the end of proof of Theorem 4.3.4, there is no loss of generality if we assume $F0 = 0$. Hence, taking into account (4.3.20) and (4.3.21) we obtain

$$\frac{\rho}{\sigma}|Fu|_{L^{p'}} \leq |Kf + g|_{L^p}|Fu|_{L^{p'}} + \frac{\rho}{\sigma} \sup_{|w|=\rho} |Fw|_{L^{p'}}.$$

Since ρ is arbitrary, we can choose it such that $\rho > \sigma|Kf + g|_{L^p}$, and from the last inequality we derive a bound for $|Fu|_{L^p}$ in terms of (f, g). This enables us to find an upper bound, for the norms of u and v in terms of f and g, if we take into account that $v = -f + Fu$, and then $u = g - Kv$.

This ends the existence part of the proof of Theorem 4.3.7.

The uniqueness of the solution of the equation (4.3.1) follows without difficulty. Indeed, if $u + KFu = h$ is verified by $u, \bar{u} \in L^p$, then $u - \bar{u} + K(Fu - F\bar{u}) = 0$. This means that $(Fu - F\bar{u}, u - \bar{u}) + (Fu - F\bar{u}, K(Fu - F\bar{u})) = 0$, and since both F and K are monotone we must have $(Fu - F\bar{u}, u - \bar{u}) = 0$ and $(Fu - F\bar{u}, K(Fu - F\bar{u})) = 0$. By Lemma 4.3.6, from $(Fu - F\bar{u}, u - \bar{u}) = 0$ we obtain $Fu = F\bar{u}$. Hence $KFu = KF\bar{u}$, or $h - u = (h - \bar{u})$, which shows that $u = \bar{u}$.

This ends the proof of Theorem 4.3.7.

Corollary. *Owing to the uniqueness of a solution to (4.3.1), we can assert the existence of the operator $(I + KF)^{-1}$, and write the solution of (4.3.1) in the form $u = (I + KF)^{-1}h$. It is worth pointing out that $(I + KF)^{-1}$ is a continuous operator on $L^p(\Omega, R^n)$.*

Indeed, assume $h \in L^p(\Omega, R^n)$ is given, and consider the equation $v + KFv = g$, with $g \in L^p(\Omega, R^n)$, and generally speaking close to h. Since K is monotone, from $u - v + K(Fu - Fv) = h - g$ we obtain

$$(Fu - Fv, u - v) \leq (Fu - Fv, h - g). \qquad (4.3.22)$$

The angle-boundedness of F leads to

$$(Fv - Fw, w - u) \leq \sigma(Fu - Fv, u - v) \leq \sigma|h - g||Fu - Fv|, \qquad (4.3.23)$$

for every $w \in L^p$. From (4.3.22) and (4.3.23) we derive, if we take into account

$$(Fv - Fu, w - u) = (Fw - Fu, w - u) + (Fv - Fw, w - u),$$

the following inequality

$$(Fv - Fu, w - u) \leq |w - u| \cdot |Fw - Fu| + \sigma |h - g| \cdot |Fu - Fv|. \quad (4.3.24)$$

In (4.3.24) we take $w = u + z$, $|z| = \lambda$, and obtain

$$\lambda |Fv - Fu| \leq \lambda \sup_{|z|=\lambda} |F(u + z) - Fu| + \sigma |h - g| \cdot |Fu - Fv|. \quad (4.3.25)$$

But F is a continuous operator from $L^p(\Omega, R^n)$ into $L^{p'}(\Omega, R^n)$, and a continuity modulus at u does exist: $|F(u + z) - Fu| \leq \omega(\lambda)$ for $|z| \leq \lambda$. Hence, (4.3.25) leads to

$$|Fu - Fv| \leq \omega(\lambda) + \frac{\sigma}{\lambda} |Fu - Fv| |h - g|. \quad (4.3.26)$$

If we now assume $|h - g| < \lambda(2\sigma)^{-1}$, then (4.3.26) yields $|Fu - Fv| < 2\omega(\lambda)$ and, since $\omega(\lambda) \to 0$ as $\lambda \to 0+$, the continuity of the map $g \to Fv = F(I + KF)^{-1}g$ is assured.

But $(I + KF)^{-1} = I - KF(I + KF)^{-1}$ and, since K is continuous, we obtain the continuity of $(I + KF)^{-1}$ from $L^p(\Omega, R^n)$ into itself.

Remark. The proof of Theorem 4.3.7 relies substantially on the condition $1 < p < \infty$, which means the reflexiveness of the underlying space is involved. There are results available using monotone operators in nonreflexive spaces: see Brézis and Browder [2], for instance.

4.4   Ultimate behavior of solutions to certain Volterra functional-differential equations

The ultimate behavior of solutions to various classes of functional equations, including integral or integrodifferential equations, has been the object of many investigations and publications. It is my aim here to illustrate the kind of ultimate (asymptotic) behavior which is similar to that encountered in ordinary differential equations.

A prototype of result in this category is due to Moser [1], and it can also be found in C. Corduneanu [4]. Some recent generalizations have been obtained by C. Corduneanu [11] and Cassago and Corduneanu [1].

To be more specific, let us consider the functional-differential equation

$$\dot{x}(t) + (Lx)(t) + f(x(t)) = g(t), \quad t \in R_+ \quad (4.4.1)$$

where x, f and g are n-dimensional vector-valued maps, and L is a linear abstract Volterra operator acting between convenient function spaces.

In the existing literature on equations involving abstract Volterra operators, there are two distinct methods used in discussing the asymptotic behavior of their solutions. The first method is based on monotonicity assumptions that allow us to compare the behavior of solutions of the Volterra equations with those of convenient ordinary differential systems. This method was initiated by Moser [1], and in the case of nonconvolution integral operators has been used by myself [4], and by Cassago and Corduneanu [1]. This method applies without difficulty to the case of equations involving abstract Volterra operators, as we shall emphasize in what follows. A second method is based on admissibility techniques and has its origin in Massera–Schäffer admissibility theory for ordinary differential equations (see Massera and Schäffer [1], [2]). In both areas, there are still a good many problems to be investigated.

The conditions under which our results will be derived are related to the first method of investigation mentioned above, and the problem of the asymptotic behavior of solutions to (4.4.1) will be reduced to the similar problem for the more common system of ordinary differential equations

$$\dot{x}(t) + f(x(t)) = 0, \quad t \in R_+, \tag{4.4.2}$$

Before we consider the asymptotic behavior of solutions to the systems (4.4.1), it is appropriate to make a few comments on the existence of solutions, verifying an initial condition of the form

$$x(0) = x^0 \in R^n. \tag{4.4.3}$$

As shown in Chapter 3, equation (4.4.1) with the initial condition (4.4.3) can be reduced to the nonlinear Volterra integral equation

$$x(t) + \int_0^t k(t,s)x(s)\,\mathrm{d}s + \int_0^t f(x(s))\,\mathrm{d}s = x^0 + \int_0^t g(s)\,\mathrm{d}s, \tag{4.4.4}$$

for which local existence can be obtained by means of standard methods. The kernel $k(t,s)$ results from the representation

$$\int_0^t (Lx)(s)\,\mathrm{d}s = \int_0^t k(t,s)x(s)\,\mathrm{d}s,$$

and does possess some regularity properties, which help in investigating (4.4.4) in relation to the local existence problem.

The hypotheses we shall assume on (4.4.1) will be of such a nature that the global existence of solutions, as well as their boundedness (on R_+), will be assured.

Let us consider now equation (4.4.1) under the following hypotheses:

(a) The linear operator L is of Volterra type, and is continuous from $C(R_+, R^n)$ – the space of continuous maps from R_+ into R^n, with the topology of uniform convergence on any compact – into $L_{\mathrm{loc}}(R_+, R^n)$, and it takes the bounded functions on R_+ into functions belonging to $M_0(R_+, R^n)$: $LBC(R_+, R^n) \subset M_0(R_+, R^n)$, i.e.

$$\int_t^{t+1} |(Lx)(s)|\,\mathrm{d}s \to 0 \quad \text{as} \quad t \to \infty, \quad \forall x \in BC(R_+, R^n). \tag{4.4.5}$$

(b) $f: R^n \to R^n$ is continuous, and such that

$$\int_0^t \langle (Lx)(s) + f(x(s)), x(s) \rangle \, ds \geq 0, \quad t \in R_+, \tag{4.4.6}$$

for any $x \in C(R_+, R^n)$.

(c) $g \in L^1(R_+, R^n)$.

The following result can be stated in relation to equation (4.4.1):

Theorem 4.4.1. *Assume that conditions* (a), (b), (c) *are satisfied for equation* (4.4.1). *Then any solution is defined on* R_+, *it is bounded there, and its limit set coincides with that of a convenient solution of the ordinary differential equation* (4.4.2).

Proof. Let $x = x(t)$ be a solution of (4.4.1) such that $x(0) = x^0 \in R^n$. This solution is defined on some interval $[0, T)$, $T \leq \infty$ (possibly, for small T only).

From (4.4.1) by scalar multiplication with $x(t)$, $t \in [0, T)$, we obtain

$$\frac{1}{2} \frac{d}{dt} |x(t)|^2 + \langle (Lx)(t) + f(x(t)), x(t) \rangle = \langle g(t), x(t) \rangle, \tag{4.4.7}$$

which implies (on the same interval)

$$\frac{1}{2} |x(t)|^2 + \int_0^t \langle (Lx)(s) + f(x(s)), x(s) \rangle \, ds = \frac{1}{2} |x^0|^2 + \int_0^t \langle g(s), x(s) \rangle \, ds. \tag{4.4.8}$$

Taking condition (b) into account, we obtain from (4.4.8) the inequality

$$|x(t)|^2 \leq |x^0|^2 + 2 \int_0^t |\langle g(s), x(s) \rangle| \, ds, \tag{4.4.9}$$

valid on $[0, T)$. If we now let $X(t) = \sup |x(s)|$, $0 \leq s \leq t < T$, then (4.4.9) yields

$$X^2(t) \leq |x^0|^2 + 2X(t) \int_0^t |g(s)| \, ds, \tag{4.4.10}$$

on $[0, T)$. By (c), (4.4.10) leads to

$$X^2(t) \leq |x^0|^2 + 2X(t) \int_0^\infty |g(s)| \, ds, \tag{4.4.11}$$

also on $[0, T)$. But (4.4.11) implies

$$X(t) \leq \int_0^\infty |g(s)| \, ds + \left\{ |x^0|^2 + \left(\int_0^\infty |g(s)| \, ds \right)^2 \right\}^{1/2}, \tag{4.4.12}$$

on $[0, T)$, from which we obtain

$$|x(t)| \leq \int_0^\infty |g(s)| \, ds + \left\{ |x^0|^2 + \left(\int_0^\infty |g(s)| \, ds \right)^2 \right\}^{1/2}, \tag{4.4.13}$$

on $[0, T)$. Since the right-hand side in (4.4.13) is a constant (with respect to t), we see that $x(t)$ must remain bounded on its maximal interval of existence. This means that $x(t)$ can be extended on the whole semi-axis R_+, and it remains bounded there. Hence, the limit set of $x(t)$ is nonempty.

In order to prove that the limit set of $x(t)$ coincides with that of a convenient solution to (4.4.2) an important property to be established for $x(t)$ is the compactness of the family $\{x(t + h); h \in R_+\}$ in the space $C(R_+, R^n)$. First, $\{x(t + h); h \in R_+\}$ is uniformly bounded on R_+ because $x(t)$ is bounded there. Second, this family of functions is equicontinuous on each compact interval of R_+. Indeed, the equicontinuity is a consequence of the uniform continuity of $x(t)$ on R_+.

From (4.4.1), taking into account the boundedness of $x(t)$ on R_+, we see that on R_+ we have $x'(t) \in L^\infty \oplus M_0$. We claim that this implies the uniform continuity of $x(t)$ on R_+. To prove this, let $x' = v + w$, with $v \in L^\infty(R_+, R^n)$ and $w \in M_0(R_+, R^n)$. Then

$$|x(t) - x(u)| = \left| \int_u^t x'(s)\, ds \right| \le K|t - u| + \left| \int_u^t |w(s)|\, ds \right|, \qquad (4.4.14)$$

and since

$$\left| \int_u^t |w(s)|\, ds \right| \le \int_\tau^{\tau+1} |w(s)|\, ds, \quad \tau = \min\{u, t\}, \quad |t - u| < 1,$$

we see that (because $w \in M_0$)

$$\left| \int_u^t |w(s)|\, ds \right| < \frac{\varepsilon}{2} \quad \text{for } u, t \ge T(\varepsilon), \quad |t - u| < 1. \qquad (4.4.15)$$

But $w \in L^1([0, T], R^n)$, and consequently

$$\left| \int_u^t |w(s)|\, ds \right| < \frac{\varepsilon}{2} \quad \text{for } u, t \in [0, T], \quad |t - u| < \delta(\varepsilon). \qquad (4.4.16)$$

From (4.4.15) and (4.4.16) we derive

$$\left| \int_u^t |w(s)|\, ds \right| < \frac{\varepsilon}{2} \quad \text{for } |t - u| < \delta(\varepsilon) < 1, \qquad (4.4.17)$$

for arbitrary $u, t \in R_+$. From (4.4.14) we easily obtain for $u, t \in R_+$

$$|x(t) - x(u)| < \varepsilon \quad \text{when } |t - u| < \min\{\delta(\varepsilon), \varepsilon/2K\},$$

which shows the uniform continuity of $x(t)$ on R_+. Hence $\{x(t + h); h \in R_+\}$ is compact in $C(R_+, R^n)$. This implies the existence of a sequence $\{h_p\} \subset R_+$, $h_p \to \infty$ as $p \to \infty$, such that

$$\lim_{p \to \infty} x(t + h_p) = z(t), \quad t \in R_+, \qquad (4.4.18)$$

uniformly on any compact interval of R_+.

We want to prove now that $z(t)$ is a solution of (4.4.2), while its limit set coincides with that of $y(t)$. Let us fix an interval $(t, t + h)$ in R_+, and integrate both sides of (4.4.1) on this interval, after changing t into $s + h_p$:

$$\int_t^{t+h} \{\dot{x}(s + h_p) + (L\dot{x})(s + h_p)\}\, ds + \int_t^{t+h} f(x(s + h_p))\, ds = \int_t^{t+h} g(s + h_p)\, ds.$$

$$(4.4.19)$$

Since $x(t + h_p) \to z(t)$ uniformly on each compact of R_+, and conditions (a), (c) allow us to state

$$\int_t^{t+h} (L\dot{x})(s + h_p)\, ds = \int_{t+h_p}^{t+h_p+h} (L\dot{x})(s)\, ds \to 0 \quad \text{as} \quad p \to \infty,$$

and

$$\int_t^{t+h} g(s + h_p)\, ds = \int_{t+h_p}^{t+h_p+h} g(s)\, ds \to 0 \quad \text{as} \quad p \to \infty,$$

(4.4.19) yields

$$z(t + h) - z(t) + \int_t^{t+h} f(z(s))\, ds = 0, \qquad (4.4.20)$$

from which equation (4.4.2) follows immediately, owing to arbitrariness of h.

Let us now denote by X the limit set of the solution $x(t)$ of equation (4.4.1), and by Z that of the solution $z(t)$ of (4.4.2). It is easy to see that $Z \subset X$. Indeed, if $z(t_k) \to \bar{z} \in Z$, taking into account that $z(t_k) = \lim x(t_k + \tau_m^k)$ as $m \to \infty$, we obtain

$$\bar{z} = \lim x(t_k + \tau_{m_k}^k) \quad \text{as} \quad k \to \infty,$$

where $\{\tau_{m_k}^k\} \subset \{\tau_m^k\}$ is chosen in such a way that

$$|z(t_k) - x(t_k + \tau_{m_k}^k)| < \frac{1}{k}.$$

Conversely, let $\bar{x} \in X$. Therefore $x(t_k) \to \bar{x}$ as $k \to \infty$, with $t_k \to \infty$. Let $z(t)$ be the solution of (4.4.2) with $z(0) = \bar{x}$. Without loss of generality, we can assume $x(t + t_k) \to z(t)$ as $k \to \infty$, uniformly on any compact interval of R_+ (the compactnss property). Let us now choose a subsequence $\{t_{k_m}\} \subset \{t_k\}$, such that $t_{k_m} > 2t_m$. Then $\bar{t}_m = t_{k_m} - t_m > t_m \to \infty$ as $m \to \infty$, and

$$|x(\bar{t}_m + t_m) - z(\bar{t}_m)| = |x(t_{k_m}) - z(t_{k_m} - t_m)| \to 0$$

as $m \to \infty$. Hence, $\bar{x} = \lim x(t_{k_m}) = \lim z(t_{k_m} - t_m) \in Z$.

This ends the proof of Theorem 4.4.1.

Remark. There are many possibilities for satisfying condition (4.4.6) in (b) of Theorem 4.4.1. For instance, one may assume that both L and f satisfy the conditions

$$\int_0^t \langle (Lx)(s), x(s) \rangle \, ds \geq 0, \quad \int_0^t \langle f(x(s)), x(s) \rangle \, ds \geq 0$$

for any continuous $x(t)$, $t \in R_+$. The last condition is practically the same as $\langle f(x), x \rangle \geq 0$ for any $x \in R^n$. The first condition states, basically, the *monotonicity* of L regarded as an operator on the space $L^2_{loc}(R_+, R^n)$.

Of course, condition (4.4.6) in (b) is satisfied when

$$\int_0^t \langle (Lx)(s) - \lambda x(s), x(s) \rangle \, ds \geq 0, \quad \lambda > 0,$$

which means that L is strictly monotone, while f is subject to the less restrictive assumption

$$\int_0^t \langle f(x(s)), x(s) \rangle \, ds \geq -\lambda \int_0^t |x(s)|^2 \, ds.$$

For instance, such a condition holds (in the scalar case) when $f(x) = -\lambda x + \alpha x^3$, with $\alpha > 0$.

It is interesting to apply Theorem 4.4.1 to the integrodifferential system

$$\dot{x}(t) + \int_0^t k(t,s)x(s) \, ds + f(x(t)) = g(t), \quad t \in R_+, \tag{4.4.21}$$

which means that in (4.4.1) we choose the operator L as follows:

$$(Lx)(t) = \int_0^t k(t,s)x(s) \, ds. \tag{4.4.22}$$

The following statement is a consequence of Theorem 4.4.1:

Corollary. *Assume that conditions* (b) *and* (c) *of Theorem* 4.4.1 *are satisfied, with L given by* (4.4.22). *Moreover, let $k(t,s)$ be a measurable matrix kernel defined for $0 \leq s \leq t < \infty$, with values in $\mathscr{L}(R^n, R^n)$, such that*

$$\int_0^t |k(t,s)| \, ds \in M_0(R_+, R). \tag{4.4.23}$$

Then the conclusion of Theorem 4.4.1 *remains valid for equation* (4.4.21).

The proof of the Corollary is immediate, if we notice the fact that condition (4.4.23) assures that any $x \in BC(R_+, R^n)$ is taken by the operator L defined by (4.4.22), into a function belonging to $M_0(R_+, R^n)$.

Let us now consider the Volterra functional-differential system

$$\dot{x}(t) + (L\dot{x})(t) + f(x(t)) = g(t); \quad t \in R_+, \tag{4.4.24}$$

in which L stands for a linear Volterra abstract operator acting on the space

$L^2_{loc}(R_+, R^n)$. The following result, similar to Theorem 4.4.1, holds and provides conditions for the existence of convergent solutions.

Theorem 4.4.2. *Consider the system (4.4.24) in which L stands for an abstract Volterra linear operator, acting continuously on $L^2_{loc}(R_+, R^n)$, and such that*

$$\int_0^t \langle (Lx)(s) + \eta x(s), x(s) \rangle \, ds \geq 0, \quad t \in R_+, \tag{4.4.25}$$

for any $x \in L^2_{loc}(R_+, R^n)$, with η real and $\eta < 1$. Moreover, L is assumed to take the space $L^2(R_+, R^n)$ into $M_0(R_+, R^n)$.

Furthermore, let $f: R^n \to R^n$ be such that

$$f(x) = \nabla U(x), \quad U \in C^{(1)}(R^n, R), \tag{4.4.26}$$

where

$$U(x) \to \infty \quad \text{as} \quad |x| \to \infty. \tag{4.4.27}$$

If $g \in L^2(R_+, R^n)$, then the system (4.4.24) has all its solutions defined on R_+, they are bounded there, and the limit set of each solution of (4.4.24) coincides with the limit set of a convenient solution of the ordinary differential system (4.4.2).

Proof. Let $x = x(t)$ be a solution of (4.4.24) with the initial condition $x(0) = x^0 \in R^n$. Such a solution does exist locally, say on some interval $[0, T)$, $T \leq \infty$. We shall prove that $x(t)$ can actually be continued to R_+, and remains bounded there.

Indeed, let us multiply scalarly both members of (4.4.24) by $\dot{x}(t)$, and integrate over $[0, t)$, $t < T$. We obtain

$$\int_0^t \langle \dot{x}(s) + (L\dot{x})(s), \dot{x}(s) \rangle \, ds + U(x(t)) = U(x^0) + \int_0^t \langle g(s), \dot{x}(s) \rangle \, ds, \tag{4.4.28}$$

if we take into account

$$\int_0^t \langle f(x(s)), \dot{x}(s) \rangle \, ds = U(x(t)) - U(x^0),$$

which is a consequence of (4.4.26). Let $\delta = 1 - \eta > 0$, and notice that (4.4.25), (4.4.28) yield the inequality

$$\delta \int_0^t |\dot{x}(s)|^2 \, ds + U(x(t)) \leq U(x^0) + \int_0^t \langle g(s), \dot{x}(s) \rangle \, ds \tag{4.4.29}$$

on the same interval $[0, t)$. Since

$$\int_0^t \langle g(s), \dot{x}(s) \rangle \, ds \leq \frac{2}{\delta} \int_0^t |g(s)|^2 \, ds + \frac{\delta}{2} \int_0^t |\dot{x}(s)|^2 \, ds,$$

we obtain from (4.4.29) the inequality

$$\frac{\delta}{2} \int_0^t |\dot{x}(s)|^2 \, ds + U(x(t)) \le U(x^0) + \int_0^\infty |g(s)|^2 \, ds. \qquad (4.4.30)$$

Because of assumption (4.4.27) on $U(x)$, (4.4.30) implies the boundedness of $x(t)$ on $[0, T)$, as well as that $\dot{x}(t)$ belongs to $L^2([0, T), R^n)$. Hence $x(t)$ is uniformly continuous on $[0, T)$, and $\lim x(t)$ as $t \uparrow T$ must exist and be finite. That means the continuability of $x(t)$ is assured beyond T, which shows that T cannot be finite. This ends the proof of the assertion that any (local) solution of (4.4.24) can be extended to a saturated solution, defined on R_+, and that this solution is bounded. Indeed, (4.4.30) implies

$$U(x(t)) \le U(x^0) + \int_0^\infty |g(x)|^2 \, ds, \quad t \in R_+,$$

which together with condition (4.4.27) lead to the boundedness of $x(t)$ on R_+.

In order to prove the last part of Theorem 4.4.2, i.e., the identity of the limit set of a solution of (4.4.24) to that of a solution of (4.4.2), we can proceed in exactly the same way as in the proof of Theorem 4.4.1.

The compactness of the family $\{x(t + h); h \in R_+\}$ with $x(t)$ a given solution of (4.4.24), is the consequence of the boundedness of x and its uniform continuity on R_+. Since $\dot{x} \in L^2(R_+, R^n)$, according to our assumption we have $L\dot{x} \in M_0(R_+, R^n)$, which implies

$$\int_t^{t+1} |(L\dot{x})(s)| \, ds \to 0 \quad \text{as } t \to \infty,$$

a type of condition that was used in the proof of Theorem 4.4.1.

This ends the proof of Theorem 4.4.2.

Remark. If $z(t) = \lim x(t + h_p)$ as $p \to \infty$, with $h_p \to \infty$ for $p \to \infty$, then $z(t)$ is a solution of (4.4.2). This solution is bounded on R_+, and its boundedness implies that of $\dot{z}(t)$. Hence, $z(t)$ is uniformly continuous on R_+. But (4.4.2) shows that $\dot{z}(t)$ is also uniformly continuous and bounded on R_+, from which one derives the uniform continuity of $\dot{z}^2(t)$ on R_+. This property, together with $\dot{z} \in L^2(R_+, R^n)$, leads to the conclusion $\dot{z}(t) \to 0$ as $t \to \infty$. From (4.4.2) we see that $f(z(t)) \to 0$ as $t \to \infty$.

If U has only a finite number of critical points (i.e., is such that f vanishes at these points), it is obvious that $f(z(t)) \to 0$ as $t \to \infty$ implies $z(t) \to \xi \in R^n$, ξ being one of the critical points (the limit set is connected).

Therefore, each solution of (4.4.24) converges (as $t \to +\infty$) to a critical point of the function U.

Corollary. *Assume the following conditions are verified with respect to equation* (4.4.21):

(1) *k is a measurable kernel from* $\Delta = \{(t,s); 0 \le s \le t < \infty\}$ *into* $\mathscr{L}(R^n, R^n)$, *such that for some* $A > 0$

$$\int_0^t |k(t,s)|^2 \, ds \le A^2, \quad \forall t \in R_+, \tag{4.4.31}$$

and for every $T > 0$

$$\int_0^T |k(t,s)|^2 \, ds \in C_0(R_+, R); \tag{4.4.32}$$

(2) *for every* $t \ge 0$

$$\lim_{h \to 0} \left[\int_t^{t+h} |k(t+h,s)|^2 \, ds + \int_0^t |k(t+h,s) - k(t,s)|^2 \, ds \right] = 0. \tag{4.4.33}$$

Then, the conclusion of Theorem 4.4.1 *holds for equation* (4.4.21).

The proof of the Corollary will follow from the fact that the Volterra integral operator

$$(Kx)(t) = \int_0^t k(t,s)x(s) \, ds, \quad t \in R_+, \tag{4.4.34}$$

is continuous from $L^2(R_+, R^n)$ into $C_0(R_+, R^n)$. If $x \in L^2(R_+, R^n)$, then (4.4.31) guarantees that $(Kx)(t)$ is defined for every $t \in R_+$, and

$$|(Kx)(t)| \le A|x|_{L^2}, \quad \forall t \in R_+. \tag{4.4.35}$$

Condition (4.4.33) yields easily the continuity of $(Kx)(t)$, as a function of $t \in R_+$. It only remains to prove that

$$Kx \in C_0(R_+, R^n), \tag{4.4.36}$$

using also (4.4.32). Indeed, for $t > T > 0$ one can write

$$\left| \int_0^t k(t,s)x(s) \, ds \right| \le \left(\int_0^T |k(t,s)|^2 \, ds \right)^{1/2} \left(\int_0^T |x(s)|^2 \, ds \right)^{1/2}$$

$$+ \left(\int_T^t |k(t,s)|^2 \, ds \right)^{1/2} \left(\int_T^t |x(s)|^2 \, ds \right)^{1/2}. \tag{4.4.37}$$

Given $\varepsilon > 0$, let us choose $T = T(\varepsilon)$ large enough such that

$$\int_T^\infty |x(s)|^2 \, ds < (\varepsilon 2^{-1} A^{-1})^2. \tag{4.4.38}$$

Therefore, for $t > T$ we have

$$\left(\int_T^t |k(t,s)|^2 \, ds\right)^{1/2} \left(\int_T^t |x(s)|^2 \, ds\right)^{1/2} < \frac{\varepsilon}{2}. \tag{4.4.39}$$

On the other hand, (4.4.32) allows us to write for $t \geq T_1(\varepsilon)$:

$$\int_0^T |k(t,s)|^2 \, ds < \left(\frac{\varepsilon}{2}|x|_{L^2}^{-1}\right)^2. \tag{4.4.40}$$

Hence, choosing $t > \max(T, T_1)$ we can write as a result of (4.4.37), (4.4.39) and (4.4.40)

$$\left|\int_0^t k(t,s)x(s) \, ds\right| < \frac{\varepsilon}{2} + \frac{\varepsilon}{2} = \varepsilon, \tag{4.4.41}$$

which shows that (4.4.36) holds.

From (4.4.35) we see that the mapping $K: L^2(R_+, R^n) \to C_0(R_+, R^n)$ is continuous. But $C_0 \subset M_0$, which ends the proof.

Remark. If one assumes $k(t,s) = k(t-s)$, where $k \in L^2(R_+, \mathscr{L}(R^n, R^n))$, then all the conditions in the Corollary are verified. Indeed, condition (4.4.31) is obviously satisfied. Condition (4.4.33) is a consequence of $k \in L^2$, while condition (4.4.32) reduces to

$$\int_t^{t+T} |k(s)|^2 \, ds \in C_0(R_+, R),$$

which again follows from $k \in L^2$.

An immediate application of Theorem 4.4.1 or 4.4.2 can be obtained if we consider *infinite delay* equations of the form

$$\dot{y}(t) + \int_{-\infty}^t k(t,s)y(s) \, ds + f(y(t)) = g(t), \quad t \in R_+, \tag{4.4.42}$$

with the initial condition

$$y(s) = \psi(s), \quad -\infty < s \leq 0. \tag{4.4.43}$$

Since because of (4.4.43), (4.4.42) can be rewritten in the form

$$\dot{y}(t) + \int_0^t k(t,s)y(s) \, ds + f(y(t)) = \bar{g}(t), \quad t \in R_+, \tag{4.4.44}$$

with the initial condition $y(0) = \psi(0)$, and with

$$\bar{g}(t) = g(t) - \int_{-\infty}^0 k(t,s)\psi(s) \, ds, \quad t \in R_+, \tag{4.4.45}$$

the only real concern is to assure the properties required for $g(t)$, keeping the same assumptions we made in Theorem 4.4.1 or Theorem 4.4.2 for other data in the equation (4.4.44).

To be more specific, let us consider the case when Theorem 4.4.1 is to be applied to (4.4.42), under condition (4.4.43).

It is obvious that to secure the condition $\bar{g}(t) \in L^1(R_+, R^n)$, it suffices to assume $g(t) \in L^1(R_+, R^n)$ and

$$\int_{-\infty}^{0} k(t,s)\psi(s)\,ds \in L^1(R_+, R^n). \qquad (4.4.46)$$

Condition (4.4.46) will be satisfied, for instance, if we assume that $k(t,s)$ is measurable on $R_+ \times R_-$, and such that

$$\int_{-\infty}^{0} |k(t,s)|\,ds \in L^1(R_+, R), \qquad (4.4.47)$$

while ψ is continuous and bounded on $(-\infty, 0]$.

In summarizing the above discussion in relation to the *infinite delay* equation (4.4.42), we can say that the ultimate behavior of its solutions is the same as in the case of the equation (4.4.2), provided the following conditions hold: k, f and g are as in Theorem 4.4.1; for $(t,s) \in R_+ \times R_-$, $k(t,s)$ is measurable and satisfies (4.4.47); the initial condition (4.4.43) is verified for some ψ which is continuous and bounded from R_- into R^n.

In concluding this section we will discuss the problem of positiveness of the operator K given by (4.4.34). This condition appears in both Theorems 4.4.1 and 4.4.2, and our purpose is to illustrate how the positiveness can be secured, by means of certain assumptions on the kernel $k(t,s)$. We follow here Gripenberg [3], and restrict our considerations to the scalar case.

Theorem 4.4.3. *Let $k(t,s)$ be a real-valued functional defined on $\Delta = \{(t,s); 0 \le s \le t < \infty\}$, satisfying the following conditions:*

(1) $k(t,s) \ge 0$ *and, for every $t > 0$, $k(t,s)$ is absolutely continuous and nondecreasing in s on $[0,t]$;*
(2) $\int_0^t k(t,s)\,ds \in L^\infty_{loc}(R_+, R)$;
(3) $k(t,0)$ *is left continuous and nonincreasing for $t > 0$;*
(4) *for almost all $s > 0$, $k_s(t,s)$ is left continuous and nonincreasing on (s, ∞).*

Then for any $T > 0$, and $y(t)$ continuous from R_+ into R, the following inequality holds:

$$\left(\int_0^T k(T,s)y(s)\,ds \right)^2 \le 4k(T,T) \int_0^T y(t)\,dt \int_0^t k(t,s)y(s)\,ds. \qquad (4.4.48)$$

Proof. Taking into account conditions (1), (2) and (3) in the statement of Theorem 4.4.3, we conclude that the functions $tk(t,0)$ and $\int_0^t (t-s)k_s(t,s)\,ds$ are locally bounded on R_+. Integration by parts yields

$$\int_0^t k(t,s)y(s)\,ds = k(t,0)\int_0^t y(u)\,du + \int_0^t k_s(t,s)\int_s^t y(u)\,du\,ds, \quad 0 \le t \le T.$$

$$(4.4.49)$$

Let $k(t,0) - k(T,0) = \alpha(t)$. On behalf of property (3), $\alpha(t)$ is a nonnegative measure: $\alpha(t) = \text{meas}[t,T)$, $t > 0$. We obtain as a result of integration by parts

$$\int_0^T y(t)k(t,0)\int_0^t y(u)\,du\,dt = \frac{1}{2}k(T,0)\left(\int_0^T y(u)\,du\right)^2 + \frac{1}{2}\int_0^T\left(\int_0^t y(u)\,du\right)^2 d\alpha(t).$$

$$(4.4.50)$$

For every s for which $k_s(t,s)$ is nondecreasing according to condition (4), we denote $k_s(t,s) = k_s(T,s) + \beta_s(t)$, with β_s being a nonnegative measure. Since $k_s(t,s)$ is a measurable function of (t,s) from conditions (1) and (4), the theorem of Fubini can be applied and, using integration by parts, we obtain

$$\int_0^T y(t)\int_0^t k_s(t,s)\int_s^t y(u)\,du\,ds\,dt = \frac{1}{2}\int_0^T k_s(T,s)\left(\int_s^T y(u)\,du\right)^2 ds$$

$$+ \frac{1}{2}\int_0^T\left(\int_s^T\left(\int_s^t y(u)\,du\right)^2 d\beta_s(t)\right)ds.$$

$$(4.4.51)$$

Based on (4.4.49) we have

$$\left(\int_0^T k(T,s)y(s)\,ds\right)^2 \le 2[k[T,0]]^2\left(\int_0^T y(u)\,du\right)^2$$

$$+ 2\left(\int_0^T k_s(T,s)\int_s^T y(u)\,du\,ds\right)^2, \qquad (4.4.52)$$

and by means of Holder's inequality and condition (1) we obtain

$$\left(\int_0^T k_s(T,s)\int_s^T y(u)\,du\,ds\right)^2 \le \int_0^T k_s(T,s)\,ds\int_0^T k_s(T,s)\left(\int_s^T y(u)\,du\right)^2 ds.$$

Combining this last inequality with (4.4.49)–(4.4.52), we obtain (4.4.48) if we also take into account

$$0 \le \int_0^T k_s(T,s)\,ds = k(T,T) - k(T,0) \le k(T,T).$$

This ends the proof of Theorem 4.4.3.

In concluding this section, let us notice that in the case of convolution kernels, $k(t,s) = k(t-s)$, the positiveness conditions on $k(t)$ can be expressed in terms of its Fourier transform $\tilde{k}(is)$. For some illustrations of this see, for instance, C. Corduneanu [4].

4.5   Hammerstein integrodifferential equations and boundary value problems

In this section we shall consider some Hammerstein integral equations arising in connection with boundary value problems for ordinary differential equations, or Volterra functional differential equations. The existence of solutions will be obtained by again using results on monotone operators, or fixed point theorems.

As seen in Section 1.1, quite general boundary value problems for ordinary differential equations can be reduced to integral equations of Hammerstein type. In this section we shall consider first classical two-point boundary value problems for second-order differential equations of the form

$$x''(t) + f(t, x(t), x'(t)) = h(t), \quad t \in [0, T], \qquad (4.5.1)$$

under boundary value conditions

$$x(0) = a, \quad x(T) = b, \qquad (4.5.2)$$

or

$$x(0) = a, \quad x'(T) = b. \qquad (4.5.3)$$

Before we state the conditions under which existence results can be obtained, notice that by a simple substitution, namely $x = y + (b - a)T^{-1}t + a$, the problem (4.5.1), (4.5.2) reduces to

$$y''(t) + f(t, y(t) + (b - a)T^{-1}t + a, y'(t) + (b - a)T^{-1}) = h(t), \quad t \in [0, T], \qquad (4.5.4)$$

with homogeneous boundary value conditions

$$y(0) = 0, \quad y(T) = 0. \qquad (4.5.5)$$

If we introduce the Green's function

$$G(t, s) = \begin{cases} \dfrac{1}{T}(t - T)s, & 0 \le s \le t \le T, \\[2mm] \dfrac{1}{T}(s - T)t, & 0 \le t \le s \le T, \end{cases} \qquad (4.5.6)$$

then for any $y(t)$ satisfying (4.5.5), and such that $y''(t) \in L^2([0, T], R)$, we can write

$$y(t) = \int_0^T G(t, s) y''(s) \, ds,$$

$$y'(t) = \int_0^T G_t(t, s) y''(s) \, ds,$$

where $G_t(t, s)$ denotes the partial derivative of G with respect to t. Notice that this derivative exists in the square $0 \le s, t \le T$, and is piecewise continuous there.

Equation (4.5.4) can now be rewritten as an integrodifferential equation,

$$y(t) = \int_0^T G(t,s)[h(s) - f(s,y(s) + (b-a)T^{-1}s + a, y'(s) + (b-a)T^{-1})]\,ds.$$

(4.5.7)

Moreover, from the formulas above we derive on $[0, T]$ the inequalities

$$|y(t)| \le \left[\sqrt{\frac{T^3}{48}}\right]|y''|_{L^2}, \quad |y'(t)| \le \left[\sqrt{\frac{T}{3}}\right]|y''|_{L^2}.$$

(4.5.8)

Indeed, using the Schwartz inequality we obtain

$$|y(t)| \le \left\{\int_0^T G^2(t,s)\,ds\right\}^{1/2}|y''|_{L^2},$$

and taking into account

$$\int_0^T G^2(t,s)\,ds = \frac{1}{3T}t^2(T-t)^2,$$

the first inequality in (4.5.8) follows immediately. One proceeds similarly with the second one.

Without fear of confusion we can denote by the same letter G the linear integral operator generated by the Green's function $G(t,s)$, and by G' the one generated by $G_t(t,s)$. Both are obviously acting from $L^2([0, T], R)$ into $C([0, T], R) \subset L^2([0, T], R)$. Then we can write $y = Gy''$ and $y' = G'y''$, for $y'' \in L^2([0, T], R)$, provided $y(0) = y(T) = 0$.

We shall now consider $y'' = z$ as the new variable in $L^2([0, T], R^n)$, and notice that (4.5.4) can be rewritten in the form

$$z + f(t, Gz + (b-a)T^{-1}t + a, G'z + (b-a)T^{-1}) = h(t).$$

(4.5.9)

If we assume that $f(t, u, v)$ is a continuous map from $[0, T] \times R \times R$ into R, then it follows that for each $z \in L^2([0, T], R)$ the left hand side of (4.5.4) is in $L^2([0, T], R)$, i.e., the operator given by

$$(Az)(t) = z(t) + f(t, (Gz)(t) + (b-a)T^{-1}t + a, (G'z)(t) + (b-a)T^{-1})$$

(4.5.10)

is acting on $L^2([0, T], R)$. Equation (4.5.4) is, in fact,

$$Az = h,$$

(4.5.11)

and, if we assume $h \in L^2([0, T], R)$, our problem reduces to proving that the operator A given by (4.5.10) is onto $L^2([0, T], R)$. While this may not necessarily be easy to prove directly, we can come up with a construction that will allow us to apply the Corollary to Theorem 2.4.5.

Let us consider, for given $\rho > 0$, the following set:

$$\Omega_\rho = \{(t,u,v); 0 \le t \le T, |u| \le \rho + \max(|a|,|b|), |v| \le \rho + |b-a|T^{-1}\}.$$

Let us also consider a continuous map $f_\rho(t, u, v)$ from $[0, T] \times R \times R$ into R, such that $f_\rho(t, u, v) \equiv f(t, u, v)$ for $(t, u, v) \in \Omega_\rho$, and

$$|f_\rho(t, u, v)| \leq \sup_{\Omega_\rho} |f(t, u, v)| = \phi(\rho). \tag{4.5.12}$$

To obtain $f_\rho(t, u, v)$ one can proceed in many ways. For instance, for $(t, U, V) \bar{\in} \Omega$ one can define $f_\rho(t, U, V) = f(t, u, v)$ where $(t, u, v) \in \Omega$ is the closest point to (t, U, V) on the ray joining (t, U, V) to $(t, 0, 0)$.

Let us now consider the equation analogous to (4.5.11) namely

$$A_\rho z = h, \quad h \in L^2([0, T], R), \tag{4.5.13}$$

where the operator A_ρ is defined by means of f_ρ in the same way A is defined in terms of f.

It is easy to show that $A_\rho: L^2 \to L^2$ is a *coercive* operator. Indeed, one has

$$\langle A_\rho z, z \rangle = |z|^2 + \langle f_\rho, z \rangle \geq |z|^2 - \phi(\rho)|z| = (|z| - \phi(\rho))|z|,$$

and since $|z| - \phi(\rho) \to \infty$ as $|z| \to \infty$, the assertion is proven.

Moreover, the operator A_ρ is weakly continuous because $A_\rho(z + y) - A_\rho z$ tends to zero in the weak topology of L^2, as $y \to 0$ in the same topology. Taking into account the continuity of f_ρ, the above statement is easily checked. Indeed, for any $w \in L^2$ one has

$$\langle A_\rho(z + y) - A_\rho z, w \rangle = \langle y, w \rangle + \langle f_\rho(t, (Gz)(t) + (Gy)(t)$$
$$+ (b - a)T^{-1}t + a, (G'z)(t) + (G'y)(t)$$
$$+ (b - a)T^{-1}) - f_\rho(t, (Gz)(t)$$
$$+ (b - a)T^{-1}t + a, (G'z)(t) + (b - a)T^{-1}), w \rangle.$$

By definition, $\langle y, w \rangle \to 0$ as $y \to \theta$ weakly, for any $w \in L^2$. Applying the Cauchy–Schwartz inequality to the second scalar product, we find that it suffices to show that

$$f_\rho(t, (Gz)(t) + (Gy)(t) \cdots) \to f_\rho(t, (Gz)(t) \cdots)$$

in L^2, when $y \to \theta$ weakly. It is obvious that the dominated Lebesque's criterion can be applied since f_ρ is bounded. On the other hand, for every $t \in [0, T]$, $(Gy)(t) \to 0$ and $(G'y) \to 0$ as $y \to \theta$ weakly (because $(Gy)(t) = \langle G(t, \cdot), y(\cdot) \rangle$). This shows the correctness of the statement.

Consequently, equation (4.5.13) has a solution for every $h \in L^2([0, T], R)$.

In order to apply Corollary 2.4.6 and obtain equation (4.5.11), we need to check the validity of the second condition in the statement. Hence, let us assume $|h| \leq H =$ const. Taking into account the definition of A_ρ and (4.5.13), we obtain $|z|^2 + \langle f_\rho(\cdots), z \rangle = \langle h, z \rangle$ from which we derive the estimate

$$|z| \le H + \phi(\rho)\sqrt{T}, \qquad (4.5.14)$$

with $\phi(\rho)$ defined by (4.5.12).

Since (4.5.8) holds for any y vanishing at 0 and T, and such that $y'' \in L^2([0, T], R)$, we obtain, keeping in mind that $z = y''$ and (4.5.14),

$$|y|_1 = \max\{\sup|y|, \sup|y'|\} \le \max\left\{\sqrt{\frac{T}{3}}, \sqrt{\frac{T^3}{48}}\right\}[H + \phi(\rho)\sqrt{T}]. \quad (4.5.15)$$

In our underlying space $L^2([0, T], R)$ we now introduce the new norm

$$|z|_s = |Gz|_1, \quad z \in L^2([0, T], R), \qquad (4.5.16)$$

and notice that, according to (4.5.15),

$$|z|_s \le \max\left\{\sqrt{\frac{T}{3}}, \sqrt{\frac{T^3}{48}}\right\}[H + \phi(\rho)\sqrt{T}]$$

for the solution z of $A_\rho z = h$. Therefore, if we can determine $\rho > 0$ such that

$$\max\left\{\sqrt{\frac{T}{3}}, \sqrt{\frac{T^3}{48}}\right\}[H + \phi(\rho)\sqrt{T}] \le \rho, \qquad (4.5.17)$$

then Corollary 2.4.6 allows us to conclude that equation (4.5.11) is solvable in $L^2([0, T], R)$, for each h, such that $|h| \le H$.

We can now state the following existence result for equation (4.5.11), which is equivalent to the Hammerstein integrodifferential equation (4.5.7), or to the boundary value problem (4.5.1), (4.5.2).

Theorem 4.5.1. *Consider the boundary value problem (4.5.1), (4.5.2), under the following assumptions on the data:*

(1) *$f(t, u, v)$ is a continuous map from $[0, T] \times R \times R$ into R, where $T > 0$;*
(2) *the real numbers a, b, T and $H > 0$ are such that inequality (4.5.17), with $\phi(\rho)$ given by (4.5.12), has at least one positive solution in ρ;*
(3) *$h \in L^2([0, T], R)$.*

Then there exists at least one solution $x(t)$ for (4.5.1), (4.5.2), with $x'(t)$ absolutely continuous on $[0, T]$ and $x''(t) \in L^2([0, T], R)$, provided $|h|_{L^2} \le H$. This statement is equivalent to the solvability of equation (4.5.7) in the space $W_{2,2}^0([0, T], R)$, or to the solvability of equation (4.5.11) in the space $L^2([0, T], R)$.

Remark 1. It is obvious that the result in Theorem 4.5.1 can be extended to the case when $x(t)$ takes values in R^n instead of R. Only superficial adjustments are required to obtain the proof in the vector-valued case.

Remark 2.   A key condition in Theorem 4.5.1 is, of course, the inequality (4.5.11). It is evident that this condition imposes a certain restriction on the growth of $\phi(\rho)$. The following special case is, none the less, signficant.

Assume that the function $f(t, u, v)$ is continuous and bounded on $[0, T] \times R \times R$. Then we can take $\phi(\rho) \equiv \text{Const.} > 0$, and (4.5.17) is verified by any $\rho > 0$ sufficiently large, and for any $H > 0$. The result obtained under this assumption generalizes a similar one in the case $f(t, u, v) \equiv f(t, u)$, $h(t) \equiv 0$ obtained by Caccioppoli in the 1930s [1].

Another acceptable assumption on $\phi(\rho)$ is

$$\liminf_{\rho \to \infty} \frac{\phi(\rho)}{\rho} < \min\left\{\frac{\sqrt{3}}{T}, \frac{4\sqrt{3}}{T^2}\right\}. \tag{4.5.18}$$

If this condition holds, then (4.5.17) is solvable in ρ, for any $H > 0$. Condition (4.5.18) is verified if $\phi(\rho)$ has linear growth, and T is chosen sufficiently small. It has been known since Picard that (4.5.1), (4.5.2), with $h \equiv 0$ and f Lipschitz continuous, has a solution, provided the length of the interval $[0, T]$ is small enough.

Let us now consider the problem (4.5.1), (4.5.3). The discussion is very much like the one conducted above, and we shall limit ourselves to the most basic steps of the procedure.

First, the substitution to be performed, in order to reduce the boundary value conditions to the homogeneous ones, is $x(t) = y(t) + bt + a$.

Second, the Green's function associated with the boundary value conditions $y(0) = 0$, $y'(T) = 0$ is

$$G(t, s) = \begin{cases} s, & 0 \le s \le t, \\ t, & t \le s \le T. \end{cases} \tag{4.5.19}$$

Then

$$y(t) = \int_0^T G(t, s) y''(s) \, ds,$$

$$y'(t) = \int_0^T G_t(t, s) y''(s) \, ds,$$

for any $y(t)$ satisfying the homogeneous boundary value conditions, and such that $y''(t) \in L^2([0, T], R)$.

Third, the problem (4.5.1), (4.5.3) is equivalent to the Hammerstein integrodifferential equation (4.5.7), in which $G(t, s)$ is given by (4.5.19). The equivalence should be understood within the class of continuous functions.

Fourth, the following estimates hold for any $y(t)$ satisfying $y(0) = 0$, $y'(T) = 0$, such that $y''(t) \in L^2([0, T], R)$:

$$|y(t)| \le \left[\sqrt{\frac{T^3}{3}}\right] |y''|_{L^2}, \quad |y'(t)| \le [\sqrt{T}] |y''|_{L^2}. \tag{4.5.20}$$

Fifth, in defining the operator analogous to (4.5.10), the following formula should be used instead of (4.5.10):

$$(Az)(t) = z(t) + f(t,(Gz)(t) + bt + a,(G'z)(t) + b), \qquad (4.5.21)$$

with G standing for the integral operator generated by the Green's function (4.5.19).

Sixth, the set Ω_ρ to be used in constructing the function f_ρ and defining $\phi(\rho)$ is given by

$$\Omega_\rho = \{(t,u,v); 0 \le t \le T, |u| \le \rho + \max(|a|,|a + bT|), |v| \le \rho + |b|\}. \qquad (4.5.22)$$

Seventh, the inequality (4.5.17) must be replaced by

$$\max\left\{\sqrt{T}, \sqrt{\frac{T^3}{3}}\right\}[H + \phi(s)\sqrt{T}] \le \rho, \qquad (4.5.23)$$

and it must be solvable for some $\rho > 0$.

The conclusion of the above sketched discussion in regard to the boundary value problem (4.5.1), (4.5.3) can be summarized as follows:

Theorem 4.5.2. *Consider the boundary value problem* (4.5.1), (4.5.3), *under the following assumptions:*

(1) *and* (3), *as in Theorem* 4.5.1;
(2) *the real numbers a, b, T, and H > 0 are such that inequality* (4.5.23), *in which $\phi(\rho)$ is given by* (4.5.12), *and Ω_ρ by* (4.5.22), *has at least one positive solution.*

Then, there exists at least one solution $x(t)$ for (4.5.1), (4.5.3), *with $x'(t)$ absolutely continuous on* $[0,T]$, *and $x''(t) \in L^2([0,T], R)$, as soon as $|h|_{L^2} \le H$. This statement is equivalent to the solvability of the equation $y(t) = \int_0^T G(t,s) \times [h(s) - f(s,y(s) + bs + a, y'(s) + b)]\,ds$ with $G(t,s)$ given by* (4.5.19), *in the space of functions $y(t)$ satisfying $y(0) = 0$, $y'(T) = 0$, and such that $y''(t) \in L^2([0,T], R)$, or to the solvability of* (4.5.11), *with A given by* (4.5.21), *in the space $L^2([0,T], R)$.*

Remark. The Cauchy problem $x(0) = x$, $x'(0) = b$ for (4.5.1) can be treated by the same method. The substitution to be performed in order to reduce the initial data to the homogeneous one is still $x(t) = y(t) + bt + a$. Instead of equation (4.5.7) or its equivalent, one has to use the equation obtained by integrating equation (4.5.1) twice:

$$x(t) = a + bt + \int_0^t (t - s)[h(s) - f(s,x(s),x'(s))]\,ds.$$

The equivalent equation for $y(t)$ is

$$y(t) = \int_0^t (t - s)[h(s) - f(s,y(s) + bt + a, y'(s) + b)]\,ds, \qquad (4.5.24)$$

and instead of a Hammerstein–Fredholm equation like (4.5.7), we obtain a Hammerstein–Volterra integrodifferential equation. Equation (4.5.24) has to be considered in the space of functions satisfying $y(0) = y'(0) = 0$, and such that $y''(t) \in L^2([0, T], R)$.

Another proof of Theorem 4.5.1. Let us go back now to equation (4.5.7), with $G(t, s)$ given by (4.5.6), and investigate the existence problem in $W_{2,2}^0([0, T], R)$ by means of the fixed point method. We shall see that we can obtain the same result as in Theorem 4.5.1, by using the Schauder fixed point theorem (see Section 2.4).

First, notice that any solution of (4.5.7) also satisfies $y''(t) \in L^2([0, T], R)$. Therefore, it is enough to deal with (4.5.7) in the space $C^{(1)}([0, T], R)$. And, to be more specific, we should say that we can deal with the subspace of $C^{(1)}([0, T], R)$ consisting of those elements that vanish at 0 and T.

The norm in $C^{(1)}([0, T], R)$ will be defined as in (4.5.15), and we look for a number $\rho > 0$, such that the ball of radius ρ, centered at the origin of $C^{(1)}([0, T], R)$, is taken into itself by the operator

$$(Uy)(t) = \int_0^T G(t, s)[h(s) - f(s, y(s) + (b - a)T^{-1}t + a, y'(s) + (b - a)T^{-1})]\,ds.$$

(4.5.25)

According to (4.5.8) and (4.5.14), if we assume $|y(t)|$, $|y'(t)| \leq \rho$ on $[0, T]$ we obtain from (4.5.12)

$$|(Uy)(t)|, |(Uy)'(t)| \leq \max\left\{\sqrt{\frac{T}{3}}, \sqrt{\frac{T^3}{48}}\right\}[H + \phi(\rho)\sqrt{T}],$$

provided $|h|_{L^2} \leq H$. Consequently, the condition that U take the ball of radius ρ in $C^{(1)}([0, T], R)$, centered at the origin, into itself is precisely (4.5.17).

It suffices to show that U is continuous and takes the ball of radius ρ into a relatively compact subset of the ball. The continuity of U in the topology of $C^{(1)}([0, T], R)$ follows easily from (4.5.25), and the similar equation obtained by differentiation

$$(Uy)'(t) = \int_0^T G_t(t, s)[h(s) - f(\cdots)]\,ds,$$

in which f is continuous. The compactness of the image of the ball by U follows also immediately from $(Uy)''(t) = h(t) - f(\cdots) \in L^2([0, T], R)$, an upper bound for the right hand side (in L^2) being $H + \phi(\rho)\sqrt{T}$.

Consequently, the operator U given by (4.5.25) satisfies the conditions required by Schauder's fixed point theorem, which means that (4.5.7) has at least one solution under the hypotheses of Theorem 4.5.1.

This second proof of Theorem 4.5.1 is more in the spirit of using integral equations for solving boundary value problems. It also shows that monotone operator techniques and classical fixed point theorems can lead to the same conclusion.

We shall now consider the linear functional-differential equation

$$\dot{x}(t) = (Lx)(t) + f(t), \quad t \in [0, T], \tag{4.5.26}$$

where L is a linear continuous Volterra operator on $L^2([0, T], R^n)$, and $f \in L^2([0, T], R^n)$. Such equations have been dealt with in Section 3.3 and, as shown there, the solution of (4.5.26) which satisfies the initial condition $x(0) = x^0$ is given by the variation of parameters formula

$$x(t) = X(t, 0)x^0 + \int_0^t X(t, s)f(s) \, ds. \tag{4.5.27}$$

The construction of the kernel $X(t, s)$ has also been discussed in Section 3.3.

Let us assume now that we want to find a solution of (4.5.26) such that the following two-point boundary value condition holds:

$$Ax(0) + Bx(T) = c, \tag{4.5.28}$$

where A and B are n by n matrices with real entries, i.e., $A, B \in \mathscr{L}(R^n, R^n)$, and $c \in R^n$.

If we substitute (4.5.27), for $t = 0$ and $t = T$, in (4.5.28), then the following equation is obtained for x^0:

$$[A + BX(T, 0)]x^0 = c - B \int_0^T X(T, s)f(s) \, ds. \tag{4.5.29}$$

It is clear that the problem (4.5.26), (4.5.28) will have a unique solution for any $c \in R^n$, if and only if

$$\det[A + BX(T, 0)] \neq 0. \tag{4.5.30}$$

Assuming condition (4.5.30), we have

$$x^0 = [A + BX(T, 0)]^{-1} \left[c - B \int_0^T X(T, s)f(s) \, ds \right], \tag{4.5.31}$$

which together with (4.5.27) provides the unique solution of the boundary value problem (4.5.26), (4.5.28). It is useful to notice that the solution of this boundary value problem can be represented by means of the formula

$$x(t) = x_0(t) + \int_0^T \tilde{X}(t, s)f(s) \, ds, \tag{4.5.32}$$

where

$$x_0(t) = X(t, 0)[A + BX(T, 0)]^{-1}c, \tag{4.5.33}$$

and

$$\tilde{X}(t,s) = \begin{cases} X(t,s) - X(t,0)[A + BX(T,0)]^{-1}BX(T,s), & 0 \le s \le t, \\ -X(t,0)[A + BX(T,0)]^{-1}BX(T,s), & t \le s \le T. \end{cases} \quad (4.5.34)$$

Formula (4.5.32) plays a basic role, not only because it provides the unique solution of problem (4.5.26), (4.5.28), under assumption (4.5.30), but because it also allows us to reduce nonlinear boundary value problems to Hammerstein integral equations.

Let us substitute for (4.5.26) the nonlinear functional differential equation

$$\dot{x}(t) = (Lx)(t) + (fx)(t), \quad t \in [0, T], \quad (4.5.35)$$

and look for a solution that also verifies the boundary value condition (4.5.28). We assume, as above, that L is a linear continuous Volterra operator on $L^2([0, T], R^n)$, while f is a map, not necessarily of Volterra type, between two convenient function spaces. Then (4.5.35), (4.5.28) are equivalent to the Hammerstein integral equation

$$x(t) = x_0(t) + \int_0^T \tilde{X}(t,s)(fx)(s)\, ds. \quad (4.5.36)$$

In the classical theory of Hammerstein equations, f is usually an operator of Niemytzki type. While this remains a valid choice, it is obviously not necessary to limit our investigation to this particular case.

We shall now concentrate our attention on the Hammerstein equation (4.5.36) in order to obtain an existence result. Such a result will imply, under conditions that have been partially specified, the existence of at least one solution to the boundary value problem (4.5.35), (4.5.28).

Let us point out first that the kernel $\tilde{X}(t,s)$ appearing in (4.5.36) is constructed by means of the formula (4.5.34) in terms of the kernel $X(t,s)$ that occurs in the variation of parameters formula (4.5.27). On the other hand, $X(t,s)$ was constructed in Section 4.3 by means of the formula

$$X(t,s) = I + \int_s^t \tilde{k}(t,u)\, du, \quad (4.5.37)$$

where $\tilde{k}(t,u)$ represents the resolvent kernel associated with the kernel $k(t,s)$ in the representation of $\int_0^t (Lx)(s)\, ds$:

$$\int_0^t (Lx)(s)\, ds = \int_0^t k(t,s)x(s)\, ds. \quad (4.5.38)$$

Formula (4.5.38) holds for any $x \in L^2([0, T], R^n)$ and $t \in [0, T]$. Since the left hand side in (4.5.38) is an absolutely continuous function on $[0, T]$, and hence in $L^2([0, T], R)$, $k(t, s)$ satisfies the Hilbert–Schmidt condition

$$\int_0^T dt \int_0^t |k(t, s)|^2 ds < +\infty.$$

This condition on $k(t, s)$ guarantees the existence of the resolvent kernel $R(t, s)$ (see Section 1.5), belonging to the same Hilbert–Schmidt class.

The above mentioned facts enable us to conclude that $X(t, s)$ and $\tilde{X}(t, s)$ possess certain useful properties that will allow us to investigate equation (4.5.36) without difficulty. First, from (4.5.37) we see that $X(t, s)$ is defined for every t, and every s, in $[0, T]$. Second, $X(t, s)$ is absolutely continuous in s, for every fixed $t \in [0, T]$. Third, the map $t \to X(t; \cdot)$ is L^2-continuous, as is the map $t \to \tilde{X}(t; \cdot)$. In particular

$$\sup_{t \in [0, T]} \int_0^T |\tilde{X}(t, s)|^2 ds = \gamma^2 < \infty. \tag{4.5.39}$$

With regard to the operator f involved in (4.5.36) we can assume, for instance, that it is continuous from $C([0, T], R^n)$ into $L^2([0, T], R^n)$. But because any solution (if any) of (4.5.36) is absolutely continuous, the space $L^2([0, T], R^n)$ does not seem to be naturally involved, and without serious loss of generality we can assume f to be a continuous operator from $C([0, T], R^n)$ into itself. This will enable us to apply Schauder's fixed point theorem in order to prove the existence result for (4.5.36), using the space $C([0, T], R^n)$ as underlying space.

A growth condition on f, with respect to $|x|_C = \sup|x(t)|$, $t \in [0, T]$, has to be imposed. We notice that the continuity of f, which in general is not a linear operator, does not imply its boundedness on a ball $|x|_C \leq \rho$. Therefore, it appears quite natural to assume

$$\sup_{|x|_C \leq \rho} |fx| = \phi(\rho) < \infty, \tag{4.5.40}$$

with $\phi(\rho)$ a positive nondecreasing function for $\rho > 0$.

Equation (4.5.36) can be written as $x(t) = (Ux)(t)$, where

$$(Ux)(t) = x_0(t) + \int_0^T \tilde{X}(t, s)(fx)(s) ds, \tag{4.5.41}$$

and since the right hand side of (4.5.41) is nothing other than the right hand side of (4.5.27), with $f(s)$ replaced by $(fx)(s)$, and x^0 is given by (4.5.33), we obtain the result that $x \to Ux$ is an operator from $C([0, T], R^n)$ into itself.

The continuity of U on $C([0, T], R^n)$ follows easily from the continuity of f, and from the continuity of the integral operator

$$x \to \int_0^T \tilde{X}(t, s)x(s) ds.$$

More precisely, one has

$$|Ux - Uy|_C \leq (\gamma\sqrt{T})|fx - fy|_C,$$

which implies the continuity of U on $C([0, T], R^n)$.

The compactness of the operator U follows from the following considerations. Let $v = Ux$ for $x \in C([0, T], R^n)$, with $|x|_C \leq \rho$. From (4.5.39)–(4.5.41) we obtain

$$|v|_C \leq |x_0(t)|_C + (\gamma\sqrt{T})\phi(\rho), \qquad (4.5.42)$$

which also implies the boundedness of $|v|_{L^2}$. Hence $|Lv|_{L^2}$ remains bounded when $|x|_C \leq \rho$. Consequently, the set $\{v'; v = Ux, |x|_C \leq \rho\}$ is bounded in $L^2([0, T], R^n)$. This property immediately implies the equicontinuity of the set $\{v; v = Ux, |x|_C \leq \rho\}$ on $[0, T]$. Indeed, one has

$$|v(t) - v(s)|^2 \leq |t - s| \int_0^T |v'(u)|^2 \, du.$$

It remains to assure the inclusion condition $|Ux|_C \leq \rho$ for $|x|_C \leq \rho$, for some $\rho > 0$. If we take (4.5.42) into account, the condition of the inclusion is

$$|x_0(t)|_C + (\gamma\sqrt{T})\phi(\rho) \leq \rho. \qquad (4.5.43)$$

Summarizing the discussion conducted above in regard to the boundary value problem (4.5.35), (4.5.28), we can formulate the following result.

Theorem 4.5.3. *Assume the following conditions are satisfied in respect to the boundary value problem* (4.5.35), (4.5.28):

(1) *L is a linear continuous Volterra operator on $L^2([0, T], R^n)$;*

(2) *f is a continuous map of $C([0, T], R^n)$ into itself, such that the inequality (4.5.43) has a positive solution in ρ, with $\phi(\rho)$ given by (4.5.40);*

(3) *the condition (4.5.30) holds.*

Then there exists at least one solution $x(t)$ of our problem, satisfying (4.5.35) a.e. on $[0, T]$, and such that $|x|_C \leq \rho$.

Remark 1. Condition (4.5.43) can be viewed as a condition limiting the growth of f. A stronger form of this condition is, for instance,

$$\liminf_{\rho \to \infty} \frac{\phi(\rho)}{\rho} = 0. \qquad (4.5.44)$$

In other words, if the order of growth of f is dominated by the linear growth, then (4.5.43) always has positive solutions.

Remark 2. In the proof of continuity of U, we obtained the inequality $|Ux - Uy|_C \leq (\gamma\sqrt{T})|fx - fy|_C$. It is obvious that if f satisfies a Lipschitz condition $|fx - fy|_C \leq \lambda|x - y|_C$, with λ small enough, the operator U is a contraction on $C([0, T], R^n)$. Therefore, the problem (4.5.35), (4.5.28) has a unique solution.

Remark 3.   Instead of assuming $f: C([0, T], R^n) \to C([0, T], R^n)$, one could assume $f: L^2([0, T], R^n) \to L^2([0, T], R^n)$. Equation (4.5.36) can be treated in the same way as in Theorem 4.5.3, using the space $L^2([0, T], R^n)$ as an underlying space. Of course, the solution will always be considered in the Carathéodory sense (i.e., a.e. on $[0, T]$).

Remark 4.   Different kinds of boundary value conditions could be dealt with. For instance, if we substitute for (4.5.28) the condition

$$\int_0^T A(t)x(t)\,dt = b,$$

with $A \in L^2([0, T], \mathscr{L}(R^n, R^n))$ and $b \in R^n$, we also obtain a linear equation (system) for the coordinates of x^0, and we are led to a condition similar to (4.5.30) for the solvability of the linear system in x^0.

4.6   Periodic and almost periodic solutions to some integrodifferential equations

We shall first consider equations of the form

$$\dot{x}(t) = \int_R [dA(s)]x(t - s) + f(t), \quad t \in R, \tag{4.6.1}$$

where x and f take values in $\mathscr{C}^n$, and A is an n by n matrix whose entries are real valued functions with bounded variation on R. As usual, we will assume that A is left continuous.

Equation (4.6.1) has well-known special cases, such as

$$\dot{x}(t) = Ax(t) + f(t), \tag{4.6.2}$$

in which A is any n by n matrix with real entries (which means that linear ordinary equations with constant coefficients belong to the class denoted by (4.6.1)). To obtain (4.6.2) from (4.6.1) one has to assume $A(s) \equiv 0$ for $s \leq 0$, and $A(s) = A = (a_{ij})$ for $s > 0$. Another noteworthy special case of (4.6.1) corresponds to the choice $A(s) = 0$ for $s < 0$, which means that (4.6.1) reduces to the following integro differential equation with delayed argument:

$$\dot{x}(t) = \int_0^\infty [dA(s)]x(t - s) + f(t), \tag{4.6.3}$$

representing a rather general form for infinite delay equations. From (4.6.3) we can obtain as special cases systems of the form

$$\dot{x}(t) = \sum_{k=1}^{N} A_k x(t - t_k) + f(t), \tag{4.6.4}$$

which are more commonly encountered in the literature on delay equations. In (4.6.4), the A_k, $k = 1, 2, \ldots, N$, stand for arbitrary n by n matrices with real entries, while t_k, $k = 1, \ldots, N$, are distinct nonnegative numbers.

We shall seek *periodic solutions*, of period $\omega > 0$, for (4.6.1), assuming that $f(t) \in L^2([0, \omega], \mathscr{C}^n)$, $t \in [0, \omega]$, and satisfies the periodicity condition $f(t + \omega) = f(t)$ a.e. for $t \in R$. In particular, $f(t)$ could be chosen as any continuous periodic function of period ω.

Since periodic or almost periodic functions are uniquely determined by the corresponding Fourier series, it is natural – especially if we keep in mind the linearity of the equation – to try to construct such series for the periodic solutions of (4.6.1), if any.

Suppose that

$$f(t) \sim \sum_{k \in Z} f_k \exp\left(\frac{2k\pi i}{\omega} t\right). \tag{4.6.5}$$

Since $f \in L^2([0, \omega], \mathscr{C}^n)$, one has the so-called Bessel–Parseval relationship

$$\sum_{k \in Z} |f_k|^2 = \frac{1}{\omega} \int_0^\omega |f(t)|^2 \, dt. \tag{4.6.6}$$

Let us seek now the solution of (4.6.1) as a Fourier series, namely

$$x(t) = \sum_{k \in Z} x_k \exp\left(\frac{2k\pi i}{\omega} t\right). \tag{4.6.7}$$

We certainly want the series in the right-hand side of (4.6.7) to converge in some sense. And, to be practical, we should look for the best available kind of convergence. Nevertheless, it should be pointed out that if we have the Fourier series of a periodic/almost periodic function, the function can be reconstituted. Moreover, there are constructive procedures to obtain the function if its Fourier series is available (see, for instance, Edwards [2]). Cesaro's summability procedure is an example. Substituting $x(t)$ and $\dot{x}(t)$ in (4.6.1) one has

$$i\omega_k x_k = \left(\int_R [dA(s)] \exp(-i\omega_k s) \right) x_k + f_k, \quad k \in Z, \tag{4.6.8}$$

where $\omega_k = 2k\pi/\omega$, $k \in Z$. These can be rewritten as

$$[i\omega_k I - \tilde{\mathscr{A}}(i\omega_k)] x_k = f_k, \quad k \in Z, \tag{4.6.9}$$

where

$$\tilde{\mathscr{A}}(\text{is}) = \int_R [\mathrm{d}A(t)] \exp(-\text{is}t) \tag{4.6.10}$$

is the Fourier–Stieltjes transform of A. It does exist owing to our assumption on the matrix $A(t)$.

If we make the assumption

$$\det[\text{is}I - \tilde{\mathscr{A}}(\text{is})] \neq 0, \quad s \in R, \tag{4.6.10}'$$

then each system (4.6.9) has unique solution (for any $\omega > 0$!). But condition (4.6.10)$'$ requires too much for fixed $\omega > 0$. It is, indeed, sufficient to impose the weaker condition

$$\det[i\omega_k I - \tilde{\mathscr{A}}(i\omega_k)] \neq 0, \quad k \in Z, \tag{4.6.10}''$$

in order to derive the existence of a unique x_k, $k \in Z$. Moreover, it appears that even condition (4.6.10)$''$ requires too much from $A(t)$, in order to secure the above property. This becomes clear if we prove the following

Lemma 4.6.1. *If $s \in R$ satisfies*

$$\det[\text{is}I - \tilde{\mathscr{A}}(\text{is})] = 0, \tag{4.6.11}$$

then necessarily

$$|s| \leq \int_R |\mathrm{d}A(t)| = \gamma < \infty. \tag{4.6.12}$$

Proof. If condition (4.6.11) holds for some $s \in R$, then the linear system $\text{is}\xi = \tilde{\mathscr{A}}(\text{is})\xi$ has nontrivial solutions $\xi \in \mathscr{C}^n$. This implies $|s||\xi| \leq |\tilde{\mathscr{A}}(\text{is})||\xi|$, where $|\xi|$ denotes the Euclidean norm in $\mathscr{C}^n$, and $|\tilde{\mathscr{A}}(\text{is})|$ is the matrix norm induced by the Euclidean norm for vectors. Therefore, keeping in mind that $|\xi| \neq 0$, we obtain $|s| \leq |\tilde{\mathscr{A}}(\text{is})|$. On the other hand, from (4.6.10) one derives

$$|\tilde{\mathscr{A}}(\text{is})| \leq \int_0^\infty |\mathrm{d}A(t)|,$$

which means that (4.6.12) is satisfied.

Remark. The main consequence we can derive from the Lemma 4.6.1 is the fact that the system (4.6.8) may fail to produce a unique solution x_k only in the case $|\omega_k| \leq \gamma$, which means $2|k|\pi \leq \omega\gamma$ (i.e., there are only finitely many values for $k \in Z$, for which (4.6.8) might not be uniquely solvable). Of course, (4.6.8) fails to produce a unique solution only when $s = \omega_k$, with $|k| \leq \omega\gamma/(2\pi)$, satisfies (4.6.11). One can also say that ω_k is a characteristic root of the equation (4.6.1).

Before we state the main result on the existence of periodic solutions for (4.6.1), let us notice the fact that there exists a unique nonnegative integer p, such that

$$\omega_p \leq \gamma < \omega_{p+1}, \qquad (4.6.13)$$

where $\omega_k = 2k\pi/\omega$, $k \in Z$.

Theorem 4.6.2. *Consider the equation* (4.6.1) *in which* $f \in L^2([0, \omega], \mathscr{C}^n)$, $f(t + \omega) = f(t)$ *a.e. on R, and* $A(t)$ *is a matrix of type n by n whose entries are functions with bounded variation on R, continuous from the left.*

If condition (4.6.10)″ *holds for* $|k| \leq p$, *where* p *is determined by* (4.6.13), *then there exists a unique periodic solution* $x(t)$ *of* (4.6.1) *satisfying this equation a.e. The Fourier series of the solution is*

$$\sum_{k \in Z} [i\omega_k I - \mathscr{A}(i\omega_k)]^{-1} f_k \exp(i\omega_k t), \qquad (4.6.14)$$

and it converges uniformly and absolutely to $x(t)$.

Proof. The series (4.6.14) can be constructed if the hypotheses of Theorem 4.6.2 are verified, with condition (4.6.10) $|k| \leq p$, playing the central role.

The uniform and absolute convergence of series (4.6.14) (on R) follows immediately from the estimate

$$|[i\omega_k I - \mathscr{A}(i\omega_k)]^{-1} f_k| \leq A_0(\omega_k^{-2} + |f_k|^2), \qquad (4.6.15)$$

which holds for $|k| > p + 1$, with $A_0 > 0$ conveniently chosen. This is always possible, because $\mathscr{A}(i\omega_k)/\omega_k \to 0$ as $|k| \to \infty$, and (4.6.6) holds.

If we denote by $x(t)$ the sum of the series (4.6.14), it only remains to show that $x(t)$ is indeed a solution of (4.6.1). We shall achieve this conclusion by comparing the Fourier series of both sides of equation (4.6.1), after substituting (4.6.14) for $x(t)$.

The series obtained by formal differentiation of (4.6.14) has the coefficients $i\omega_k[i\omega_k I - \mathscr{A}(i\omega_k)]^{-1} f_k = b_k$, $k \in Z$. It is easily seen that $\sum_{k \in Z} |b_k|^2 < \infty$, and therefore there exists a unique element in $L^2([0, \omega], \mathscr{C}^n)$ whose former coefficients are exactly b_k, $k \in Z$. As we know, this function can only be $x'(t)$.

On the other hand, the integral in the right hand side of (4.6.1) does exist for $x(t)$ constructed above, and we can easily calculate (because of the uniform convergence) the Fourier coefficients of the convolution product. We have

$$\int_R [dA(s)]x(t - s) = \int_R [dA(s)] \sum_{k \in Z} x_k \exp\{i\omega_k(t - s)\}$$

$$= \sum_{k \in Z} \left(\int_R [dA(s)] \exp(-i\omega_k s) \right) x_k \exp(i\omega_k t)$$

$$= \sum_{k \in Z} \mathscr{A}(i\omega_k) x_k \exp(i\omega_k t),$$

with x_k given by (4.6.9). Hence, the Fourier coefficients of the right hand side of

(4.6.1) are

$$\mathscr{A}(i\omega_k)[i\omega_k I - \mathscr{A}(i\omega_k)]^{-1}f_k + f_k$$

$$= [\mathscr{A}(i\omega_k) + i\omega_k I - \mathscr{A}(i\omega_k)][i\omega_k I - \mathscr{A}(i\omega_k)]^{-1}f_k$$

$$= i\omega_k[i\omega_k I - \mathscr{A}(i\omega_k)]^{-1}f_k = b_k.$$

Since b_k are the Fourier coefficients of $x'(t)$, Theorem 4.6.2 is thereby proven.

Remark 1. The estimate (4.6.12) for the solutions of the (characteristic) equation (4.6.11) is certainly not the best possible in all cases (think, for instance, of the case when (4.6.1) reduces to (4.6.2)). Consequently, the condition (4.6.10)″ for $|k| \le p$ might contain some superfluous assumptions (i.e., the inequality is automatically verified when $i\omega_k$ does not belong to the smallest interval $[-iT, iT]$ which contains all the roots of (4.6.11)).

Remark 2. Of course, a very interesting situation with regard to the existence of periodic solutions of (4.6.1) is the one in which condition (4.6.10)″ is violated for some ks with $|k| \le p$. In such a case, the system (4.6.9) is either deprived of solutions, or it has more than one solution. Accordingly, the system (4.6.1) does not possess periodic solutions of period ω, or it has more than one periodic solution of period ω.

We postpone the answer to this problem until we show that system (4.6.1) is equivalent in the class of periodic functions to another system in which the interval of integration is finite (equal to the period).

Remark 3. Under the assumptions of Theorem 4.6.2 we easily derive from (4.6.9) an inequality of the form $|x_k| \le \bar{A}|f_k|$, $k \in Z$, with $\bar{A} > 0$ depending only on $A(t)$. If we take into account Parseval's equality for functions in $L^2([0, \omega], \mathscr{C}^n)$, then we obtain

$$|x|_{L^2} \le \bar{A}|f|_{L^2}, \qquad (4.6.16)$$

where x stands for the (unique) periodic solution, of period ω, to (4.6.1).

If instead of the linear system (4.6.1) we consider the nonlinear system

$$\dot{x}(t) = \int_R [dA(s)]x(t - s) + f(t; x), \qquad (4.6.17)$$

with $f(t; x) = (fx)(t)$ an operator such that $f: L^2([0, \omega], \mathscr{C}^n) \to L^2([0, \omega], \mathscr{C}^n)$ and verifies a Lipschitz condition of the form

$$|fx - fy|_{L^2} \le \lambda|x - y|_{L^2},$$

then (4.6.16) shows that for $\lambda\bar{A} < 1$, the system (4.6.17) has a unique periodic solution of period ω.

We leave it to the reader to carry out the details of the proof of the above statement (contraction mapping).

We shall now consider the system (4.6.1) from the point of view of transforming it into another integrodifferential system, in which the range of integration is finite. This transformation is aimed only at the investigation of periodic solutions.

If $x(t)$ is a periodic solution of (4.6.1), with period ω, then for every $k \in Z$ we have

$$\int_{k\omega}^{(k+1)\omega} [dA(s)]x(t-s) = \int_0^\omega [dA(s+k\omega)]x(t-s)$$

$$= \int_0^\omega \{d[A(s+k\omega) - A(k\omega)]\}x(t-s), \quad t \in R.$$

Let

$$B(s) = \sum_{k \in Z} [A(s+k\omega) - A(k\omega)], \quad 0 \le s \le \omega. \tag{4.6.18}$$

Owing to our assumptions on $A(s)$, there results the uniform and absolute convergence of the series in the right hand side of (4.6.18), which implies that $B(s)$ is continuous from the left on $[0, \omega]$, and of bounded variation there.

From the above considerations we see that (4.6.1) can be written as

$$\dot{x}(t) = \int_0^\omega [dB(s)]x(t-s) + f(t), \quad t \in [0, \omega], \tag{4.6.19}$$

where $x(t)$ denotes any periodic solution of (4.6.1), of period ω.

Conversely, if we now start with a periodic solution, of period ω, for (4.6.19), in which $B(s)$ is defined by (4.6.18), then this periodic solution will also satisfy the system (4.6.1).

The equivalence of the system (4.6.1) and (4.6.19), with respect to the existence of periodic solutions of period ω is thereby established. Of course, those systems might have solutions which are not common, and even periodic ones, provided the period is different from ω.

More general systems than (4.6.19) have been investigated by many authors. For instance, systems of the form

$$\dot{x}(t) = \int_0^\omega [d_s B(t,s)]x(s)\,ds + f(t), \quad t \in [0, \omega], \tag{4.6.20}$$

are investigated in Schwabik et al. [1], under constraints such as

$$Mx(0) + \int_0^\omega L(t)x'(t)\,dt = r, \tag{4.6.21}$$

where M and $L(t)$ are matrices, and r is a vector. Choosing these elements conveniently, we can write (4.6.21) in the form $x(0) = x(\omega)$, which characterizes the periodic solutions of (4.6.20), of period ω.

As shown in Schwabik *et al.* [1], under appropriate conditions which are verified in the special case of the system (4.6.19), the problem (4.6.20), (4.6.21) with $r = \theta$ has an absolutely continuous solution for a given $f \in L^2([0, \omega], \mathscr{C}^n)$, if and only if the orthogonality condition

$$\int_0^\omega y^*(t)f(t)\,\mathrm{d}t = 0$$

holds for any solution $y^*(t)$ of the adjoint problem

$$y^*(t) + \int_0^\omega y^*(s)B(s, t)\,\mathrm{d}s + \lambda^*L(t) = 0,$$

$$\lambda^*M + \int_0^\omega y^*(s)B(s, 0)\,\mathrm{d}s = 0,$$

where λ is a convenient finite-dimensional vector. For the periodic case, $M = 0$, L is an aribtrary n by n nonsingular matrix, and $r = \theta$. The conditions for the adjoint problem become

$$y^*(t) + \int_0^\omega y^*(s)B(s, t)\,\mathrm{d}t = 0, \qquad \int_0^\omega y^*(s)B(s, 0)\,\mathrm{d}s = 0.$$

As we can see, the solutions $y^*(t)$ of the adjoint problem need not be periodic. In some special cases (see Halanay [1]), when the adjoint equation is also a differential functional equation, the solutions $y^*(t)$ involved in condition (4.6.22) are periodic, of the same period as f. This seems to be a more appropriate form for the Fredholm alternative in the case of systems of the form (4.6.20).

Let us now consider equation (4.6.1), and investigate *almost-periodic solutions* under the basic assumption that $f(t)$ is almost periodic in some sense soon to be specified. It is well known that the almost periodicity concept of H. Bohr has been generalized in several interesting ways, and we shall confine our discussion to the case when $f(t)$ is almost periodic in the Stepanov sense. This concept of almost periodicity has been discussed in some detail in Section 2.2, and we refer the reader to that section for more information.

Since $S(R, \mathscr{C}^n) \subset M(R, \mathscr{C}^n)$, we can restrict our considerations to the case of solutions of (4.6.1) which belong to the space $M(R, \mathscr{C}^n)$.

The following problem is a variant of the classical Bohr–Neugebauer problem for ordinary differential equations (see C. Corduneanu [3], for instance): let $x(t) \in M(R, \mathscr{C}^n)$ be a solution of (4.6.1), in which $f(t) \in S(R, \mathscr{C}^n)$. Show that $x(t) \in S(R, \mathscr{C}^n)$. In other words, does the boundedness of the solution in the norm of M imply almost periodicity in Stepanov's sense?

Before we get into details with the investigation of the above formulated problem, we shall notice that it is not really more general to require $x(t) \in M$, than to require the usual boundedness of x: $x \in BC(R, \mathscr{C}^n)$.

Indeed, if $x(t) \in M(R, \mathscr{C}^n)$ is a solution of (4.6.1), then we derive from that equation $\dot{x}(t) \in M(R, \mathscr{C}^n)$, because $A * x \in M$ (see Section 2.2). Hence, both x and $\dot{x}$ are bounded in the norm of M, which implies that $\dot{x}(t) + x(t) = h(t) \in M(R, \mathscr{C}^n)$. But the only solution of $\dot{x} + x = h$ which belongs to M is

$$x(t) = \int_{-\infty}^{t} e^{-(t-s)} h(s)\, ds = (e * h)(t),$$

where $e(t) = \exp(-t)$ for $t \geq 0$, $e(t) = 0$ for $t < 0$; and, as seen in Section 2.2, this is bounded on R.

Going back to (4.6.1) we find that $\dot{x} = A * x + f \in BC(R, \mathscr{C}^n) \cup S(R, \mathscr{C}^n)$. Consequently, $x(t)$ is uniformly continuous on R, and this implies the (Bohr) almost periodicity of $x(t)$.

Therefore, we do not loose generality if we reformulate the above problem in the classical form: show that any bounded solution (on R) of the equation (4.6.1) is Bohr almost periodic.

We shall now reduce the investigation of the problem (Bohr–Neugebauer) formulated above to the investigation of the similar problem in relation to an integral equation (also of convolution type). Let us formulate this result as

Lemma 4.6.3. *Consider the system* (4.6.1) *with A as described above, and $f \in S(R, \mathscr{C}^n)$. Then any bounded (on R) solution of* (4.6.1) *is also a solution of the integral equation*

$$x = a * x + \bar{f}, \tag{4.6.22}$$

where

$$a(t) = e(t)I + (e * A)(t), \tag{4.6.23}$$

and

$$\bar{f}(t) = (e * f)(t). \tag{4.6.24}$$

Proof. If x is a bounded (on R) solution of (4.6.1), i.e. $x \in BC(R, \mathscr{C}^n)$, then x' can be represented as the sum of a function in $BC(R, \mathscr{C}^n)$, and another function in $S(R, \mathscr{C}^n)$. This is easily seen from (4.6.1), taking into account the fact that $A * x \in BC(R, \mathscr{C}^n)$ when $x \in BC(R, \mathscr{C}^n)$. This means that $x' \in M(R, \mathscr{C}^n)$, since both $BC(R, \mathscr{C}^n)$ and $S(R, \mathscr{C}^n)$ are subsets of $M(R, \mathscr{C}^n)$. Hence, the convolution $e * x'$ makes sense, and an integration by parts yields

$$(e * x')(t) = x(t) - (e * x)(t). \tag{4.6.25}$$

If we now take the convolution product of both sides of equation (4.6.1) by $e(t)$, and take into account (4.6.23)–(4.6.25), then we obtain the integral equation (4.6.22).

This ends the proof of Lemma 4.6.3.

Remark 1. As noticed in the proof of the Lemma 4.6.3, the derivative x' of a bounded solution of (4.6.1) can be represented as a sum of a term in $BC(R, \mathscr{C}^n)$, and another in $S(R, \mathscr{C}^n)$. This implies the uniform continuity of the solution x, on the entire real axis. See Section 2.2 for details.

Remark 2. In what follows we shall need a relationship between the Fourier transform of $\delta I - a$ and the (matrix-valued) function

$$D(i\omega) = i\omega I - \tilde{\mathscr{A}}(i\omega). \tag{4.6.26}$$

This is motivated by the fact that the equation (4.6.22) can be written as $(\delta I - a) * x = f$. Since

$$\widetilde{(\delta I - a)}(i\omega) = I - (1 + i\omega)^{-1}[\tilde{\mathscr{A}}(i\omega) - I],$$

by substituting $\tilde{\mathscr{A}}(i\omega)$ from (4.6.26) we obtain

$$\widetilde{(\delta I - a)}(i\omega) = (1 + i\omega)^{-1}D(i\omega). \tag{4.6.27}$$

Therefore, $\widetilde{(\delta I - a)}(i\omega)$ is a singular matrix if and only if $D(i\omega)$ is singular.

From now on, our attention will be concentrated on equation (4.6.22). We want to show that any bounded (on R) solution of (4.6.22), with $\bar{f}(t)$ almost periodic in the sense of Bohr and $a(t)$ given by (4.6.23), is Bohr almost periodic.

Actually, the only fact that remains to be established is the uniform continuity on R of any bounded solution of (4.6.22). Then, a spectral characterization of almost periodic functions can be used in order to obtain the almost periodicity of any bounded solution $x(t)$.

We notice first that under our assumptions one has $|a| \in L^1(R, R)$, while $\bar{f}$ is Bohr almost periodic from R into $\mathscr{C}^n$. From elementary considerations concerning Banach algebras of Fourier transforms (see, for instance, C. Corduneanu [4]), one derives the existence of a kernel b with the properties $|b| \in L^1(R, R)$, and

$$[I + \tilde{b}(i\omega)][I - \tilde{a}(i\omega)] = I \tag{4.6.28}$$

for $|\omega| \geq \Omega > 0$. We have to keep in mind that $|\tilde{a}(i\omega)| \to 0$ as $|\omega| \to \infty$ (Riemann–Lebesque property of the transform).

Now let η be a scalar L^1-function whose Fourier transform $\tilde{\eta}$ has compact support, and such that $\tilde{\eta}(i\omega) \equiv 1$ for $|\omega| \leq \Omega + 1$. As usual, δ will denote the Dirac distribution: $\delta * \phi = \phi$. If we take the convolution product of both sides of (4.6.22) by $\delta - \eta$, we obtain

$$(I\delta - a) * (x - \eta * x) = \bar{f} - \eta * \bar{f}. \qquad (4.6.29)$$

The (distribution) Fourier transform of $x - \eta * x$ vanishes on $(-\Omega - 1, \Omega + 1)$, and taking (4.6.28) into account we have

$$(I\delta + b) * (I\delta - a) * (x - \eta * x) = x - \eta * x.$$

Hence, taking the convolution product of both sides in (4.6.29) by $(I\delta + b)$, we obtain

$$x - \eta * x = \bar{f} - \eta * \bar{f} + b * \bar{f} - b * \eta * \bar{f}. \qquad (4.6.30)$$

The right-hand side of (4.6.30) is uniformly continuous, and hence $x - \eta * x$ is also uniformly continuous on R. Since the Fourier transform of $\eta * x$ has compact support, $\eta * x$ is uniformly continuous on R. Consequently, x is uniformly continuous.

The property characterizing the almost periodic functions can be stated as follows: If $x: R \to \mathscr{C}^n$ is bounded and uniformly continuous, then its (Bohr) almost periodicity is equivalent to the property that its (distribution) Fourier transform is almost periodic, except perhaps for a countable set.

For details concerning this property, see Staffans [7] (a reference in book form does not seem to be available).

Summing up the discussion conducted above, we can formulate the following result on almost periodicity of solutions to equation (4.6.1).

Theorem 4.6.4. *Consider equation* (4.6.1) *with $A(t)$ consisting of real entries which are left continuous functions and with bounded variation on R. Assume further that* $\det D(i\omega) = 0$ *only for a countable set of values of ω. If $f: R \to \mathscr{C}^n$ is Stepanov almost periodic and $x: R \to \mathscr{C}^n$ is a solution of* (4.6.1) *which belongs to $M(R, \mathscr{C}^n)$, then x is necessarily bounded on R, and Bohr almost periodic.*

4.7   Integral inclusions of Volterra–Hammerstein type

The inclusions we shall consider in this section are of the form

$$x(t) \in (VFx)(t) + g(t), \quad t \in [t_0, T], \qquad (4.7.1)$$

where x and g are n-vectors, V stands for the linear Volterra operator

$$(Vh)(t) = \int_{t_0}^{t} v(t, s) h(s) \, ds, \qquad (4.7.2)$$

and F stands for a multivalued map generated by the map $(t, x) \to f(t, x) \in 2^{R^n}$ by means of the formula

$$(Fx)(t) = f(t, x(t)), \quad t \in [t_0, T]. \qquad (4.7.3)$$

More precise conditions on the functions involved in the inclusion (4.7.1) will be given subsequently. We only want now to point out that (4.7.1) can also be written in the form

$$x(t) \in \int_{t_0}^{t} v(t,s)f(s,x(s)) \, ds + g(t), \quad t \in [t_0, T], \tag{4.7.4}$$

where

$$\int_{t_0}^{t} v(t,s)f(s,x(s)) \, ds = \left\{ y(t): y(t) = \int_{t_0}^{t} v(t,s)h(s) \, ds, h(s) \in f(s,x(s)) \right\}, \tag{4.7.5}$$

while h stands for a measurable function, and therefore the equalities or inclusions should be understood as valid almost everywhere.

We will show that, under adequate conditions on the data, there exist solutions $x(t)$ to (4.7.1), their set is connected and compact, and the dependence of this set with respect to the 'initial' g is upper semicontinuous in a natural metric.

Let us recall first a few concepts related to the multivalued maps, as well as the concept of distance between subsets of a metric space (Hausdorff).

If (X, ρ) is a metric space, and $A, B \subset X$, then we first define the function

$$\beta(A, B) = \sup\{\rho(a, B), a \in A\} = \inf\{\varepsilon, A \in B^{\varepsilon}\}, \tag{4.7.6}$$

where B^{ε} stands for the ε-neighborhood of the set B. The equivalence of the two definitions is an elementary matter.

Then we define the Hausdorff metric by

$$\alpha(A, B) = \max\{\beta(A, B), \beta(B, A)\}. \tag{4.7.7}$$

If P is a multivalued map from the metric space X into another metric space Y, more exactly

$$P: X \to 2^{Y},$$

then P is called *upper semicontinuous* at the point $x \in X$, if the following property holds: for any sequence $\{x_k\} \subset X$ which converges in X, $x_k \to x$ as $k \to \infty$, we have $\beta(Px_k, Px) \to 0$, as $k \to \infty$.

A similar definition can be formulated for the *lower semicontinuity* of P at x: for any sequence $\{x_k\} \subset X$ which converges in X, $x_k \to x$ as $k \to \infty$, we have $\beta(Px, Px_k) \to 0$, as $k \to \infty$.

The *continuity* of P at x is defined as the simultaneous validity of both upper and lower semicontinuities. It is obvious that the continuity means $\alpha(Px_k, Px) \to 0$, for any sequence $\{x_k\} \subset X$, such that $x_k \to x$ as $k \to 0$.

The continuity or semicontinuity on a set means such a property holds at any point of that set.

The multivalued map $P: X \to 2^{Y}$ is *closed* if for any sequence $\{x_k\} \subset X$, such that $x_k \to x$ as $k \to \infty$, and any sequence $\{y_k\} \subset Y$, with $y_k \subset Px_k$, such that

$y_k \to y$, then $y \in P_x$. This property is obviously equivalent to the closure of the graph of P in $X \times Y$.

The definition of *compactness* of the multivalued map P can be formulated as follows: for any bounded set $M \subset X$, the image $PM = \{\bigcup Px; x \in M\} \subset Y$ is compact.

A multivalued map P, which is both *closed* and *compact*, is *upper semicontinuous*. Indeed, if $\{x_k\} \subset X$ is such that $x_k \to x$ in X, we need to show that $\beta(Px_k, Px) \to 0$. Assuming this is not true, we can determine a sequence $\{k_m\}$ of natural numbers, $k_m \to \infty$, such that $\beta(Px_{k_m}, Px) \geq \varepsilon_0 > 0$ for some ε_0. We now choose $y_{k_m} \in Px_{k_m}$ such that $\rho(y_{k_m}, Px) \geq \frac{1}{2}\varepsilon_0 > 0$. Since the set $\bigcup Px_{k_m}$ is compact in Y, we can assume without loss of generality that $y_{k_m} \to y$ as $m \to \infty$. But the closedness of P implies $y \in Px$, which contradicts the inequalities $\rho(y_{k_m}, Px) \geq \frac{1}{2}\varepsilon_0 > 0$.

Many other properties can be established for multivalued maps. An example is the following one, whose proof can be found in Berge [1].

P_1. Let $P: X \to 2^Y$ a multivalued map which is upper semicontinuous, and such that Px is compact and connected for every $x \in K \subset X$, with K connected. Then PK is also connected.

Another property which we shall use in this section can be found in Kuratowski [1] and can be stated as follows.

P_2. Let $A_k \subset X, k \geq 1$, be connected, and assume $\alpha(A_k, A) \to 0$ as $k \to \infty$. Then A is also connected.

If Y is a Banach space, the multivalued operator $P: X \to 2^Y$ is called *weakly closed* if the following property holds: for any sequence $x_k \to x$ in X, and any sequence $y_k \to y$ weakly in Y, with $y_k \in Px_k, k \geq 1$, one has $y \in Px$.

This concept will be useful in the construction of the solution of the inclusion (4.7.1).

Let us now state the basic assumptions on the functions involved in the inclusion (4.7.1).

(a) $f: [t_0, T] \times R^n \to 2^{R^n}$ takes only nonempty convex compact values;

(b) $f(t, \cdot)$ is upper semicontinuous on R^n;

(c) $f(\cdot, x)$ possesses a measurable selection;

(d) $|f(t, x)| \leq v(t)$ on $[t_0, T] \times R^n$, where $v \in L^1([t_0, T], R)$;

(e) $v(t, s)$ is measurable for $t_0 \leq s \leq t \leq T$, and $k(t) = \text{ess-sup}|v(t, s)|, t_0 \leq s \leq t$ is bounded on $[t_0, T]$;

(f) the map $t \to v(t, \cdot)$, from $[t_0, T]$ into $L^\infty([t_0, T], \mathscr{L}(R^n, R^n))$ is continuous.

With regard to the condition (f), we point out that, as usual for Volterra kernels, $v(t, s) = 0$ for $t_0 \leq t < s \leq T$.

Still with regard to v, we notice that for every $x \in L^1([t_0, T], R^n)$ we have

$$\left| \int_{t_0}^{t} v(t, s)x(s)\,ds - \int_{t_0}^{\tau} v(\tau, s)x(s)\,ds \right| \le \sup k(t) \left| \int_{\tau}^{t} |x(s)|\,ds \right|$$

$$+ |v(t, \cdot) - v(\tau, \cdot)|_{L^\infty} \cdot \int_{t_0}^{T} |x(s)|\,ds, \quad (4.7.8)$$

which shows that the integral operator V, given by (4.7.2), takes the space $L^1([t_0, T], R^n)$ into the space $C([t_0, T], R^n)$. Its continuity follows from the estimate

$$\left| \int_{t_0}^{t} v(t, s)x(s)\,ds \right| \le \sup k(t) \int_{t_0}^{T} |x(s)|\,ds, \quad t \in [t_0, T]. \quad (4.7.9)$$

The following property for the operator V will be needed in what follows.

Lemma 4.7.1. *Under assumptions (e) and (f), the operator V has the property: if $h_k \to h$ weakly in L^1, as $k \to \infty$, then there exists a subsequence $\{h_{k_m}\} \subset \{h_k\}$ such that $Vh_{k_m} \to Vh$ in C, as $m \to \infty$.*

Proof. Since $\{h_k\} \subset L^1$ is weakly compact, the Dunford–Pettis theorem says that $\{Vh_k\} \subset C$ is compact. Indeed, the estimate (4.7.8), taken for $x = h_k$, shows the equicontinuity of the sequence $\{Vh_k\}$, because

$$\left| \int_{\tau}^{t} |h_k(s)|\,ds \right| < \varepsilon \text{ for } |t - \tau| < \delta(\varepsilon), \quad k \ge 1.$$

Therefore, from the sequence $\{Vh_k\}$ one can extract a subsequence which converges in $C([t_0, T], R^n)$. The next property relates to the multivalued map f, or Niemytzki operator F given by (4.7.3).

Lemma 4.7.2. *Under assumptions (a), (b), (c), (d), the operator $F: S \to 2^{L^1}$ is weakly closed.*

Proof. Assume $x_k \to x$ in $S = S([t_0, T], R^n)$, i.e., x_k converges in measure to x, and let $h_k \to h$ weakly in L^1, with $h_k \in Fx_k$, $h \ge 1$. Without loss of generality we can assume $x_k(s) \to x(s)$ a.e. on $[t_0, T]$, replacing $\{x_k\}$ by a convenient subsequence, if necessary. Let $\varepsilon > 0$ be arbitrary, and consider s such that $x_k(s) \to x(s)$ as $k \to \infty$. Then, in view of condition (b), we can state $h_k(s) \in f^\varepsilon(s, x(s))$, for $k \ge K(s, \varepsilon)$. At this point, in order to conclude $h(s) \in f(s, x(s))$, we will apply a well-known result on weakly convergent sequences (see Dunford and Schwartz [1], p. 422). We have $h(s) \in f^\varepsilon(s, x(s))$ for every $\varepsilon > 0$, which implies $h(s) \in f(s, x(s))$, because the set $f(s, x(s)) \subset R^n$ is compact. Since this is true for almost all $s \in [t_0, T]$, we conclude $h \in Fx$, which proves Lemma 4.7.2.

Corollary. *If $x_k \to x$ in S as $k \to \infty$, and $Fx_k \neq \emptyset$, $k > 1$, then $Fx \neq \emptyset$. In particular, $Fx \neq \emptyset$ for any $x \in S$, because one can choose for given x a sequence x_k of step functions such that $x_k \to x$ in S, and rely on property* (c).

Indeed, FS is weakly compact from (d) and the Dunford–Pettis theorem, and using Lemma 4.7.2 one easily obtains the conclusion.

It is also true that Fx is convex in L^1 for any $x \in S$.

A few remarks are necessary about the product VF, which is obviously a map from S into 2^C. It is shown that VFx is convex for any $x \in S$. Then, based on the weak compactness of the set FS, we conclude that the set VFS is compact in C. Finally, the operator VF is closed. If $x_k \to x$ in S, as $k \to \infty$, and $z_k \to z$ in C, with $z_k = Vh_k \in VFx_k$, then from the weak compactness of FS and the weak closedness of F we can infer the existence of a subsequence $\{k_m\}$ of positive integers, $k_m \to \infty$ as $m \to \infty$, such that $h_{k_m} \to h \in Fx$, as $m \to \infty$, weakly in L^1. If we take Lemma 4.7.1 into account, we obtain $z = Vh \in VFx$, which proves the closedness of the operator VF.

Let us now introduce the operator

$$T(x, g) = VFx + g, \qquad (4.7.10)$$

which is, in fact, the right hand side of the inclusion (4.7.1). The following property can be easily obtained based on the assumptions of this section, and is useful subsequently.

Lemma 4.7.3. *Under assumptions* (a), (b), (c), (d), (e), (f), *and* $g \in C([t_0, T], R^n)$, *the following conditions are satisfied:*

(1) $T(x, g) \subset C$ *for any* $x \in S$, $g \in C$, *and is nonempty and convex;*
(2) *for any compact set* $M \subset C$, *the set* $T(S, M)$ *is compact in* C;
(3) *the map* $(x, g) \to T(x, g)$ *is closed.*

Proof. Condition (1) is obvious. Condition (2) is immediately obtained by means of Arzela's criterion of compactness in C. In particular, assumption (d) yields the equicontinuity. With regard to condition (3) in the Lemma 4.7.3, it is the consequence of the closedness of the operator (multivalued map) VF established above.

In what follows, we shall need a few facts related to the operator

$$(V_{ab}x)(t) = \int_a^t v(t, s)x(s) \, ds, \quad t \in [a, b] \subset [t_0, T], \qquad (4.7.11)$$

and to the multivalued map

$$f_{ab}(t, x) = f(t, x), \quad t \in [a, b] \subset [t_0, T], \, x \in R^n, \qquad (4.7.12)$$

and $f_{ab}(t, x) = 0$ elsewhere. The meaning of $F_{ab}x$ is then obvious. In accordance with previous notation, we have $V_{t_0T} = V$, and $F_{t_0T} = F$. Let us point out that the case $a > b$ is not excluded. Then in (4.7.11) we will consider $b \leq t \leq s \leq a$.

If we take into account (4.7.11) and (4.7.12), we see that the operators V_{ab} and F_{ab} have the properties we have established already for V and F. Moreover, the following result holds.

Lemma 4.7.4. *Let the function h be the weak limit of the sequence $\{h_k\}$, $h_k \in F_{a_k b_k} x_k$, $k = 1, 2, \ldots$, where $x_k \to x$ in S, $a_k \to a$ and $b_k \to b$, as $k \to \infty$. Then $h \in F_{ab}x$.*

Proof. We assume $a < b$. Let $\varepsilon > 0$ arbitrary, and $\varepsilon_0 > 0$ such that $a + \varepsilon_0 < b - \varepsilon_0$. If we take into account conditions (a) and (b) for $f(t, x)$, and proceed as in the proof of Lemma 4.7.2, we obtain for almost all $s \in [a + \varepsilon_0, b - \varepsilon_0]$, $h(s) \in f^\varepsilon(s, x(s))$. It means that, almost everywhere on $[a, b]$, $h(s) \in f(s, x(s))$. Of course, for $s \in [t_0, T] \backslash [a, b]$ we have $h(s) = 0$.

Lemma 4.7.4 is thereby proven.

We now start a construction which is a generalization of the Kneser's procedure in the theory of differential equations without uniqueness (see C. Corduneanu [6], for instance). This construction will enable us to obtain the basic result of this section.

Let D_j be a partition of the interval $[t_0, T]$, say $t_0 < t_1 < t_2 < \cdots < t_j = T$. Consider the multivalued maps A_i and $B_{z,i}$, $i = 0, 1, 2, \ldots, j - 1$, $z \in C([t_0, T], R^n)$,

$$A_i: C([t_0, t_i], R^n) \to 2^{C([t_0, t_{i+1}], R^n)},$$

$$B_{z,i}: [t_i, t_{i+1}] \to 2^{C([t_0, t_{i+1}], R^n)},$$

given by:

$$A_i x = \{(V_{at_{i+1}} h)(\cdot) + \phi(\cdot); h \in F_{t_i t_{i+1}} y\}, \qquad (4.7.13)$$

where $y \equiv x(t_i)$, $\phi(t) = x(t)$ for $t \in [t_0, t_i]$, and $\phi(t) = g(t) - g(t_i) + x(t_i)$ for $t \in [t_i, t_{i+1}]$;

$$B_{z,i}(\tau) = \{(V_{at_{i+1}} h)(\cdot) + \psi(\cdot); h \in F_{\tau t_{i+1}} y\}, \qquad (4.7.14)$$

where $y \equiv z(t_i)$, $\psi(t) = z(t)$ for $t \in [a, \tau]$, $\psi(t) = g(t) - g(\tau) + z(\tau)$, for $t \in [\tau, t_{i+1}]$.

The following statements are true: for any $x \in C([t_0, t_i], R^n)$ and any $\tau \in [t_i, t_{i+1}]$ the sets $A_i x$ and $B_{z,i}(\tau)$ are nonempty, and convex; the set $A_i M$, with $M \subset C([t_0, t_i], R^n)$ compact, is also compact; $B_{z,i}([t_i, t_{i+1}])$ is compact; the maps $x \to A_i x$ and $\tau \to B_{z,i}(\tau)$ are closed.

The definition formulas (4.7.13) and (4.7.14) lead easily to the properties stated above. Lemma 4.7.4 is required to obtain the closedness of the multivalued maps $\tau \to B_{z,i}(\tau)$, $i = 1, 2, \ldots, j - 1$.

Moreover, from the compactness and the closedness of the maps A_i and $B_{z,i}$ we derive the fact that both are upper semicontinuous.

Let

$$\bar{B}_{z,i}(\tau) = A_{j-1} \cdot A_{j-2} \cdots A_{i+1} \cdot B_{z,i}(\tau), \qquad (4.7.15)$$

and, for any $t \in [t_i, t_{i+1}]$, $i = 0, 1, \ldots, j - 1$,

$$B_z(t) = \bar{B}_{z,i}(t), \qquad (4.7.16)$$

and define the multivalued maps

$$\bar{B}_{z,i} \colon [t_i, t_{i+1}] \to 2^C, \quad B_z \colon [t_0, T] \to 2^C. \qquad (4.7.17)$$

From the definitions above one derives immediately that any $x \in B_z(\tau)$ can be represented in the form

$$x(\cdot) = (Vh)(\cdot) + \phi(\cdot), \qquad (4.7.18)$$

where $\phi(t) = z(t)$ for $t \in [t_0, \tau]$, $\phi(t) = g(t) - g(\tau) + z(\tau)$ for $t \in [\tau, T]$, and $h \in F_{\tau T} y$, with $y(t) \equiv x(t_i)$, $t \in [t_i, t_{i+1}]$, $i = 0, 1, 2, \ldots, j - 1$.

We can now state the last lemma we need for the final existence result.

Lemma 4.7.5. (1) *The set $B_z(\tau)$, $\tau \in [t_0, T]$, defined by (4.7.16) is nonempty and connected*; (2) *for any compact set $M \subset C([t_0, T], R^n)$, the set $B_M([t_0, T]) = \bigcup_{z \in M} B_z([t_0, T])$ is compact*; (3) *the map $(\tau, z) \to B_z(\tau)$ is closed.*

Proof. The connectedness of the set $B_z(\tau)$ is a consequence of the definition and of property P_1 stated above, taking into account the properties of the maps A_i and $B_{z,i}$. The fact that $B_z(\tau)$ is nonempty is obvious (see (4.7.13) and (4.7.14)). From (4.7.18) we derive that $B_M([t_0, T])$ is compact, if M is. So, the property (2) in Lemma 4.7.5 is verified. It remains to check the validity of (3). Let $z_k \to z$ in C, $\tau_k \to \tau$, and $x_k \to x$ as $k \to \infty$, where $x_k \in B_{z_k}(\tau_k)$, $k = 1, 2, \ldots$. Let us construct the step functions y_k, $k = 1, 2, \ldots$, and y, such that $y_k(t) = x_k(t_i)$ for $t \in [t_i, t_{i+1})$, and $y(t) = x(t_i)$. The points t_i belong to the partition D_j, $j = 0, 1, 2, \ldots, j - 1$. Let us also construct the functions ϕ_k and ϕ, by putting $\phi_k(t) = z_k(t)$ for $t \in [t_0, \tau_k)$, $\phi_k(t) = g(t) - g(\tau_k) + z_k(\tau_k)$ for $t \in [\tau_k, T]$, $\phi(t) = z(t)$ for $t \in [t_0, \tau)$, $\phi(t) = g(t) - g(\tau) + z(\tau)$ on $[\tau, b]$. If we take (4.7.18) into account, we can write $x_k(\cdot) = (Vh_k)(\cdot) + \phi(\cdot)$, with $h_k \in F_{\tau_k T} y_k$. But in the space S we have $y_k \to y$, and in the space C we have $\phi_k \to \phi$, as $k \to \infty$. From property (3) in Lemma 4.7.3, we conclude $x(\cdot) = (Vh)(\cdot) + \phi(\cdot)$ letting $k \to \infty$, where $h \in F_{\tau T} y$ is the weak limit in L^1 of a weakly convergent subsequence of $\{h_k\}$. This means $x \in B_z(\tau)$, which proves the closedness of the map $(\tau, z) \to B_z(\tau)$.

Remark 1. It can be shown, also based on Arzela's criterion, that for every compact set $M \subset C([t_0, T], R^n)$, the set $\bigcup B_M([t_0, T], D)$, where the union is taken with respect to all partitions D of $[t_0, T]$, is also compact.

Remark 2. If in the proof of Lemma 4.7.5, part (3), instead of $z_k \to z$, $\tau_k \to \tau$, and $x_k \to x$ where $x_k \in B_{z_k}(\tau_k)$, $k = 1, 2, \ldots$, we assume $z_k \to z$, $\tau_k \to \tau$, and $x_k \to x$, where $x_k \in B_{z_k}(\tau_k, D_{j_k})$, $k = 1, 2, \ldots$, with $\{D_{j_k}\}$ a sequence of partitions of $[t_0, T]$, such that $\delta(D_{j_k}) = \max\{|t_{i+1} - t_i|; t_i, t_{i+1} \in D_{j_k}\} \to 0$ as $k \to \infty$, then in accordance with the construction of step functions $\{y_k\}$, this sequence converges pointwise to x on $[t_0, T]$. From (4.7.18), the limit function x satisfies the inclusion

$$x(t) \in (VF_{\tau T}x)(\tau) + \phi(t), \quad t \in [t_0, T],$$

where $\phi(t) = z(t)$ for $t \in [t_0, \tau)$, and $\phi(t) = g(t) - g(\tau) + z(\tau)$ for $t \in [\tau, T]$. It is clear that if z satisfies (4.7.1), then x in (4.7.18) is also a solution of (4.7.1).

Let us denote by $H(g)$ the set of all (continuous) solutions of (4.7.1). Then the following basic result holds.

Theorem 4.7.1. *Assume the conditions (a), (b), (c), (d), (e), (f) are satisfied. Then for every $g \in C([t_0, T], R^n)$, the set $H(g)$ is nonempty, connected and compact. For every compact set $M \subset C([t_0, T], R^n)$, the set $H(M)$ of all solutions of (4.7.1) with arbitrary $g \in M$ is also compact. The map $g \to H(g)$ is closed (and, therefore, semicontinuous).*

Proof. In order to show that $H(g) \neq \varnothing$ for every $g \in C([t_0, T], R^n)$, we will rely on Remark 2 to Lemma 4.7.5. Let us choose the sequence $\{x_k\}$, $x_k \to x$, such that $x_k \in B_{z_k}(\tau_k, D_k)$, $k = 1, 2, \ldots$, with $\tau_k = t_0$, $z_k(\cdot) \equiv g(t_0)$, the sequence of partitions $\{D_k\}$ of $[t_0, T]$ being such that $\delta(D_k) \to 0$ as $k \to \infty$. It can easily be seen that the sequence $\{x_k\}$ is compact in C. Hence, the limit of any convergent subsequence, say $\{x_{k_m}\}$, is a solution of the inclusion (4.7.1). This proves that $H(g) \neq \varnothing$.

The compactness of the set $H(M)$, with $M \subset C$ compact, can be obtained from property (2) in the statement of Lemma 4.7.3.

The closedness of the map $g \to H(g)$ follows without difficulty from property (3) in Lemma 4.7.3. As noticed at the beginning of this section, the closedness and the compactness of $H(M)$ for every compact M, imply the upper semicontinuity.

It remains only to show that $H(g)$ is connected. Let us consider the sequence of sets in C

$$E_k = \bigcup_{x \in H(g)} B_x([t_0, T], D_k), \quad k = 1, 2, \ldots,$$

with $\delta(D_k) \to 0$ as $k \to \infty$. Property P_1 stated at the beginning of this section and Lemma 4.7.5 allow us to conclude that E_k are connected sets, if we also take into account the fact that $x_1, x_2 \in H(g)$ implies $B_{x_1}(t_0; D_k) = B_{x_2}(t_0; D_k)$. According to Remark 1 to Lemma 4.7.5, we conclude $\beta(E_k, H(g)) \to 0$ as $k \to \infty$. But $H(g) \subset E_k$, $k \geq 1$, which means that $\beta(H(g), E_k) \to 0$ as $k \to \infty$. Hence, $\alpha(E_k, H(g)) \to 0$ as $k \to \infty$, which means that $H(g)$ is connected (see Kuratowski [1]).

Theorem 4.7.1 is thus proved.

Bibliographical notes

Section 4.1 is written mostly following Cushing [1], [2]. Cushing's theory of admissibility is formulated in a very general framework and generalizes the theory of admissibility as developed by Massera and Schäffer [1] for ordinary differential equations, as well as the theory developed for integral equations (see, for instance, C. Corduneanu [4]). I found it satisfactory to present this theory in the case of abstract operators, and then apply it to the case of integral operators. See also C. Corduneanu [7] for further comments and applications.

In Section 4.2 I included some recent results due to Staffans [5]. The construction of the resolvent kernel in the nonconvolution case is a rather difficult problem and progress is to be expected in the future.

Section 4.3 is devoted to the investigation of certain Hammerstein integral equations on measure spaces. The results provided were obtained by Brezis and Browder [1], [2], [3]. These results are mainly based on monotonicity conditions. Related results are available in the literature, and I mention here the following references: Browder [2], Guo [1], [2], Pascali [1], Pascali and Sburlan [1], Petryshyn [1], Vainberg [1].

In Section 4.4 the asymptotic behavior of solutions to certain Volterra integro-differential equations is investigated, using a method that has been devised by Moser [1]. While the class of equations to which this method can be applied is not a very general one, the main advantage of obtaining such results lies in the fact that the limit set of a solution of the Volterra integrodifferential equation coincides with the limit set of a solution of an ordinary differential equation. A monotonicity condition is usually assumed, together with other hypotheses that guarantee the property of the limiting system to be an ordinary differential system (for which our knowledge of the asymptotic behavior is better known). The results in this section are due to C. Corduneanu [10], and Cassago Jr and Corduneanu [1]. Theorem 4.4.3 on the monotonicity of linear Volterra operators is due to Gripenberg [3]. The topics discussed in this section are certainly open to further development. In particular, the case when the limiting equation does not reduce to an ordinary differential equation must be considered.

Section 4.5 deals with some integrodifferential equations of Hammerstein type, as they result from the theory of boundary value problems for ordinary differential equations (two-point boundary value problem). The results pertaining to this problem were obtained in 1984 by Gaponenko [1]. Such results have been obtained by various methods during the last 70–80 years, and an early valuable reference is Caccioppoli [1]. I followed the approach of Gaponenko based on monotone operators, but then also provided a proof based on the fixed point technique (as it appears in Caccioppoli's paper). The remaining results are concerned with boundary value problems for differential functional equations involving abstract Volterra operators. The interesting part is that boundary value problems for such equations also lead to Fredholm equations. These beginnings of the theory of boundary value problems for equations with abstract Volterra operators will certainly be developed in the future.

In Section 4.6, the problems considered relate to certain convolution equations, and the main concern is the existence of periodic or almost periodic solutions. In the first part of this section, results on the existence of periodic solutions are given following C. Corduneanu [13], while in the second part a discussion of the Bohr–Neugebauer problem is conducted following Staffans [7]. There are numerous problems still open in this area, and equations as simple as $x(t) = \int k(t - s)r(s)x(s)\,ds + f(t)$ have not been investigated yet. Pertinent references are Cromer [1], [2].

Section 4.7 is dedicated to the presentation of a few results related to integral inclusions, as they appear in the references Bulgakov [1], and Bulgakov and Lyapin [1], [2]. The necessity of studying such mathematical objects was made clear in Section 1.2, in which we collected a few examples of integral and related equations occurring in applications (see Glashoff and Sprekels [1], [2]).

Further references related to topics discussed in this chapter are: B. Alfawicka [1], Amann [1], [2], [3], [4], Appell [1], Bantas [1], Brykalov [1], [2], Bushell [1], Bykov and Ruzikulov [1], A. Corduneanu [1], Fedorenko [1], C. N. Friedman [1], Gillot [1], Hochstadt [1], Hönig [1], Jordan and Wheeler [2], [3], Karakostas [2], Karakostas, Sficas and Staikos [1], Langenhop [1], Leitman and Mizel [1], Levin [2], [3], Luca [1], [2], Marocco [1], Mossaheb [1], [2], Nashed and Wong [1], Peichl [1], Prüss [3], Puljaev [1]. Rama Mohana Rao and Sivasundaram [1], Ramm [3], Reich [2], Schumacher [1] Staffans [1], [4], Svec [1], Vinokurov and Smolin [1], [2], Zabreyko and Mayorova [1].

5

Integral equations in abstract spaces

The theory of integral equations in Banach/Hilbert spaces is more recent than the theory of integral equations in finite-dimensional spaces. Practically, the theory of integral equations in infinite-dimensional spaces, involving not necessarily bounded operators, has come to life during the 1970s and 1980s. Most of the significant contributions were made in the 1980s, especially with regard to the various applications found for such equations in applied science (primarily, in continuum mechanics).

Several approaches can be encountered in the literature on this subject: fixed point methods in various function spaces, in relation to the existence problem; use of the concept of measure of noncompactness; semigroup theory; construction of the resolvent kernel, and its applications; transform theory methods. Classical types of Volterra equations, various classes of integro-partial differential equations, and to some extent abstract Volterra equations (i.e., involving an abstract nonanticipative operator) have been investigated by many authors. This kind of research illustrates the growing role of Volterra equations as mathematical models for evolution phenomena.

To date, there are only a few references in book form, related to Volterra equations in infinite-dimensional spaces: Hönig [1], Prüss [1], and Renardy, Hrusa and Nohel [1]. This last book particularly illustrates semigroup methods in the mathematical theory of viscoelasticity, but has also a good many basic theoretical results.

The aim, in this chapter, is to provide results relating to the above mentioned directions of research, and illustrate the methods that have already been successfully applied to further the investigation in this area.

As is probably the case with all chapters of this book, we shall succeed in covering only a rather limited number of results on the theory of integral equations in spaces of infinite dimension: it is hoped, among some of the most significant available in the literature.

5.1   Equations with bounded operators

It is known that even continuous nonlinear operators in spaces of infinite dimension need not be bounded (i.e., take bounded sets into bounded sets). Such operators might be very useful in various problems if we can attach in some convenient fashion a semigroup of (bounded) operators by means of which existence and behavior results can be obtained (in both linear and nonlinear cases).

In this section we shall examine only integral equations or Volterra functional equations involving bounded operators (or special cases, such as compact operators). The case of equations involving unbounded operators will be discussed in subsequent sections of this chapter.

Despite the fact that most of the equations with bounded operators have not found as significant applications as those involving unbounded operators, their analysis is nevertheless instructive.

Let us start with the most elementary result in this regard, which bears some resemblance to the results obtained in Section 1.3.

Theorem 5.1.1.   *Consider the integral equation*

$$x(t) = f(t) + \int_0^t k(t, s, x(s))\,ds, \quad t \in [0, T] \tag{5.1.1}$$

where x, f and k take their values in a Banach space X. Assume that the following conditions hold:

(1) *$f(t)$ is strongly continuous from $[0, T]$ into X (i.e., continuous in the norm of X);*

(2) *$k(t, s, x)$ is strongly continuous in (t, s, x) on the set*

$$0 \leq t \leq T, \quad 0 \leq s \leq t, \quad |x - f(t)| \leq r, \tag{D}$$

where r is a positive number;

(3) *k satisfies on D a Lipschitz condition*

$$|k(t, s, x) - k(t, s, y)| \leq L|x - y|. \tag{5.1.2}$$

Then, there exists a unique (strongly) continuous solution $x = x(t)$ of (5.1.1), defined on the interval

$$0 \leq t \leq \delta, \quad \delta = \min\{T, M^{-1}r\}, \tag{5.1.3}$$

where $M = \sup_D |k(t, s, x)|$.

Proof. The proof can be easily carried out by the method of successive approximations. Let us obtain the existence and uniqueness by contraction mapping, considering the operator T defined by

$$(Tx)(t) = f(t) + \int_0^t k(t, s, x(s)) \, ds, \quad t \in [0, \delta], \tag{5.1.4}$$

on the space $C([0, \delta], X)$. More precisely, T is defined only on the ball $S_r \subset C([0, \delta], X)$ of radius r, centered at f.

Instead of using the usual supremum norm of $C([0, \delta], X)$, we shall use an equivalent norm, namely

$$|x|_1 = \sup\{|x(t)| \exp(-Lt); t \in [0, \delta]\}. \tag{5.1.5}$$

Let us point out the fact that we always have $M < +\infty$, because (5.1.2) implies

$$|k(t, s, x)| \le L|x - f(t)| + |k(t, s, f(t))|, \tag{5.1.6}$$

and the right hand side in (5.1.6) is obviously bounded on D.

From (5.1.4) we obtain

$$(Tx)(t) - (Ty)(t) = \int_0^t \{k(t, s, x(s)) - k(t, s, y(s))\} \, ds \tag{5.1.7}$$

for any $x, y \in S_r$. Based on (5.1.2) we obtain from (5.1.7)

$$|Tx - Ty|_1 \le |x - y|_1 \int_0^t L \exp(-Ls) \, ds, \tag{5.1.8}$$

or

$$|Tx - Ty|_1 \le [1 - \exp(-L\delta)]|x - y|_1, \tag{5.1.9}$$

if we notice that $1 - \exp(-Lt) \le 1 - \exp(-L\delta)$ for $t \in [0, \delta]$.

Since $1 - \exp(-L\delta) < 1$, we obtain from (5.1.9) that the operator T is a contraction mapping on S_r.

On the other hand, $x \in S_r$ implies $Tx \in S_r$. This property is easily obtained from (5.1.4) which leads to

$$|Tx - f| \le M\delta \le r.$$

The above considerations conclude the proof of Theorem 5.1.1.

Remark 1. If D is substituted by the set

$$0 \le t \le T, \quad 0 \le s \le t, \quad x \in X, \tag{D_∞}$$

and $k(t, s, x)$ is (strongly) continuous on D_∞, while (5.1.2) holds in D_∞, then obviously r can be taken as large as we want, and $\delta = T$. In other words, existence and uniqueness are guaranted on the interval $[0, T]$.

Remark 2. In particular, the linear equation

$$x(t) = f(t) + \int_0^t k(t, s) x(s) \, ds, \tag{5.1.10}$$

in which $k(t, s)$ is a map from the set $0 \le s \le t \le T$ into the space $\mathcal{L}(X, X)$ of linear bounded operators on X, continuous in the uniform topology of $\mathscr{L}(X, X)$, obviously satisfies the condition specified in Remark 1. Therefore, equation (5.1.10) has a unique continuous solution $x(t)$ on $[0, T]$. A resolvent kernel $\tilde{k}(t, s)$ can be constructed, following the same steps as in the finite-dimensional case.

Remark 3.   The norm $|\cdot|_1$ defined by (5.1.5), was first used by A. Bielecki (see C. Corduneanu [11] for developments related to this topic), and only has the role of improving the length of the interval $[0, \delta]$ on which the existence is assured.

Let us now consider the abstract Volterra equation

$$x(t) = (Vx)(t), \quad t \in [0, T], \tag{5.1.11}$$

where $x: [0, T] \to X$ is a map belonging to some function space $E([0, T], X)$, and $V: E \to E$ is a Volterra operator, in general nonlinear.

Equation (5.1.11) was discussed in Sections 3.2–3.4, for the case when $X = R^n$. It is possible to generalize some of the results obtained in those sections, to the case when X is an infinite-dimensional Banach space.

For instance, if $E([0, T], X) = C([0, T], X)$, the conditions to be imposed on the operator V are the same as in the finite-dimensional setting (see Theorem 3.2.1). More precisely, the following result holds.

Theorem 5.1.2.   *Consider equation* (5.1.11) *under the following conditions*:

(1) V *is a Volterra operator from the space* $C([0, T], X)$ *into itself*;
(2) V *is continuous and compact on the ball* $S_r \subset C([0, T], X)$ *of radius* $r > 0$, *centered at* $x^0 \in X \subset C$;
(3) x^0 *is the fixed initial value of the operator* V.

Then, there exists $\delta > 0$, $\delta \leq T$, *such that* (5.1.11) *has at least one solution on* $[0, \delta]$.

The details of the proof are left to the reader. The Schauder–Tychonoff fixed point theorem applies exactly as in the finite-dimensional case (see Theorem 3.2.1).

We notice that a good many existence results, encountered in the literature, represent special cases of Theorem 5.1.2 (particularly, the results concerning existence of continuous solutions).

On the other hand, if we are interested in existence results involving spaces of measurable functions (in which case Lebesque's spaces must consist of Bochner integrable maps from $[0, T]$ into X), then Theorem 3.4.1 should be taken as a model. We are not going to elaborate on this type of result, because of its close resemblance to the finite-dimensional case.

A somewhat different type of result could be obtained using weak topologies in the spaces $L^p([0, T], X)$, $1 \leq p < \infty$ (and accordingly, weak continuity, weak compactness, etc.). Apparently, this kind of result has not yet been dealt with in the case of integral equations in Banach spaces. All necessary tools seem to be available, including the variant of the Dunford–Pettis theorem on compactness, provided X is a reflexive Banach space (see Theorem 2.2.3 and the comments related to it).

Let us now consider equation (5.1.11), under the assumptions of Theorem 3.2.1, with the difference that instead of the finite-dimensional space R^n, we deal with an arbitrary Banach space X. Let $x^0 \in X$ be the fixed initial value of the operator V.

A function $x \in C([0, T], x)$ will be called an ε-*approximate solution* of the equation (5.1.11), if it satisfies the conditions

$$x(0) = x^0, \quad |x - Vx|_C < \varepsilon. \tag{5.1.12}$$

It is obvious that the restriction of x at an interval $[0, \delta] \subset [0, T]$, also represents an ε-approximate solution of (5.1.11) on the interval $[0, \delta]$.

For Volterra equations like (5.1.11), it is possible to construct approximate solutions following Tonelli's procedure that we used in Section 3.1 (the second proof of Theorem 3.1.1), in the case of classical Volterra operators.

Following Szufla [1], we shall construct approximate solutions for equation (5.1.11), in the case of Banach spaces. The fact that we possess a method of constructing such approximate solutions has some implications for the numerical approach for this class of abstract equations.

First, notice that the compactness of the operator V implies the existence of a modulus of continuity, common to all Vx with x belonging to a bounded set in $C([0, T], X)$. More precisely, if we consider the ball $S_r \subset C$, centered at x^0, then for all $x \in S_r$ we shall have

$$|(Vx)(t) - (Vx)(s)| \le \omega_r(|t - s|), \tag{5.1.13}$$

provided $t, s \in [0, T]$, The function ω_r is such that $\omega_r(0) = 0$, is positive for $r > 0$, and is continuous from the right at the origin (i.e., $\omega_r(u) \to 0$ as $u \to 0+$).

Let us now consider a partition of the interval $[0, T]$ into $n(n \ge 1)$ equal subintervals, and denote $h = n^{-1}T$. The following construction provides an approximate solution $x_n(t)$ for (5.1.11), depending on n:

$$\left.\begin{array}{ll} x_n(t) = x^0, & 0 \le t \le h, \\ x_n(t) = (Vx_n)(t - h), & h < t \le nh = T. \end{array}\right\} \tag{5.1.14}$$

The second equation in (5.1.14) allows us to determine $x_n(t)$, in $n - 1$ steps equal to h, on the interval $[h, T]$. It is enough to point out that the construction indicated in (5.1.14) is possible because V is a Volterra operator, and we can always work with restrictions to subintervals $[0, \delta] \subset [0, T]$ (see Section 2.3). It is also appropriate to notice that $x_n(t)$, as constructed in (5.1.14), is a continuous function on the interval $[0, T]$. Since V takes continuous functions into continuous ones, even when we restrict our functions to subintervals $[0, \delta] \subset [0, T]$, it remains only to show that $x_n(t)$ is continuous at each $t = kh$, $k = 1$, 2, ..., n. This fact easily follows from (5.1.14), which allows the simultaneous construction of both $x_n(t)$ and $(Vx_n)(t)$, as soon as $x_n(t)$ is given on $[0, h]$. We proceed with $x_n(t)$ one step ahead of $(Vx_n)(t)$.

The problem we need to discuss now is whether $x_n(t)$ really does represent an approximate solution, and what the degree of accuracy is as far as equation (5.1.11) is concerned. We obtain from (5.1.14)

$$|x_n(t) - (Vx_n)(t)| = |x^0 - (Vx_n)(t)| = |(Vx_n)(0) - (Vx_n)(t)|, \quad 0 \le t \le h.$$

Hence, taking into account that $x_n(t) = x^0$ on $[0, h]$, we obtain from above

$$|x_n(t) - (Vx_n)(t)| \le \omega_r(h), \quad 0 \le t \le h. \tag{5.1.15}$$

Furthermore, (5.1.14) leads also to

$$|x_n(t) - (Vx_n)(t)| = |(Vx_n)(t - h) - (Vx_n)(t)| \le \omega_r(h)$$

on the interval $[h, T]$, provided $x_n(t)$ does not exit the ball of radius r in X, centered at x^0. Assuming this property takes place, we obtain from above

$$|x_n(t) - (Vx_n)(t)| \le \omega_r(h), \quad h \le t \le T. \tag{5.1.16}$$

Combining (5.1.15) and (5.1.16), we obtain the estimate

$$|x_n(t) - (Vx_n)(t)| \le \omega_r(h), \quad 0 \le t \le T, \tag{5.1.17}$$

under the assumption that $x_n(t)$, as constructed in (5.1.14), does not exit the ball of radius r in X, centered at x^0.

It is clear that the right hand side in (5.1.17) can be made arbitrarily small, provided n is chosen sufficiently large.

On the other hand, if we want to estimate the number of steps in (5.1.14) that we can make in order to keep $x_n(t)$ in the ball of radius r, centered at x^0, then we have to notice that

$$x^0 - x_n(t) = \sum_{j=1}^{k-1} [(Vx_n)((j - 1)h) - (Vx_n)(jh)]$$

$$+ [(Vx_n)((k - 1)h) - (Vx_n)(t - h)],$$

and assuming $t \in ((k - 1)h, kh]$ we obtain from (5.1.13)

$$|x^0 - x_n(t)| \le k\omega_r(h). \tag{5.1.18}$$

The condition (5.1.18) leads to a necessary condition for $x_n(t)$ to belong on $[0, kh]$ to the ball of radius r in X, centered at x^0:

$$k\omega_r(h) \le r. \tag{5.1.18}'$$

Conversely, if (5.1.18)$'$ holds, the first k steps in (5.1.14) are possible, and the estimate (5.1.17) for the approximate solution $x_n(t)$ occurs on the interval $[0, kh]$.

The reader can easily check the validity of this assertion, based on the representation for $x^0 - x_n(t)$ we have used above.

We shall go back now to equation (5.1.1), which we will discuss under the assumption that the underlying Banach space X is separable. The result we shall provide is due to Orlicz and Szufla [1], [2]. It states the existence of a solution to (5.1.1), belonging to an Orlicz space, under conditions involving an integral inequality of the type encountered in comparison techniques (i.e., a scalar integral inequality which is associated to the given equation, and implies certain properties for solutions).

We shall denote by $L_\phi([0, T], X)$ the Orlicz space of all strongly (Bochner) measurable functions $u: [0, T] \to X$, such that

$$|u|_\phi = \inf \left\{ r > 0: \int_0^T \phi(|u(s)|/r)\,ds < \infty \right\}. \qquad (5.1.19)$$

The function ϕ is assumed to be an N-function (see Krasnoselskii and Rutickii [1]). It is known that $L_\phi([0, T], X)$ is a Banach space. The closure in $L_\phi([0, T], X)$ of the set of all bounded functions is again a Banach space, with the norm (5.1.19), and it will be denoted by $E_\phi([0, T], X)$.

Let us recall the definition of the *measure of noncompactness*. If (S, ρ) is a metric space and $A \subset S$, then the measure of noncompactness of the set A, denoted by $\beta(A)$, is the infimum of these $\varepsilon > 0$, with the property that A has a finite coverage by balls of radius ε. Of course, for noncompact A one has $\beta(A) > 0$.

Theorem 5.1.2. *Assume the following conditions are satisfied for equation* (5.1.1) *and the Orlicz space* $L_\phi([0, T], X)$:

(a) *the map* $(t, s, x) \to k(t, s, x)$, *from* $\Delta \times X$ *into* X, *where* $\Delta = \{(t, s): 0 \le s \le t \le T\}$, *is continuous in* x *for almost all* $(t, s) \in \Delta$, *and for every* $x \in X$ *is strongly measurable in* (t, s);

(b) $|k(t, s, x)| \le K(t, s)[b(s) + H(|x|)]$ *on* $\Delta \times X$, *with a nonnegative* $K \in E_M(\Delta, R)$, $b \in L_N([0, T], R)$, *and* H *a nonnegative nondecreasing continuous function on* R_+;

(c) $N: R_+ \to$ *satisfies* $N(uv) \le \lambda N(u)N(v)$ *for some* $\lambda > 0$, *and* $u, v \ge u_0 \ge 0$;

(d) ϕ *is an* N-*function, and there exist* $\alpha, \gamma, u_0 \ge 0$ *with the property*

$$N(\alpha H(u)) \le \gamma \phi(u) \le \gamma M(u), \quad u \ge u_0, \qquad (5.1.20)$$

where N *and* M *are complementary* N-*functions*;

(e) *for every bounded set* $A \subset X$ *one has*

$$\beta(k(t, s, A)) \le h(t, s, \beta(A)), \qquad (5.1.21)$$

for almost all $(t, s) \in \Delta$, *where* $h(t, s, u)$ *is nonnegative on* $\Delta \times R_+$, *and such that* $\int_0^t h(t, s, u(s))\,ds$ *exists almost everywhere on* $[0, T]$, *for any* $u \in L^1([0, T], R)$;

(f) *for any $\delta > 0$, $\delta \leq T$, $u = 0$ is the only nonnegative integrable function on* $[0, \delta]$, *satisfying almost everywhere the comparison inequality*

$$u(t) \leq \int_0^t h(t, s, u(s)) \, ds; \qquad (5.1.22)$$

(g) *if we let*

$$(Kx)(t) = \int_0^t k(t, s, x(s)) \, ds, \quad t \in [0, T], \qquad (5.1.23)$$

then

$$\lim_{h \to 0} \int_0^T |(Kx)(t + h) - (Kx)(t)| \, dt = 0, \qquad (5.1.24)$$

uniformly with respect to x in the unit ball of $E_\phi([0, T], X)$.

Then, *for any $f \in E_\phi([0, T], X)$, there exists an interval $[0, a]$, $a \leq T$, such that the set of all solutions of equation (5.1.1), belonging to the unit ball of $E_\phi([0, a], X)$ is nonempty and compact.*

The proof of Theorem 5.1.2 involves a good deal of technicalities and auxiliary results. We refer the reader to the papers of Orlicz and Szufla [1], [2], where the proof is provided, and where other similar results are given.

Remark. In (5.1.24), the functions involved are assumed to be extended from $[0, T]$ to R, by letting them equal zero outside $[0, T]$. The condition (5.1.24) is obviously one of the conditions appearing in the criterion of compactness for the operator K, in the space $L^1([0, T], X)$, when $X = R^n$ (see Section 2.3). Of course, we cannot make any conclusions about the compactness of K, given by (5.1.23), in the Orlicz space involved.

5.2 Equations with unbounded operators in Hilbert spaces

We have usually dealt with integral operators that are bounded on the spaces they are defined on. But there are significant examples of linear integral operators that are unbounded. For instance, if $k(t, s)$ is a real-valued measurable function on $[a, b] \times [a, b]$, such that $k(t, s) = k(s, t)$ a.e., and

$$\int_a^b |k(t, s)|^2 \, ds < \infty \qquad (5.2.1)$$

then the integral operator

$$(Kx)(t) = \int_a^b k(t, s) x(s) \, ds \qquad (5.2.2)$$

has – in general – a dense domain of definition in $L^2([a, b), R)$, and is unbounded.

The integral operator K, given by (5.2.2), under assumption (5.2.1), is called a *Carleman operator*. Basic references relating to the theory of Carleman operators are: Carleman [1], von Neumann [1], and Korotkov [1].

A necessary and sufficient condition for the operator K, densely defined in $L^2([a, b], R)$ and with symmetric kernel $k(t, s)$, to be bounded on $L^2([a, b], R)$ is the existence of a positive measurable function $A: [a, b] \to R$, and a positive constant C, such that

$$\int_a^b |k(t, s)| A(s) \, ds \le CA(t) \quad \text{a.e. on } [a, b]. \tag{5.2.3}$$

Then $|K| \le C$.

The proof of this result can be found in Korotkov [1] (Theorem 4.1). It is useful because it does characterize the bounded operators (and, consequently, the unbounded ones).

If the symmetry of $k(t, s)$ is dropped, then the Hilbert–Schmidt condition

$$\int_a^b \int_a^b |k(t, s)|^2 \, dt \, ds < +\infty \tag{5.2.4}$$

assures the continuity (boundedness) of K on $L^2([a, b], R)$, and as we know (see Section 1.4), K is also compact. For details on these matters see Halmos and Sunder [1], or Korotkov [1].

It is easy to produce examples of linear integral operators which are unbounded. For instance, if $k(t, s)$ is measurable from $R_+ \times R_+$ into R, and such that

$$\int_0^\infty |k(t, s)|^2 \, ds \le M < \infty \quad \text{a.e. on } R_+, \tag{5.2.5}$$

then the integral operator

$$(Kx)(t) = \int_0^\infty k(t, s)x(s) \, ds \tag{5.2.6}$$

is bounded from $L^2(R_+, R)$ into $L^\infty(R_+, R)$. Since $L^1(R_+, R) \cap L^2(R_+, R)$ is dense in both $L^1(R_+, R)$ and $L^2(R_+, R)$, it happens that K, defined by (5.2.6), is densely defined on $L^1(R_+, R)$, with values in $L^\infty(R_+, R)$. If $k(t, s)$ is such that

$$\underset{s \in R_+}{\text{ess-sup}} \, |k(t, s)| = \infty \quad \text{a.e. on } R_+, \tag{5.2.7}$$

then we cannot expect K to be bounded from $L^1(R_+, R)$ into $L^\infty(R_+, R)$. Indeed, we can choose $k(t, s) \equiv k(s)$ on R_+, with $k(s)$ unbounded (essentially), but in $L^2(R_+, R)$. It is necessary to keep in mind that the dual of $L^1(R_+, R)$ is $L^\infty(R_+, R)$.

In order to deal with integral equations such as

$$\int_0^T k(t, s)x(s) \, ds = (Fx)(t) \quad \text{a.e. on } R_+, \tag{5.2.8}$$

in which F is an operator acting between convenient function spaces, and the integral operator on the left hand side is unbounded, we shall follow Amann [5], who considered the abstract equation

$$Ax = Fx, \quad x \in H. \tag{5.2.9}$$

In (5.2.9), A stands for a linear self-adjoint operator $A: D_A \subset H \to H$, and $F: H \to H$ is an operator which admits a symmetric Gateaux derivative F'. By H we mean a real Hilbert space.

In what follows, we attempt to provide a generalization of the main result in Section 1.5. This abstract version of the Hammerstein–Dolph theory of nonlinear integral equations, obtained by Amann [5], covers many particular situations, including partial differential equations, ordinary differential equations (classical or abstract), and other types of nonlinear equations.

The following lemma is helpful with regard to the main existence result of this section.

Lemma 5.2.1. *Assume the real Hilbert space H is the direct sum of its closed subspaces H_1 and H_2. If $P_i: H \to H_i$, $i = 1, 2$, are the projectors of H on H_i, and $R = P_1 - P_2$, then the following properties hold:*

(a) *R is a continuous automorphism of the space H;*
(b) *if $A: D_A \subset H \to H$ is a linear self adjoint operator with $P_i D_A \subset D_A$, $i = 1, 2$, the restriction of R to D_A is a bijection;*
(c) *the linear operator $L = AR$ is closed, and $D_L = D_{L^*} = D_A$.*

Proof. (a) Let $\{x_j\} \subset H$ be a sequence such that $x_j \to x$, and $Rx_j \to y$. Then, taking into account $P_1 x_j = 2^{-1}(x_j + Rx_j)$ and $P_2 x_j = 2^{-1}(x_j - Rx_j)$, one obtains $P_1 x_j \to 2^{-1}(x + y)$ and $P_2 x_j \to 2^{-1}(x - y)$, as $j \to \infty$. But both P_1 and P_2 are closed, which implies $P_1 x = 2^{-1}(x + y)$, and $P_2 x = 2^{-1}(x - y)$. These imply $Rx = P_1 x - P_2 x = y$, which means that R is closed. The closed graph theorem implies the continuity of R. Since $R = P_1 - P_2$, it means that $Rx = \theta$ implies $P_1 x = P_2 x = \theta$, i.e., $x = \theta$. Hence, R is a bijection. It is onto H because $R^2 = P_1 + P_2 = I$ (the unit operator on H).

(b) Since D_A is invariant for both projectors P_1 and P_2, the bijection property is obtained without difficulty as in (a).

(c) Since R is continuous, $L = AR$ is closed (because A is closed). It is obvious that $D_L = D_A$. If $y \in D_A$, then $\langle Lx, y \rangle = \langle x, R^*Ay \rangle$, $\forall x \in D_A$. This implies $y \in D_{A^*}$. On the other hand, if $y \in D_{A^*}$, we have

$$|\langle Lx, y \rangle| = |\langle ARx, y \rangle| \le \alpha |x| \le \alpha |R^{-1}||Rx|,$$

for a convenient $\alpha > 0$, and all $x \in D_L = D_A$. Hence, since the restriction of R to D_A is a bijection onto D_A,

$$|\langle Az, y \rangle| \leq \alpha |R^{-1}| |z|, \quad \forall z \in D_A.$$

This shows that $y \in D_A = D_L$. Therefore, $D_L = D_{L^*}$, which proves (c).

Before we state existence result for equation (5.2.9), let us recall that for two symmetric linear operator on H, say B and C, the relationship $B \geq C$ means that $\langle (B - C)x, x \rangle \geq 0$, for any $x \in H$.

Theorem 5.2.2. *Consider equation (5.2.9), and assume that the following conditions are verified:*

(1) *$A: D_A \subset H \rightarrow H$ is a linear self-adjoint operator;*

(2) *H is the direct sum of two closed subspaces H_1 and H_2, and D_A is invariant for both projectors P_1, P_2, $P_i: H \rightarrow H_i$, $i = 1, 2$;*

(3) *$F: H \rightarrow H$ has a symmetric weak Gateaux derivative $F'(x)$, at each $x \in H$;*

(4) *there exist two linear symmetric operators B and C, such that*

$$B \leq F'(x) \leq C, \quad \forall x \in H; \tag{5.2.10}$$

(5) *there exists $\gamma > 0$ such that*

$$\left.\begin{array}{l} \langle (A - B)x, x \rangle \leq -\gamma |x|^2, \quad \forall x \in D_A \cap H_1, \\ \langle (A - C)x, x \rangle \leq \gamma |x|^2, \quad \forall x \in D_A \cap H_2. \end{array}\right\} \tag{5.2.11}$$

Then there exists a unique solution $x^* \in D_A$ of equation (5.2.9), and the estimate

$$\max\{|x^*|, |Rx^*|\} \leq 2\gamma^{-1} |F(\theta)| \tag{5.2.12}$$

holds.

Proof. We will consider first the operator $M = (A - F)R = L - FR$, from D_L into H, and show that the inequality

$$\langle Mx - My, x - y \rangle \geq \frac{\gamma}{2} \max\{|x - y|^2, |Rx - Ry| |x - y|\} \tag{5.2.13}$$

holds for all $x, y \in D_L$.

Indeed, let $x, y \in D_L$ be fixed. Then, there exists $t \in (0, 1)$ with the property

$$\langle FRx - FRy, x - y \rangle = \langle DR(x - y), x - y \rangle, \tag{5.2.14}$$

with

$$D = F'(Ry + t(Rx - Ry)) \tag{5.2.15}$$

a linear operator on H, which is self-adjoint according to assumption (3). Hence, $A - D$ is also self-adjoint on H. Therefore

$$\langle Mx - My, x - y \rangle = \langle AR(x - y), x - y \rangle - \langle FRx - FRy, x - y \rangle$$

$$= (A - D)[P_1(x - y) - P_2(x - y)], P_1(x - y) + P_2(x - y) \rangle$$

$$= \langle (A - D)P_1(x - y), P_1(x - y) \rangle$$

$$- \langle (A - D)P_2(x - y), P_2(x - y) \rangle$$

$$\geq \langle (A - C)P_1(x - y), P_1(x - y) \rangle$$

$$- \langle (A - B)P_2(x - y), P_2(x - y) \rangle$$

$$\geq \gamma(|P_1(x - y)|^2 + |P_2(x - y)|^2).$$

But we have for every $z \in H$

$$|P_1 z \pm P_2 z|^2 = |P_1 z|^2 \pm 2\langle P_1 z_1 P_2 z \rangle + |P_2 z|^2$$

$$\leq |P_1 z|^2 + 2|P_1 z||P_2 z| + |P_2 z|^2 \leq 2(|P_1 z|^2 + |P_2 z|^2),$$

and taking into account $P_1 + P_2 = I$ and $P_1 - P_2 = R$, the inequality (5.2.13) is proven.

Now, we notice that equation (5.2.9) is equivalent to the equation $Mx = \theta$, where $M = L - FR$. Indeed, we have $M = (A - F)R$ and, since R is a bijection, the assertion follows.

The inequality (5.2.13) for the operator M shows that M is strongly monotone. More precisely, (5.2.13) does imply for $x = z$ and $y = \theta$,

$$\langle Mz, z \rangle \geq \left[\frac{\gamma}{2}|z| - |M\theta| \right]|z|, \quad \forall z \in D_L, \tag{5.2.16}$$

which also proves the coerciveness of M.

Therefore, $M: D_L \subset H \to H$ is monotone and coercive (not continuous!). According to a result of Browder [1] (Theorem 16), there exists $x^* \in D_L$, such that $M(x^*) = \theta$. This means, as seen above, that $Ax^* = Fx^*$.

The uniqueness is an immediate consequence of the strong monotonicity of M: if $Mx = My$ in (5.2.13), then $x = y$.

Finally, to obtain the estimate (5.2.12) we use (5.2.16) for $z = x^*$, and notice that $M\theta = F\theta$. Moreover, if we let $y^* = Rx^*$ and keep in mind the relationship $R^2 = I$, then we obtain $x^* = Ry^*$, and again use (5.2.16).

Theorem 5.2.2 is thereby proven.

Remark 1. Browder's result used in the proof is in line with those stated at the end of Section 2.4 in relation to the range of monotone operators. Unlike the results quoted in Section 2.4, the result under discussion does not require the operator M to satisfy strong restrictions.

Remark 2. The conditions (5.2.11) are conditions related to the spectrum of the operator A, which being a symmetric operator satisfies $\sigma(A) \subset R$. More precisely, these conditions imply that $\sigma(A)$ has gaps on R. From (5.2.10) we see that $C \geq B$, which combined with (5.2.11) leads to the above conclusion. That this situation (existence of gaps) really occurs in applications follows from discussions related to Schroedinger's operator (see, for instance, Reed and Simon [1]).

Remark 3. Sometimes, instead of conditions (5.2.11), one encounters the following type of condition: there exist two real numbers α, β, with $\alpha < \beta$, such that $[\alpha, \beta] \subset \rho(A)$, the resolvent set of the operator A, and

$$\alpha \leq \frac{\langle Fx - Fy, x - y \rangle}{|x - y|^2} \leq \beta,$$

for all $x, y \in H$, $x \neq y$. This is the typical condition on nonlinearity which we used, in the scalar case, for Hammerstein equations in Section 1.5. In that particular situation, α and β represent two consecutive eigenvalues of the kernel. In Theorem 5.2.2, there is no condition on the nature of the spectrum of A outside $[\alpha, \beta]$.

In concluding this section, we shall apply Theorem 5.2.2 to the case of linear integrodifferential equations of the form

$$-\frac{d}{dt}\left[p(t)\frac{dx}{dt}\right] + q(t)x = \int_a^b k(t,s)x(s)\,ds + f(t), \tag{5.2.17}$$

where $p(t)$, $p'(t)$, $q(t)$ are continuous functions on $[a,b]$, with values in R, and $p(t) > 0$. The function $k(t,s)$ is assumed symmetric and measurable on $[a,b] \times [a,b]$, and such that Hilbert–Schmidt condition (5.2.4) takes place. The function $f(t)$ belongs to the space $L^2([a,b], R)$.

The following boundary value conditions will be associated with the equation (5.2.17):

$$\alpha, x(a) + x_2\dot{x}(a) = 0, \quad \beta_1 x(b) + \beta_2\dot{x}(b) = 0, \tag{5.2.18}$$

where $|\alpha_1| + |\alpha_2| > 0$, and $|\beta_1| + |\beta_2| > 0$.

One more basic assumption is needed, concerning the (linear) integral operator K in the right hand side of equation (5.2.17) (see (5.2.2)). Before we can formulate it, let us specify in more detail how this problem can be treated in the framework provided by Theorem 5.2.2.

First, the Hilbert space H is chosen to be the $L^2([a,b], R)$ space.

Second, the operator $A: D_A \subset H \to H$ is given by

$$(Ax)(t) = -\frac{d}{dt}\left[p(t)\frac{dx}{dt}\right] + q(t)x, \tag{5.2.19}$$

where D_A consists of all $x \in L^2([a,b], R)$ which are continuously differentiable, possess a second-order derivative belonging to $L^2([a,b], R)$ and verify the boundary value conditions (5.2.18). It is easily seen that D_A is dense in $L^2([a,b], R)$. It is well known (see, for instance, C. Corduneanu [6]) that A is a self-adjoint operator which admits a compact inverse, if we assume that the problem $Ax = 0$, (5.2.18) has only the trivial solution $x = 0$. Moreover, this inverse can be represented as an integral operator whose kernel – the Green's function – is symmetric and continuous on $[a, b] \times [a, b]$. The eigenfunctions of A and A^{-1} are the same, and they provide a complete orthogonal system in $L^2([a, b], R)$, say $\{\psi_j\}, j \geq 1$, with $\langle \psi_j, \psi_k \rangle = 0$ for $j \neq k$. The eigenvalues of A, say $\{\lambda_j\}, j \geq 1$, are the same as those of the integral operator A^{-1}: $\psi_j = \lambda_j A^{-1} \psi_j, j \geq 1$ (according to the traditional notation, which we adopted in Chapter 1). It is known that we can label them in such a way that $\lambda_j < \lambda_{j+1}$. Then $\lambda_j \to \infty$ as $j \to \infty$, and only a finite number of λ_js can be negative.

Since the Gateaux derivative of the right hand side in (5.2.17) is the symmetric operator generated by the kernel $k(t, s)$, i.e., the operator K given by (5.2.2), we will assume that the following conditions are satisfied.

If $\mu < \nu$ are two real numbers such that for some natural m one has

$$\lambda_m < \mu < \nu < \lambda_{m+i}, \tag{5.2.20}$$

then we define H_1 to be the (finite-dimensional) subspace generated by ψ_1, $\psi_2, \ldots, \psi_m$, while H_2 is its orthogonal complement in $L^2([a, b], R)$. Of course, we can say that H_2 is the subspace of $L^2([a, b], R)$ generated by the elements $\psi_j, j \geq m + 1$.

The operator K, given by (5.2.2), will be required to satisfy the conditions

$$\mu |x|^2 \leq \langle Kx, x \rangle, \quad \forall x \in H_1 \cap D_A, \tag{5.2.20}$$

$$\langle Kx, x \rangle \leq \nu |x|^2, \quad \forall x \in H_2 \cap D_A. \tag{5.2.21}$$

One can easily check that (5.2.20) and (5.2.21) are basically the same as conditions (5.2.11) of Theorem 5.2.2. The number γ in (5.2.11) can be chosen as any positive number such that $\gamma < \min\{\mu - \lambda_m, \lambda_{m+1} - \nu\}$.

The conclusion of the above discussion is that (5.2.17) has a unique solution in D_A, if the problem $Ax = 0$, (5.2.18) has only trivial solution, and the operator K verifies (5.2.20), (5.2.21). It is worth while noticing that if the problem $Ax = 0$, (5.2.18) has nontrivial solutions, instead of the operator A defined by (5.2.19) one has to consider an operator of the form $A - \varepsilon I$ which is invertible.

Finally, let us point out that this result cannot be derived from the theory of Hammerstein equations as presented in Section 1.5.

5.3   The semigroup method for integrodifferential equations

We shall consider in this section some integrodifferential equations of the form

$$x'(t) = Ax(t) + \int_0^t B(t - s)x(s)\,ds + f(t), \quad t \in R_+, \tag{5.3.1}$$

under the initial condition

$$x(0) = x^0 \in D_A \subset X,   \tag{5.3.2}$$

where X stands for a Banach space (real or complex). The operator A is, in general, unbounded on X, and the main hypothesis will be that A generates a semigroup in X. Then, under convenient assumptions on $B(t)$, one can attach a semigroup $S(t)$, $t \in R_+$ to (5.3.1), which allows us to represent, roughly speaking, the solution of (5.3.1), (5.3.2) by means of the variation of constants formula

$$x(t) = S(t)x^0 + \int_0^t S(t - s)f(s)\,ds, \quad t \in R_+.   \tag{5.3.3}$$

On the other hand, if a *resolvent operator* $R(t)$, $t \in R_+$, can be constructed for (5.3.1), then a similar formula to (5.3.3) holds

$$x(t) = R(t)x^0 + \int_0^t R(t - s)f(s)\,ds, \quad t \in R_+.   \tag{5.3.4}$$

If we compare (5.3.3) and (5.3.4), taking into account the arbitrariness of $x^0 \in D_A$ and $f \in E$ = a function space, we conclude that $S(t)$, $t \in R_+$, and $R(t)$, $t \in R_+$, must be the same. Therefore, looking for a semigroup of continuous operators that describes the family of solutions to (5.3.1), and looking for a resolvent operator attached to (5.3.1) represent, more or less, the same thing.

Following Desch and Schappacher [1], we shall prove that from the construction of a semigroup we can always derive the existence of the resolvent operator. At the same time, by means of a convenient example, we will show that this procedure has its limitations, and we cannot count on it as a 'universal' method. Nevertheless, this is an efficient method in obtaining exponentially bounded resolvent operators.

The following hypotheses will be admitted with regard to the operators A and $B(t)$, $t \in R_+$, appearing in (5.3.1).

(a) A is the infinitesimal generator of a semigroup of bounded linear operators acting in X. Since A is closed, D_A can be organized as a Banach space with the graph norm: $x \to |x| + |Ax|$. This Banach space will be denoted in the sequel by $(Y, |\cdot|_Y)$.

(b) $\{B(t); t \in R_+\}$ is a family of bounded linear operators from Y into X.

(c) For every $x \in Y$, the function $(Bx)(t) = B(t)x$ is Bochner measurable (from R_+ into X).

(d) There exists $\omega \geq 0$, such that

$$e^{-\omega t}|B(t)| \in L^1(R_+, R),   \tag{5.3.5}$$

where $|B(t)|$ means the usual norm of the (bounded) linear operator $B(t)$ from Y into X.

These hypotheses will be somewhat strengthened subsequently, in order to obtain existence results.

Let us now define the *resolvent operator* corresponding to (5.3.1). A family $\{R(t); t \in R_+\}$ of bounded linear operators on X is called a *resolvent operator* for (5.3.1), if the following conditions are satisfied.

R_1. $R(0) = I$ = the unit operator of X.

R_2. For any $x \in X$, the map $t \to R(t)x$ is continuous on R_+.

R_3. For any $x \in Y$, the map $t \to R(t)x$ belongs to $C(R_+, Y) \cap C^{(1)}(R_+, X)$, and verifies

$$R'(t)x = AR(t)x + \int_0^t B(t - s)R(s)x \, ds. \qquad (5.3.6)$$

R_4. For any $x \in Y$, the following equation holds on R_+:

$$R'(t)x = R(t)Ax + \int_0^t R(t - s)B(s)x \, ds. \qquad (5.3.7)$$

The following lemma will be needed for our discussion on the connection between the semigroup of operators attached to (5.3.1), and the resolvent operator corresponding to the same equation.

Lemma 5.3.1. *Assume that* $\{R(t); t \in R_+\}$ *is a family of bounded linear operators on X, satisfying R_1 and R_2, and such that*

$$|R(t)| \le Me^{\omega t}, \quad t \in R_+, \qquad (5.3.8)$$

for some constants M and ω. Then the following three conditions are equivalent:

(1) *$R(t)$ satisfies R_3;*
(2) *$R(t)$ satisfies R_4;*
(3) *the Laplace transform of $R(t)$*

$$\tilde{\mathscr{R}}(s) = \int_0^\infty R(t)e^{-ts} \, dt, \quad \text{Re} s > \omega, \qquad (5.3.9)$$

is given by

$$\tilde{\mathscr{R}}(s) = [sI - A - \tilde{\mathscr{B}}(s)]^{-1}. \qquad (5.3.10)$$

Proof. Let us notice first that the inverse operators of $sI - A - \tilde{\mathscr{B}}(s)$ does exist and is bounded for sufficiently large Res. Indeed, we can write

$$sI - A - \tilde{\mathscr{B}}(s) = [I - \tilde{\mathscr{B}}(s)(sI - A)^{-1}](sI - A), \qquad (5.3.11)$$

provided both $\tilde{\mathscr{B}}(s)$ and $(sI - A)^{-1}$ are defined. Since $(sI - A)^{-1}$, as an operator from X into Y, is uniformly bounded for sufficiently large s, while Re$s \to \infty$ implies $|\tilde{\mathscr{B}}(s)| \to 0$, we obtain

$$|\tilde{\mathscr{B}}(s)(sI - A)^{-1}| < 1, \quad \mathrm{Res} > s_0. \tag{5.3.12}$$

Therefore, the operator $I - \tilde{\mathscr{B}}(s)(sI - A)^{-1}$ has an inverse operator for sufficiently large Res. Consequently, for such s we have

$$[sI - A - \tilde{\mathscr{B}}(s)]^{-1} = (sI - A)^{-1}[I - \tilde{\mathscr{B}}(s)(sI - A)^{-1}]^{-1}, \tag{5.3.13}$$

which proves our statement.

We shall assume now that A^{-1} does exist as a bounded operator from X into Y. The general case requires the existence of $(A - \sigma I)^{-1}$, with the same property. We can reduce this situation to the special case, by substituting $e^{-\sigma t}R(t)$ for $R(t)$. The family of operators $B(t)$, $t \in R_+$, can now be represented as $B(t) = C(t)A$, with $C(t) = B(t)A^{-1}$ a family of bounded operators from Y into X. Let us now determine the family of operators $L(t)$, $t \in R_+$, from the equation

$$(\delta + L) * (\delta + C) = (\delta + C) * (\delta + L) = \delta, \tag{5.3.14}$$

which actually stands for

$$L(t) = C(t) - \int_0^t C(t - s)L(s)\,ds, \quad t \in R_+. \tag{5.3.15}$$

The existence of the family $L(t)$, $t \in R_+$, is a consequence of Theorem 5.1.1.

Let us now prove the equivalence of the properties (1), (2), (3) in the statement of Lemma 5.3.1.

First, let us show that $(2) \Rightarrow (3)$. If we take the Laplace transform in (5.3.7) we obtain

$$s\tilde{\mathscr{R}}(s)x - x = \tilde{\mathscr{R}}(s)Ax + \tilde{\mathscr{R}}(s)\tilde{\mathscr{C}}(s)Ax,$$

which can be rewritten as

$$\tilde{\mathscr{R}}(s)[sx - Ax - \tilde{\mathscr{C}}(s)Ax] = x, \quad \forall x \in Y.$$

If we now let $x = [sI - A - \tilde{\mathscr{B}}(s)]^{-1}y$ for $y \in X$, we obtain from above

$$\tilde{\mathscr{R}}(s)y = [sI - A - \tilde{\mathscr{B}}(s)]^{-1}y = x, \quad \forall x \in Y,$$

which shows that (3) is true.

Second, let us prove that $(3) \Rightarrow (2)$. We notice that (5.3.7) can be written in integral form as

$$R(t)x - x = \int_0^t \left[R(s)Ax + \int_0^s R(s - u)C(u)Ax\,du \right]ds. \tag{5.3.16}$$

Since $R(s)Ax + \int_0^s R(s - u)C(u)Ax\,du$ is a continuous function of s, (5.3.7) can be obtained from (5.3.16) by differentiation. If we now take the Laplace transform on both sides of (5.3.16), we obtain

$$\tilde{\mathscr{R}}(s)x - s^{-1}x = s^{-1}[\tilde{\mathscr{R}}(s)Ax + \tilde{\mathscr{R}}(s)\tilde{\mathscr{C}}(s)Ax],$$

which easily follows from (5.3.10).

We will now prove that $(1) \Rightarrow (3)$. If in (5.3.6) we integrate both sides from 0 to t, and take the Laplace transform, we obtain

$$\tilde{\mathscr{R}}(s)x - s^{-1}x = [I + \tilde{\mathscr{C}}(s)]A[s^{-1}\tilde{\mathscr{R}}(s)x], \qquad (5.3.17)$$

noticing that the integrated version of (5.3.6) can be written as $(\delta + L) *$ $(Rx - x) = A(I * Rx)$, where L is as defined above, and $(\delta + L) * (\delta + C) = \delta$. From (5.3.17) we obtain easily $(sI - A - \tilde{\mathscr{C}}(s)A)\tilde{\mathscr{R}}(s)x = x$, for any $x \in Y$, which implies (5.3.10), i.e., property (3).

The last statement to be proven is $(3) \Rightarrow (1)$. Since we have shown that $(3) \Rightarrow (2)$, we know that (5.3.7) holds, and therefore $R'(t)x$ is exponentially bounded on R_+, for every $x \in Y$. We notice that $A^{-1}(\delta + L) * R'x = Rx$, because it is the same as $(\delta + L) * R'x = ARx$, and this means $R'x = (\delta + C) * ARx$, which coincides with (5.3.6). The Laplace transform of $A^{-1}(\delta + L) * R'x - Rx$ is

$$A^{-1}[I + \tilde{\mathscr{L}}(s)]\{s[sI - A - \tilde{\mathscr{B}}(s)]^{-1}x - x\} - [sI - A - \tilde{\mathscr{B}}(s)]^{-1}x$$

$$= A^{-1}[I + \tilde{\mathscr{L}}(s)]\{s[sI - A - \tilde{\mathscr{B}}(s)]^{-1}x$$

$$\quad - x - [I + \tilde{\mathscr{C}}(s)]A[sI - A - \tilde{\mathscr{B}}(s)]^{-1}x\}$$

$$= A^{-1}[I + \tilde{\mathscr{L}}(s)]\{[sI - A - \tilde{\mathscr{B}}(s)][sI - A - \tilde{\mathscr{B}}(s)]^{-1}x - x\} = 0,$$

which concludes the proof of Lemma 5.3.1.

We shall now prove that a semigroup of operators can be attached to equation (5.3.1), under the initial condition (5.3.2). This construction will allow us to obtain the resolvent operator associated with (5.3.1), as well as the representation formula (5.3.4).

It will be assumed that the function f in (5.3.1) belongs to a convenient function space $F = F(R_+, X)$. The pair $(x^0, f) \in Y \times F$ will be considered as the state of the system at time $t = 0$. To see how this system changes in time, let us first proceed by formal calculation. We have

$$x'(s + t) = Ax(s + t) + \int_0^s B(s - u)x(u + t)\,du + g(t, s), \qquad (5.3.18)$$

where $g(t, s)$ is given by

$$g(t, s) = \int_0^t B(s + t - u)x(u)\,du + f(s + t). \qquad (5.3.19)$$

For $s = 0$ we have from (5.3.18) $x'(t) = Ax(t) + g(t, 0)$, while (5.3.19) leads to

$$\frac{d}{dt}g(t, s) = B(s)x(t) + \int_0^t B'(t + s - u)x(u)\,du + f'(t + s)$$

$$= B(s)x(t) + \frac{d}{ds}\left[\int_0^t B(t + s - u)x(u)\,du + f(t + s)\right]$$

$$B(s)x(t) + \frac{d}{ds}g(t, s).$$

This relationship, together with (5.3.1), suggests the fact that the pair (x, g) can be regarded as the solution of the system

$$(x', g') = \mathscr{A}(x, g), \tag{5.3.20}$$

under initial condition

$$(x(0), g(0)) = (x^0, f), \tag{5.3.21}$$

where

$$\mathscr{A}(x, g) = \left(Ax + g(\cdot, 0), Bx + \frac{\mathrm{d}}{\mathrm{d}s} g \right). \tag{5.3.22}$$

In order to give a precise meaning to the above considerations, we need to specify the function space F, as well as the properties necessary for B to assure that $\mathscr{A}$ is the generator of a semigroup of operators on the product space $X \times F$. We shall choose the space F to be the space $e^{\omega t} L^1(R_+, X)$, with the natural norm

$$x \to \int_0^\infty |x(t)| e^{-\omega t} \, \mathrm{d}t.$$

Another possible choice would be the space $e^{\omega t} L^2(R_+, X)$, and even an axiomatic approach (see Miller [5]) is available.

It is useful to notice that the shift semigroup $T(\cdot)$, which is given by the formula $(T(t)f)(u) = f(t + u)$, is strongly continuous and its infinitesimal generator D_u has as domain the Sobolev space $e^{\omega t} W^{1,1}(R_+, X)$. This fact implies that the Dirac operator $\delta \colon \delta f = f(0)$, is defined on the space $e^{\omega t} W^{1,1}(R_+, X)$, and it is continuous: $e^{\omega t} W^{1,1}(R_+, X) \to X$.

The following results shows that the possibility of attaching a semigroup to (5.3.1) implies the existence of a resolvent operator.

Theorem 5.3.2. *Under the general assumptions* (a), (b), (c), (d), *let us define* $D_A = Y \times e^{\omega t} W^{1,1}(R_+, X) \subset X \times e^{\omega t} L^1(R_+, X)$, $\mathscr{A}(x, g) = (Ax + \delta g, Bx + D_u g)$. *If* $\mathscr{A}$ *generates a semigroup* $\mathscr{S}(t)$ *on* $X \times e^{\omega t} L^1(R_+, X)$, *then* (5.3.1) *admits a resolvent operator* $R(t)$, *such that* $R(t)x$ *is the first component of* $\mathscr{S}(t)(x, 0)$. *For any* $f \in e^{\omega t} L^1(R_+, R)$, *the first component of* $\mathscr{S}(t)(0, f)$ *is* $R * f$.

Proof. Let us denote by $R(t)x$ the first component of $\mathscr{S}(t)(x, 0)$. For sufficiently large Res, $\tilde{\mathscr{R}}(s)x$ is the first component of y in the pair $(y, g) = (sI - \mathscr{A})^{-1}(x, 0)$. This means that $sy - Ay - g(0) = x$, and $sg - D_u g - By = 0$. The second equation leads to

$$g = (sI - D_u)^{-1} By = \int_0^\infty e^{-ts} T(t) By \, \mathrm{d}t,$$

which means

$$g(u) = \int_0^\infty e^{-ts}B(t+u)y\,dt,$$

and therefore $g(0) = \tilde{\mathscr{B}}(s)y$. From the first equation we obtain $sy - Ay - \tilde{\mathscr{B}}(s)y = x$, or $y = [sI - A - \tilde{\mathscr{B}}(s)]^{-1}x$. According to Lemma 5.3.1, $R(t)$ is the resolvent operator for (5.3.1).

For an arbitrary $f \in e^{\omega t}L^1(R_+, X)$, let $u(t)$ be the first component of $\mathscr{S}(t)(0, f)$. Then $\tilde{u}(s)$ is the first component of the pair $(y, g) = (sI - \mathscr{A})^{-1}(0, f)$. This means $sy - Ay - g(0) = 0$, $sg - D_u g - By = f$, and therefore $g = (sI - D_u)^{-1}(By + f)$. As above,

$$g(u) = \int_0^\infty e^{-ts}[B(t+u)y + f(t+u)]\,dt,$$

and $g(0) = \tilde{\mathscr{B}}(s)y + \tilde{f}(s)$. The first equation yields $sy - Ay - \tilde{\mathscr{B}}(s)y = \tilde{f}(s)$, whence $y = \tilde{u}(s) = [sI - A - \tilde{\mathscr{B}}(s)]^{-1}\tilde{f}(s) = \tilde{\mathscr{R}}(s) \cdot \tilde{f}(s)$. As we know, the last equation is equivalent to $u = R * f$, or

$$u(t) = \int_0^t R(t-u)f(u)\,du.$$

This ends the proof of Theorem 5.3.2.

Remark 1. Since strongly continuous semigroups (see Pazy [1]) always produce exponentially bounded resolvent operators $R(t)$, $t \in R_+$, any resolvent which is not in this category cannot be obtained by the procedure indicated in Theorem 5.3.2. In fact, there are exponentially unbounded resolvent operators, as the following example illustrates.

Let $X = \ell^2(\mathscr{C})$, and define A by $Ax_n = inx_n$, $n = 1, 2, \ldots$, with domain $D_A = Y = \{(x_n) = x \in X, \sum_{n=1}^\infty n^2 x_n^2 < \infty\}$. The operator $B(t)$ is defined by $B(t)x_n = b_n(t)x_n$, $n = 1, 2, \ldots$, with

$$b_n(t) = \begin{cases} 0 & \text{when } t < k_n = (lnn)^{1/2} \\ ne^{int} & \text{when } t \geq k_n. \end{cases}$$

For each fixed $t \geq 0$, only finitely many $b_n(t)$ are not identically zero. Hence, for each $t \in R_+$, $B(t)$ is bounded from X into X. Since each $b_n(t)$ is piecewise continuous on R_+, one can easily see that general hypotheses (c) and (d) satisfied for any $\omega > 0$ (because $|B(t)| \leq 1$, as an operator from Y into X).

For the resolvent operator $R(t)$, let us try $R(t)x_n = e^{int}r_n(t)x_n$, $n \geq 1$. The following relationship is easily obtained from equation (5.3.6) or (5.3.7):

$$r_n'(t) = \begin{cases} 0 & \text{when } t < k_n, \\ \int_0^{t-k_n} nr_n(s)\,ds & \text{when } t \geq k_n. \end{cases}$$

We see that each $r_n(t)$ is positive and nondecreasing.

The following estimates are obtainable from the equations for $r_n(t)$:

$$r_n(t + 3k_n) = r_n(t) + \int_t^{t+3k_n} \int_0^{s-k_n} nr_n(u)\, du\, ds$$

$$\geq \int_{t+2k_n}^{t+3k_n} \int_{s-2k_n}^{s-k_n} nr_n(u)\, du\, ds \geq nk_n^2 r_n(t).$$

Consequently, $r_n(t + 3k_n)/r_n(t) \geq nk_n^2$, $n \geq 1$ and, since $ln(nk_n^2)/3k_n \to \infty$ as $n \to \infty$, we obtain that $R(t)$ grows faster than exponentially.

Remark 2. The requirement of Theorem 5.3.2 that the operator $\mathscr{A}$ generates a (strongly continuous) semigroup can be illustrated in various ways, under different types of assumptions. Such an illustration is provided in the following result.

Theorem 5.3.3. *Consider the operator $\mathscr{A}$ of Theorem 5.3.2, under the general assumptions* (a), (b), (c), (d). *Moreover, assume that the map $t \to e^{-\omega t}B(t)x$ is of bounded variation on R_+, for any $x \in Y$. Then, there exists a resolvent operator for the equation* (5.3.1).

Remark. Before we can provide the proof of Theorem 5.3.3, a few concepts are necessary. We will present them, in order to facilitate the reader's approach to the proof of Theorem 5.3.3.

First, the following definition: if $T(t)$, $t \in R_+$, is a linear semigroup of bounded operators on the Banach space E, then the *Favard class* associated with $T(t)$ is the set

$$\left\{ x; x \in E, \limsup_{t \downarrow 0} t^{-1}|T(t)x - x| < \infty \right\}.$$

Second, a perturbation result for semigroups of linear operators on E is needed. This result can be stated as follows (see Desch and Schappacher [2]): Let E be a Banach space, and $A: D_A \subset E \to E$ be the infinitesimal generator of a (strongly continuous) semigroup $T(t)$, $t \in R_+$. Let $B: D_A \to E$ be continuous with respect to the graph norm of A on D_A, such that the range of B belongs to the Favard class of $T(t)$. Then $A + B$ generates a semigroup on E.

Proof of Theorem 5.3.3. In view of Theorem 5.3.2, we need to prove that the operator $\mathscr{A}$ given by

$$\mathscr{A}(x, g) = (Ax + \delta g, Bx + D_u g)$$

generates a semigroup on $E = X \times e^{\omega t}L^1(R_+, X)$.

We construct first the unperturbed semigroup $\mathscr{T}(t)$, $t \in R_+$, by letting

$$\mathcal{T}(t)(x,g) = \left(S(t)x + \int_0^t S(t-s)g(s)\,ds, T(t)g \right), \qquad (5.3.23)$$

where $S(t)$, $t \in R_+$, is the semigroup generated by the operator A on X. It is easy to derive from (5.3.23) that $\mathcal{T}(t)$, $t \in R_+$, is a strongly continuous family of bounded operators on E. Indeed, one has according to the definition of $\mathcal{T}(t)$:

$$\mathcal{T}(t)\mathcal{T}(u)(x,g)$$

$$= \left(S(t)\left[S(u)x + \int_0^u S(u-v)g(v)\,dv \right] + \int_0^t S(t-v)(T(u)g)(v)\,dv, T(t)T(u)g \right)$$

$$= \left(S(t+u)x + \int_0^u S(t+u-v)g(v)\,dv + \int_0^t (S(t-v)g(u+v)\,dv, T(t+u)g \right)$$

$$= \mathcal{T}(t+u)(x,g).$$

Let $\mathcal{G}$ be the infinitesimal generator of the semigroup $\mathcal{T}(t)$, $t \in R_+$. Then $(y,f) = \mathcal{G}(x,g)$ is equivalent to the relationships

$$y = \lim_{t\downarrow 0} t^{-1}\left[S(t)x - x + \int_0^t S(t-u)g(u)\,du \right],$$

and

$$f = \lim_{t\downarrow 0} t^{-1}[T(t)g - g] = D_u g.$$

The last formula implies the continuity of g which, taken into consideration in the formula for y, implies

$$y = \lim_{t\downarrow 0} t^{-1}[S(t)x - x] + S(0)g(0) = Ax + \delta g.$$

This means that $\mathcal{G}(x,g) = (Ax + \delta g, D_u g)$, the domain of $\mathcal{G}$ being $Y \times D_{D_u}$. But $\mathcal{A} = \mathcal{G} + B$, with $\mathcal{B}$ given by $\mathcal{B}(x,g) = (0, Bx)$, the range of $\mathcal{B}$ consisting of those elements of the form $(0,f)$, where f is such that $e^{-\omega t}f(t)$ is of bounded variation on R.

In order to apply the perturbation result stated above, we need only prove that these elements belong to the Favard class for $\mathcal{T}(t)$, $t \in R_+$. To estimate $t^{-1}|\mathcal{T}(t)(0,f) - (0,f)|$, we have to estimate $t^{-1}|\int_0^t S(t-u)f(u)\,du|$ and $t^{-1}|T(t)f - f|$. The first term is bounded by $\sup|S(u)| \cdot \sup|f(u)|$, $0 \le u \le t$, which is finite because f is of bounded variation on any $[0,t]$, $t > 0$. The second term leads to

$$t^{-1}\int_0^\infty e^{-\omega u}|f(t+u) - f(u)|\,du \le t^{-1}\int_0^\infty e^{-\omega(t+u)}(e^{\omega t} - 1)|f(t+u)|\,du$$

$$+ t^{-1}\int_0^\infty |e^{-\omega(t+u)}f(t+u) - e^{-\omega u}f(u)|\,du.$$

When $t \downarrow 0$, the first term on the right hand side of the last inequality is bounded by

$$2\omega \int_0^\infty e^{-\omega u} |f(u)|\,du,$$

while the second term (see Brézis [1]) tends to the variation of $e^{-\omega t}f(t)$ on R_+, as $t \downarrow 0$. Consequently,

$$\limsup_{t \downarrow 0} t^{-1} |\mathcal{T}(t)(0, f) - (0, f)|$$

is finite, and this means that the range of $\mathcal{B}$ belongs to the Favard class of the semigroup $\mathcal{T}(t)$.

Theorem 5.3.3 is thereby proven.

Remark. If $S(t)$, $t \in R_+$, is a strongly continuous semigroup on X, and A is its infinitesimal generator, then (see Desch and Schappacher [1]) the Favard class of $S(t)$ is precisely $D_{A^{\cdots}}$.

Sometimes, the definition of the resolvent operator is modified slightly so that its existence can be shown under milder conditions. Let us briefly describe here the approach used by Prüss [1], [2] in relation to the construction of the resolvent operator for the equations of the form

$$x'(t) = Ax(t) + \int_0^t [dB(u)]x(t - u) + f(t), \tag{5.3.24}$$

under the initial condition (5.3.2).

The following hypotheses are assumed in relation to A and $B(t)$, $t \in R_+$.

(a') A is the generator of an analytic (strongly continuous) semigroup $T(t)$ in X.

(b') $B(t)$, $t \in R_+$, is a family of closable linear operators in X, such that $D_{B(t)} \supset D_A$ for all $t \in R_+$, $B(0) = 0$, and $B(t)$ is continuous and of bounded variation on any finite interval $[0, T] \subset R_+$ (as a map from R_+ into the space of bounded linear operators from Y into X).

The definition of the resolvent operator is modified as follows: the family of bounded linear operators $R(t)$, $t \in R_+$, on X is called a resolvent operator for (5.3.24) – or its homogeneous counterpart – if conditions R_1 and R_2 are verified, $R(t) D_A \subset D_A$ for all $t \geq 0$, $AR(t)x$ is continuous and $R(t)x$ is locally Lipschitz and differentiable a.e. on R_+, for any $x \in D_A$; moreover, for any $x \in D_A$, and a.e. on R_+, the resolvent equations

$$R'(t)x = AR(t)x + \int_0^t [dB(u)]R(t - u)x,$$

and

$$R'(t)x = R(t)Ax + \int_0^t R(t-u)\,dB(u)x,$$

are satisfied.

It is worth mentioning that the second equation for $R(t)$, which is similar to (5.3.7), involves the convolution integral which might not necessarily make sense as a Riemann–Stieltges integral. However, in integrated form, the second equation is

$$R(t)x - x = U(t)Ax + \int_0^t U(t-u)\,dB(u)x, \qquad (5.3.25)$$

with $U(t) = \int_0^t R(u)\,du$, and this always makes sense under our hypothesis. In what follows, (5.3.25) will be considered as a variant of the second resolvent equation.

With this modified version of the resolvent operator, the following result holds.

Theorem 5.3.4. *Consider the integrodifferential equation* (5.3.24), *under the assumptions* (a'), (b') *formulated above. Then, there exists a unique resolvent operator* $R(t)$, $t \in R_+$, *which allows us to represent the solution of* (5.3.24), (5.3.2) *by means of the formula* (5.3.4), *for any* $f(t)$ *locally integrable from* R_+ *into* X. *Moreover, the map* $t \to R(t)$ *is continuous as a map from* $(0, \infty)$ *into the space of bounded linear operators on* Y, *or on* X.

For the proof of Theorem 5.3.4, which is far more technical than the proof of Theorem 5.3.3, we refer the interested reader to the references Prüss [1], [2]. Actually, a somewhat more general result is given, together with applications to partial integrodifferential equations of the form

$$u_t(t,x) = \Delta u(t,x) + \int_0^t db(s)\Delta u(t-s,x) + f(t,x), \qquad (5.3.26)$$

with initial and boundary value conditions.

In concluding this section, we want to point out that integrodifferential equations such as those discussed above are usually called of *parabolic* type. This terminology is supported by the fact that the results obtained for these abstract equations are applicable to partial integrodifferential equations such as (5.3.26).

5.4   A nonlinear Volterra equation and the associated semigroup of operators

Let E be a real Banach space, and consider the nonlinear integral equation of Volterra type

$$x(t) = f(t) + \int_0^t g(t - s, x(s)) \, ds, \quad t \in R_+, \qquad (5.4.1)$$

where the unknown function $x: R_+ \to E$, the given term $f: R_+ \to E$, as well as the integrand $g: R_+ \times E \to E$, are assumed to satisfy certain conditions (to be specified subsequently) which will assure that it is possible to treat (5.4.1) as a nonlinear semigroup of operators in a convenient product space.

In order to associate with (5.4.1) a nonlinear semigroup of operators, we shall transform this integral equation into a functional equation which can be written in the form

$$x(t) = h + G(x_t) - G(\phi), \quad t \in R_+, \qquad (5.4.2)$$

under the initial conditions

$$x(0) = h \in E, \quad x_0 = \phi \in S, \qquad (5.4.3)$$

where G stands for a nonlinear operator acting on a convenient function space S. As usual, x_t represents the restriction of $x: (-\infty, T) \to E$ to $(-\infty, t)$, $t < T$, and our hypotheses will be formulated in such a way that $x_t \in S$, for any $t \in R_+$.

If we proceed formally, then G is defined by

$$G(\phi) = \int_{-\infty}^0 g(-s, \phi(s)) \, ds, \quad \phi \in S, \qquad (5.4.4)$$

which substituted in (5.4.2) leads to

$$x(t) = h + \int_{-\infty}^0 g(-s, x_t(s)) \, ds - \int_{-\infty}^0 g(-s, \phi(s)) \, ds, \qquad (5.4.5)$$

or

$$x(t) = h + \int_{-\infty}^0 g(-s, x(t + s)) \, ds - \int_{-\infty}^0 g(-s, \phi(s)) \, ds. \qquad (5.4.6)$$

If in the second integral above we make the substitution $t + s = u$, then we obtain from (5.4.5) the equation (5.4.1), in which

$$f(t) = h - \int_{-\infty}^0 g(-s, \phi(s)) \, ds + \int_{-\infty}^0 g(t - s, \phi(s)) \, ds. \qquad (5.4.7)$$

Conversely, if $f(t)$ given by (5.4.7) is substituted into (5.4.1), then after elementary transformations one obtains (5.4.2).

We must now provide the conditions that will legitimate the above considerations. First, we need to choose a convenient function space S, which will serve as the initial function space for the problem (5.4.2), (5.4.3).

Let us choose $S = S((-\infty, 0), E)$ to be the weighted L^p-space $L_\rho^p(R_-, E)$, $1 \le p < \infty$, with $\rho: R_- \to (0, \infty)$ nondecreasing (see C. Corduneanu [4] for properties of these spaces, when $E = R^n$). More precisely, $S = L_\rho^p$ consists of all

Bochner measurable maps from R_- into E, such that

$$\int_{-\infty}^{0} |x(t)|^p \rho(t)\,dt < +\infty, \tag{5.4.8}$$

the norm being defined as usual by

$$|x|_{p,\rho} = \left\{ \int_{-\infty}^{0} |x(t)|^p \rho(t)\,dt \right\}^{1/p}. \tag{5.4.9}$$

This notation is somewhat different from the one used in Section 2.3.

Next, we must make precise the conditions on g, such that the operator G which is formally given by (5.4.4) makes sense as an operator from L_ρ^p into E. The following hypotheses will be accepted in what follows.

(a) $g: R_+ \times E \to E$ is measurable in t, $t \in R_+$, for every fixed $x \in E$, and $g(t, 0) = 0$ for $t \in R_+$.

(b) g satisfies the generalized Lipschitz condition

$$|g(t, x) - g(t, y)| \le a(t)|x - y|, \tag{5.4.10}$$

with $a: R_+ \to R$ measurable, and such that

$$a(-t)[\rho(t)]^{1/p} \in L^q(R_-, R) \tag{5.4.11}$$

where $p \ge 1(p^{-1} + q^{-1} = 1)$. As usual, $p = 1$ means $q = \infty$. The L^q norm of the left hand side in (5.4.11) will be denoted by β, $\beta > 0$.

Lemma 5.4.1. *Under assumptions* (a) *and* (b), *the operator G defined by* (5.4.4) *maps the space $L_\rho^p(R_-, E)$ into E, and is Lipschitz continuous with constant β.*

Proof. Let $\phi, \psi \in L_\rho^p$. Then

$$|G(\phi) - G(\psi)| \le \int_{-\infty}^{0} a(-s)|\phi(s) - \psi(s)|\,ds. \tag{5.4.12}$$

For $p > 1$, (5.4.12) yields

$$|G(\phi) - G(\psi)| \le \left(\int_{-\infty}^{0} \{a(-s)[\rho(s)]^{-1/p}\}^q\,ds \right)^{1/q} \left(\int_{-\infty}^{0} |\phi(s) - \psi(s)|^p \rho(s)\,ds \right)^{1/p}$$

$$\le \beta|\phi - \psi|_{L_\rho^p}, \tag{5.4.13}$$

while for $p = 1$ we obtain directly from (5.4.12)

$$|G(\phi) - G(\psi)| \le \beta \int_{-\infty}^{0} |\phi(s) - \psi(s)|\rho(s)\,ds = \beta|\phi - \psi|_{L_\rho^1}. \tag{5.4.14}$$

(5.4.13) and (5.4.14) prove Lemma 5.4.1, if we notice that $G(0) = 0$, and choose $\psi = 0$ in both inequalities.

A better result than in Lemma 5.4.1 can be obtain if we strengthen the hypotheses (a) and (b). More precisely, the new hypotheses can be formulated as follows.

(a') $g(t, x)$ is Fréchet differentiable at each $x \in E$ for every fixed $t \in R_+$, and the derivative $g_2(t, x)$ is measurable in t, $t \in R_+$, for each fixed $x \in E$.

(b') There exists a constant $C > 0$, such that

$$|g(t, x + y) - g(t, x) - g_2(t, x)y| \leq C\rho(t)|y|^p, \qquad (5.4.15)$$

for any $x, y \in E$.

(c') There exists a measurable function $c: R_+ \to R_+$, such that

$$c(-t)[\rho(t)]^{-1/p} \in L^q(R_-, R), \qquad (5.4.16)$$

and for all $t \in R_+$, and $x, y, \bar{x}, \bar{y} \in E$,

$$|g_2(t, x)y - g_2(t, \bar{x})\bar{y}| \leq c(-t)(|x - \bar{x}| + |y - \bar{y}|), \qquad (5.4.17)$$

$$|g_2(t, x)y| \leq c(t)|y|. \qquad (5.4.18)$$

Lemma 5.4.2. *Under assumptions* (a), (b), (a'), (b'), (c'), *and* $p > 1$, *the operator* G *defined by* (5.4.4) *is Fréchet differentiable at each* $\phi \in L_\rho^p$, *and the following inequalities hold:*

$$|G'(\phi)\psi - G'(\bar{\phi})\bar{\psi}| \leq C(|\phi - \bar{\phi}| + |\psi - \bar{\psi}|), \qquad (5.4.19)$$

$$|G'(\phi)\psi| \leq C(1 + |\phi|)|\psi|. \qquad (5.4.20)$$

Proof. If $\phi, \psi \in L_\rho^p$, then (5.4.18) yields $g_2(-t, \phi(t))\psi(t) \in L^1(R_-, E)$, because

$$\int_{-\infty}^0 |g_2(-t, \phi(t))\psi(t)| \, dt \leq \int_{-\infty}^0 c(t)|\psi(t)| \, dt$$

$$\leq \left(\int_{-\infty}^0 \{c(t)[\rho(t)]^{-1/p}\}^q \, dt\right)^{1/q} \left(\int_{-\infty}^0 |\psi(t)|^p \rho(t) \, dt\right)^{1/p}.$$

Now we prove that

$$G'(\phi)\psi = \int_{-\infty}^0 g_2(-t, \phi(t))\psi(t) \, dt. \qquad (5.4.21)$$

Indeed, from (b') we have

$$|\psi|^{-1} \left| G(\phi + \psi) - G(\phi) - \int_{-\infty}^0 g_2(-t, \phi(t))\psi(t) \, dt \right|$$

$$\leq |\psi|^{-1} \int_{-\infty}^0 |g(-t, \phi(t) + \psi(t)) - g(-t, \phi(t)) - g_2(-t, \phi(t))\psi(t)| \, dt$$

$$\leq |\psi|^{-1} C \int_{-\infty}^0 |\psi(t)|^p \rho(t) \, dt \leq C|\psi|^{p-1},$$

and since $|\psi|^{p-1} \to 0$ as $|\psi| \to 0$, we conclude that $G(\phi)$ is Fréchet differentiable at any $\phi \in L_p^p$.

Furthermore, in order to derive (5.4.19) and (5.4.20), we must rely on (5.4.17) and (5.4.18). For instance, using (5.4.21) and (5.4.18) we obtain $|G'(\phi)\psi| \leq C_1 |\psi|$, with

$$C_1 = \left(\int_{-\infty}^{0} \{ c(-t) [\rho(t)]^{-1/p} \}^q \, dt \right)^{1/q}.$$

The last inequality obtained above is even stronger than (5.4.20).

This ends the proof of Lemma 5.4.2.

We will concentrate now on equation (5.4.2), under initial conditions (5.4.3), with $S = L_\rho^p(R_-, E)$, with the aim of obtaining an existence result. This existence result will enable us to define the semigroup describing the same dynamics as the Volterra equation (5.4.1).

First, let us consider the produce space $X = L_\rho^p \times E$, with the norm

$$|\{\phi, h\}| = \left(\int_{-\infty}^{0} |\phi(t)|^p \rho(t) \, dt + |h|^p \right)^{1/p}. \tag{5.4.22}$$

The projection operators π_1 and π_2 are defined by $\pi_1\{\phi, h\} = \phi$, $\pi_2\{\phi, h\} = h$, respectively.

The following existence result can be obtained for (5.4.2), (5.4.3).

Lemma 5.4.3. *Under conditions* (a), (b), *and with G defined by* (5.4.4), *the problem* (5.4.2), (5.4.3) *admits a unique solution* $x: R \to E$, *for any* $\{\phi, h\} \in X$. *Moreover, this solution is continuous on* R_+.

Proof. Let us choose $t_0 > 0$ such that $\beta(t_0 \rho(0))^{1/p} < 1$. For given $\{\phi, h\} \in X$ define

$$M = \{u \mid u; (-\infty, t_0] \to E, u_0 = \phi, u(0) = h, u \text{ restricted to } [0, t_0] \text{ is continuous}\},$$

$$\rho_M(u, v) = \sup |u(t) - v(t)|, \quad t \in [0, t_0],$$

for all $u, v \in M$. It is easy to see that (M, ρ_M) is a complete metric space (as mentioned above, ϕ and h remain fixed). Let us now define a map $K: M \to M$, by setting for $t \in [0, t_0]$

$$(Ku)_0 = \phi, \quad (Ku)(t) = h + G(u_t) - G(\phi). \tag{5.4.23}$$

Since G is Lipschitz continuous and the map $t \to u_t$ is continuous from $[0, t_0]$ into L_ρ^p, we see that $(Ku)(t)$ is a continuous function from $[0, t_0]$ into E, and $(Ku)(0) = h$. Therefore, K maps M into M. We obtain for arbitrary $u, v \in M$,

$$\rho_M(Ku, Kv) = \sup_{0 \le t \le t_0} |G(u_t) - G(v_t)|$$

$$\le \beta \sup_{0 \le t \le t_0} |u_t - v_t| = \beta \sup_{0 \le t \le t_0} \left(\int_{-\infty}^0 |u(t+s) - v(t+s)|^p \rho(s) \, ds \right)^{1/p}$$

$$\le \beta [\rho(0)]^{1/p} \sup_{0 \le t \le t_0} \left(\int_0^t |u(s) - v(s)|^p \, ds \right)^{1/p}$$

$$\le \beta [t_0 \rho(0)]^{1/p} \sup_{0 \le t \le t_0} |u(t) - v(t)| = \beta [t_0 \rho(0)]^{1/p} \rho_M(u, v).$$

This shows that K is a contraction on M, with respect to the distance ρ_M, which implies the existence of a unique fixed element (in M) for K. From (5.4.23) one reads immediately that the fixed element is a solution of the equation (5.4.2), with initial conditions (5.4.3).

Consequently, for the problem (5.4.2), (5.4.3) we have a unique continuous solution on $(-\infty, t_0]$. But t_0 does not depend on $\{\phi, h\}$, and the only restriction is $\beta(t_0[\rho(0)])^{1/p} < 1$. Therefore, if we repeat the argument above in relation to the equation (equivalent to (5.4.3))

$$y(t) = \tilde{h} + G(y_t) - G(x_{t_0}), \quad 0 \le t \le t_0,$$

with initial data

$$y_0 = x_{t_0}, \quad y(0) = \tilde{h} = h + G(x_{t_0}) - G(\phi),$$

and then define $x(t_0 + t) = y(t), 0 \le t \le t_0$, we obtain for (5.4.2), (5.4.3) a solution $x(t)$ defined on $(-\infty, 2t_0]$. The procedure can be indefinitely continued, which leads to the existence of the solution to (5.4.2), (5.4.3) on R. The continuity of this solution on R_+ is obvious from the fact that it is continuous on each interval $[kt_0, (k+1)t_0]$, $k = 0, 1, 2, \ldots$, while according to the continuation procedure it has the same value from the left and from the right, at each point kt_0. Lemma 5.4.3 is thereby proven.

Based on the existence of the solution to equation (5.4.2), with initial conditions (5.4.3), we can now construct a strongly continuous semigroup of nonlinear operators in X, according to the following scheme.

For each $\{\phi, h\} \in X$, let $x(\phi, h)(t)$ be the unique solution (as guaranteed by Lemma 5.4.3) of (5.4.2), (5.4.3). Define now the family of operators $T(t), t \in R_+$, in X, by letting

$$T(t)\{\phi, h\} = \{x_t(\phi, h), x(\phi, h)(t)\}. \tag{5.4.24}$$

Theorem 5.4.4. *The family $T(t), t \in R_+$, of operators in X, defined by (5.4.24), satisfies the following properties:*

(1) $T(0) = I$;

(2) $T(t)\{\phi, h\}$ *is continuous in t, for each fixed $\{\phi, h\} \in X$;*

(3) $T(t + s) = T(t)T(s)$, *for all* $t, s \in R_+$;

(4) *there exists* $M \geq 1$, *and* ω *real, such that*

$$|T(t)\{\phi, h\} - T(t)\{\psi, k\}| \leq Me^{\omega t}|\{\phi, h\} - \{\psi, k\}|, \qquad (5.4.25)$$

for all $t \in R_+$, *and* $\{\phi, h\}, \{\psi, k\} \in X$.

Proof. Since (1) is immediate from the definition, let us consider property (2). We notice that each component of $T(t)\{\phi, h\}$ is continuous. Indeed, the translation is continuous in L^p_ρ while $x(\phi, h)(t)$ is continuous on R_+. In order to prove (3), let $\{\phi, h\} \in X$ and $s \geq 0$. Define $y: R \to E$ by letting $y(u) = x(\phi, h)(s + u)$, for $u \in R$. For $t \geq 0$ and $u \leq 0$ we have

$$y_t(u) = y(t + u) = x(\phi, h)(t + s + u) = x_{t+s}(\phi, h)(u).$$

Then, if $t \geq 0$, we have

$$y(t) = x(\phi, h)(s + t) = h + G(x_{t+s}(\phi, h)) - G(\phi) = h + G(y_t) - G(\phi).$$

Since uniqueness of solution holds for (5.4.2), (5.4.3), we must have

$$y(t) = x(\pi_1 T(s)\{\phi, h\}, \pi_2 T(s)\{\phi, h\})(t), \quad \forall t \in R_+.$$

Thus, for $t \in R_+$, we have

$$y(t) = \pi_2 T(t + s)\{\phi, h\} = \pi_2 T(t)[\pi_1 T(s)\{\phi, h\}, \pi_2 T(s)\{\phi, h\}],$$

$$y_t = \pi_1 T(t + s)\{\phi, h\} = \pi_1 T(t)[\pi_1 T(s)\{\phi, h\}, \pi_2 T(s)\{\phi, h\}],$$

which implies $T(t + s)\{\phi, h\} = T(t)T(s)\{\phi, h\}$. Finally, in order to obtain (5.4.25) we will choose t_0 such that $t_0 < [\rho(0)]^{-1}\beta^{-p}$, and notice that for $\{\phi, h\}, \{\psi, k\} \in X$, $t \in [0, t_0]$, we have

$$|x(\phi, h)(t) - x(\psi, k)(t)|^p$$

$$= |h + G(x_t(\phi, h)) - G(\phi) - k - G(x_t(\psi, k)) + G(\psi)|^p$$

$$\leq 3^{p/q}(|h - k|^p + \beta^p|\phi - \psi|^p + \beta^p|x_t(\phi, h) - x_t(\psi, k)|^p)$$

$$= 3^{p/q}\left(|h - k|^p + \beta^p|\phi - \psi|^p + \beta^p \int_{-\infty}^0 |x(\phi, h)(t + u)\right.$$

$$\left. - x(\psi, k)(t + u)|^p \rho(u)\,du\right)$$

$$\leq 3^{p/q}\left(|h - k|^p + \beta^p|\phi - \psi|^p + \beta^p \int_{-\infty}^0 |\phi(u) - \psi(u)|^p \rho(u)\,du\right.$$

$$\left. + \beta^p\rho(0) \int_0^t |x(\phi, h)(u) - x(\psi, k)(u)|^p\,du\right)$$

$$\leq 3^{p/q}\left(|h - k|^p + 2\beta^p|\phi - \psi|^p + \beta^p\rho(0)t_0 \sup_{0 \leq u \leq t_0} |x(\phi, h)(u) - x(\psi, k)(u)|^p\right).$$

Hence, for $t \in [0, t_0]$, the above inequality implies

$$|x(\phi, h)(t) - x(\psi, k)(t)|^p \le M_1 |\{\phi, h\} - \{\psi, k\}|^p, \tag{5.4.26}$$

with $M_1 = 3^{p/q}(1 + 2\beta^p)/(1 - \beta^p \rho(0)t_0)$. But (5.4.26) actually means

$$
\begin{aligned}
|x_t(\phi, h) - x_t(\psi, k)|^p &= \int_{-\infty}^{0} |x(\phi, h)(t + u) - x(\psi, k)(t + u)|^p \rho(u)\, du \\
&\le \int_{-\infty}^{0} |\phi(u) - \psi(u)|^p \rho(u)\, du \\
&\quad + \rho(0) \int_{0}^{t} |x(\phi, h)(u) - x(\psi, k)(u)|^p\, du \\
&\le M_2 |\{\phi, h\} - \{\psi, k\}|^p,
\end{aligned}
$$

with $M_2 = 1 + \rho(0)t_0 M_1^p$. From (5.4.24) and the inequality just obtained, we have

$$|T(t)\{\phi, h\} - T(t)\{\psi, k\}| \le M |\{\phi, h\} - \{\psi, k\}|, \quad \text{for } t \in [0, t_0], \{\phi, h\}, \{\psi, k\} \in X, \tag{5.4.27}$$

with $M > 0$ independent of t (it does depend on t_0, but, once t_0 is fixed as shown above, M becomes dependent of the semigroup only). Now let us consider an arbitrary $t \in R_+$, which can be represented as $t = nt_0 + s, 0 \le s \le t_0$. One then has

$$
\begin{aligned}
|T(t)\{\phi, h\} - T(t)\{\psi, k\}| &= |T(s)T(nt_0)\{\phi, h\} - T(s)T(nt_0)\{\psi, k\}| \\
&\le MM^n |\{\phi, h\} - \{\psi, k\}| \le MM^{t/t_0} |\{\phi, h\} - \{\psi, k\}| \\
&= Me^{\omega t} |\{\phi, h\} - \{\psi, k\}|,
\end{aligned}
$$

where $\omega = t_0^{-1}\ln M$, i.e., (5.4.25).

This ends the proof of Theorem 5.4.4.

Remark 1. From (5.4.25) one easily obtains the continuous dependence of the solution with respect to initial data $\{\phi, h\} \in X$, on any finite interval.

Remark 2. From the proof, it is clear that $M > 1$ and $\omega > 0$. For nonlinear semigroups, it is particularly desirable to have $M = 1$ (contraction semigroups, when also $\omega \le 0$), the case in which a general semigroup theory is available. See Webb [1], [2] for details on this, under the basic assumption that (5.4.1) can be differentiated and transformed into an integrodifferential equation.

In the remaining part of this section we shall be concerned with the problem of finding the *infinitesimal generator* of the nonlinear semigroup constructed above, as well as with further properties of the semigroup. It is generally known that the infinitesimal generator determines the basic properties of the semigroup.

If one applies the general definition of the infinitesimal generator to the case of the nonlinear semigroup defined by (5.4.24), then it yields

$$A\{\phi, h\} = \lim_{t \downarrow 0} t^{-1}(T(t)\{\phi, h\} - \{\phi, h\}), \tag{5.4.28}$$

with D_A determined by the existence of the limit in (5.4.28).
More precisely, the following result can be obtained.

Theorem 5.4.5. *The infinitesimal generator A of the semigroup $T(t)$, $t \in R_+$, defined by (5.4.24) is completely described by*

$$A\{\phi, h\} = \left\{ \phi', \lim_{t \downarrow 0} t^{-1}[G(x_t(\phi, h)) - G(\phi)] \right\} \tag{5.4.29}$$

with D_A given by

$$D_A = \Bigl\{ (\phi, h) \in X; \phi \in AC_{\mathrm{loc}}(R_-, E), \phi' \in L_p^p, h = \phi(0), $$

$$\text{and } \lim_{t \downarrow 0} t^{-1}[G(x_t(\phi, h)) - G(\phi)] \text{ exists} \Bigr\}, \tag{5.4.30}$$

if G in (5.4.2) is Lipschitz continuous.

Proof. Assume $\{\phi, h\} \in D_A$. We have

$$\left| t^{-1}[T(t)\{\phi, h\} - \{\phi, h\}] - \left\{ \phi', \lim_{t \downarrow 0} t^{-1}[G(x_t(\phi, h)) - G(\phi)] \right\} \right|^p$$

$$= \int_{-\infty}^{0} |t^{-1}[x(\phi, h)(t + u) - \phi(u)] - \phi'(u)|^p \rho(u)\, du$$

$$+ \left| t^{-1}[G(x_t(\phi, h)) - G(\phi)] - \lim_{t \downarrow 0} t^{-1}[G(x_t(\phi, h)) - G(\phi)] \right|^p. \tag{5.4.31}$$

We will show that the first term on the right hand side of (5.4.31) tends to zero as $t \downarrow 0$. Using the property that ϕ is the indefinite integral of its derivative ϕ' (see, for instance, Hille and Phillips [1]), we obtain

$$\int_{-\infty}^{0} |t^{-1}[x(\phi, h)(t + u) - \phi(u)] - \phi'(u)|^p \rho(u)\, du$$

$$= \int_{-\infty}^{-t} \left| t^{-1} \int_{u}^{t+u} [\phi'(s) - \phi'(u)]\, du \right|^p \rho(u)\, du$$

$$+ \int_{-t}^{0} |t^{-1}[x(\phi, h)(t + u) - \phi(u)] - \phi'(u)|^p \rho(u)\, du.$$

Let us now estimate the first integral in the right hand side of (5.4.32), which we will denote by I_1. The second one will be denoted by I_2. As usual, $p^{-1} + q^{-1} = 1$.

We then have

$$I_1 \le t^{-p} \int_{-\infty}^{-t} \left(t^{p/q} \int_u^{t+u} |\phi'(s) - \phi'(u)|^p \, ds \right) \rho(u) \, du$$

$$= t^{-1} \int_{-\infty}^{-t} \left(\int_0^t |\phi'(s+u) - \phi'(u)|^p \, ds \right) \rho(u) \, du$$

$$= t^{-1} \int_0^t \int_{-\infty}^{-t} |\phi'(s+u) - \phi'(u)|^p \rho(u) \, du \, ds$$

$$\le \sup_{0 \le u \le t} \int_{-\infty}^{-t} |\phi'(s+u) - \phi'(u)|^p \rho(u) \, du.$$

The estimate obtained above shows that

$$\lim_{t \downarrow 0} I_1 = 0, \tag{5.4.33}$$

because the translation is continuous in L_q^p. Taking into account $h = \phi(0)$, we have

$$I_2^{1/p} = \left(\int_{-t}^0 |t^{-1}[\phi(0) - G(\phi) + G(x_{t+u}(\phi, h)) - \phi(u)] - \phi'(u)|^p \rho(u) \, du \right)^{1/p}$$

$$\le \left(\int_{-t}^0 |t^{-1}[\phi(0) - \phi(u)]|^p \rho(u) \, du \right)^{1/p}$$

$$+ \left(\int_{-t}^0 |t^{-1}[G(\phi) - G(x_{t+u}(\phi, h))]|^p \rho(u) \, du \right)^{1/p}$$

$$+ \left(\int_{-t}^0 |\phi'(u)|^p \rho(u) \, du \right)^{1/p} = J_1^{1/p} + J_2^{1/p} + J_3^{1/p}.$$

We shall now estimate J_1:

$$J_1 \le t^{-p} \int_{-t}^0 \left((-u)^{p/q} \int_u^0 |\phi'(s)|^p \, ds \right) \rho(u) \, du$$

$$\le t^{-1} \int_{-t}^0 \int_u^0 |\phi'(s)|^p \rho(u) \, ds \, du$$

$$\le (\rho(0)/\rho(-t)) \int_{-t}^0 |\phi'(s)|^p \rho(s) \, ds.$$

It is now obvious that $J_1 \to 0$ as $t \downarrow 0$. For the same reason, $J_3 \to 0$ as $t \downarrow 0$. In J_2, we notice that $\lim t^{-1}[G(\phi) - G(x_{t+u}(\phi, h))]$ does exist as $t \downarrow 0$, and consequently $J_2 \to 0$ as $t \downarrow 0$. Taking all these findings into account, we obtain

$$\lim_{t \downarrow 0} I_2 = 0. \tag{5.4.34}$$

From (5.4.33) and (5.4.34), we see that the pair $\{\phi, h\} \in D_A$, with D_A given by (5.4.30), belongs to the domain of the infinitesimal generator.

Now let us assume that

$$\lim_{t \downarrow 0} t^{-1}[T(t)\{\phi, h\} - \{\phi, h\}] = \{\psi, k\}. \tag{5.4.35}$$

We then have (see again Hille and Phillips [1], p. 535).

$$\lim_{t \downarrow 0} |\pi_1 t^{-1}[T(t)\{\phi, h\} - \{\phi, h\}] - \psi| = 0. \tag{5.4.36}$$

On any interval (a, b), $-\infty < a < b < 0$, and for $t \in [0, -b]$, we have

$$\int_b^{b+t} \phi(u)\,du - \int_a^{a+t} \phi(u)\,du = \int_a^b [\phi(t+u) - \phi(u)]\,du$$

$$= \int_a^b [x(\phi, h)(t+u) - x(\phi, h)(u)]\,du$$

$$= \int_a^b [\pi_1 T(t)\{\phi, h\}(u) - \phi(u)]\,du.$$

If we take (5.4.36) into account, we have

$$\lim_{t \downarrow 0} t^{-1}\left(\int_b^{b+t} \phi(u)\,du - \int_a^{a+t} \phi(u)\,du \right) = \int_a^b \psi(u)\,du, \tag{5.4.37}$$

because

$$\int_a^b |t^{-1}[\pi_1 T(t)\{\phi, h\}(u) - \phi(u)] - \psi(u)|\,du$$

$$\leq \left(\int_a^b \rho(u)^{-q/p}\,du \right)^{1/q} \left(\int_a^b |t^{-1}[\pi_1 T(t)\{\phi, h\}(u) - \phi(u)] - \psi(u)|^p \rho(u)\,du \right)^{1/p},$$

and the second factor tends to zero as $t \downarrow 0$.

As we know, almost everywhere on R_-

$$\lim_{t \downarrow 0} t^{-1} \int_s^{s+t} \phi(u)\,du = \phi(s). \tag{5.4.38}$$

Let $a \in R_-$ be chosen among the s satisfying (5.4.38). Then (5.4.37) yields

$$\phi(b) - \phi(a) = \int_a^b \psi(u)\,du. \tag{5.4.39}$$

(5.4.39) leads to the conclusion that, except for a set of measure zero, $\phi(u)$ can be identified with a locally absolutely continuous function whose derivative is equal to $\psi \in L_\rho^p$ almost everywhere on R_-. Furthermore, as $t \downarrow 0$, $t^{-1}[G(x_t(\phi, h)) - G(\phi)] = t^{-1}[x(\phi, h)(t) - h] = \pi_2 t^{-1}[T(t)\{\phi, h\} - \{\phi, h\}] \to k$ in E. In order to conclude $\{\phi, h\} \in D_A$, it remains to show that $\phi(0) = h$. Assume

$\phi(0) \neq h$. Then

$$\left| t^{-1}[T(t)\{\phi, h\} - \{\phi, h\}] - \left\{ \phi', \lim_{t\downarrow 0} t^{-1}[G(x_t(\phi, h)) - G(\phi)] \right\} \right|$$

$$\geq \left(\int_{-t}^{0} |t^{-1}[\pi_1 T(t)\{\phi, h\}(u) - \phi(u)] - \phi'(u)|^p \rho(u)\, du \right)^{1/p}$$

$$= \left(\int_{-t}^{0} |t^{-1}[x(\phi, h)(t + u) - \phi(u)] - \phi'(u)|^p \rho(u)\, du \right)^{1/p}$$

$$= \left(\int_{-t}^{0} |t^{-1}[h + G(x_{t+u}(\phi, h)) - G(\phi) - \phi(u)] - \phi'(u)|^p \rho(u)\, du \right)^{1/p}$$

$$= \left[\int_{-t}^{0} \left| t^{-1} \left[h - \phi(0) - \int_{0}^{u} \phi'(s)\, ds + G(x_{t+u}(\phi_1 h)) - G(\phi) \right] - \phi'(u) \right|^p \rho(u)\, du \right]^{1/p}$$

$$\geq \left(\int_{-t}^{0} |t^{-1}[h - \phi(0)]|^p \rho(u)\, du \right)^{1/p} - \left(\int_{-t}^{0} \left| t^{-1} \int_{0}^{u} \phi'(s)\, ds \right|^p \rho(u)\, du \right)^{1/p}$$

$$- \left(\int_{-t}^{0} |t^{-1}[G(x_{t+u}(\phi, h)) - G(\phi)]|^p \rho(u)\, du \right)^{1/p} - \left(\int_{-t}^{0} |\phi'(u)|^p \rho(u)\, du \right)^{1/p}$$

$$= K_1^{1/p} - K_2^{1/p} - K_3^{1/p} - K_4^{1/p}.$$

Using the same arguments as in the case of J_1, J_2, J_3, we see easily that $K_i \to 0$ as $t \downarrow 0$, $i = 2, 3, 4$. Since, obviously, K_1 does not approach zero as $t \downarrow 0$, we have from the inequality obtained above that

$$t^{-1}[T(t)\{\phi, h\} - \{\phi, h\}] \nrightarrow \left\{ \phi', \lim_{t\downarrow 0} t^{-1}[G(x_t(\phi, h)) - G(\phi)] \right\}$$

as $t \downarrow 0$, which constitutes a contraction.

Theorem 5.4.5 is thereby proven.

Corollary. If we impose the condition that G is Fréchet differentiable on E, then the infinitesimal generator A of the semigroup (5.4.24) is given by

$$A\{\phi, h\} = \{\phi', G'(\phi)\phi'\}. \tag{5.4.40}$$

The proof follows from

$$\lim_{t\downarrow 0} t^{-1}[G(x_t(\phi, h)) - G(\phi)] = G'(\phi)\phi'. \tag{5.4.41}$$

if we keep in mind (5.4.29). But $\{\phi, h\} \in D_A$ means that

$$\lim_{t\downarrow 0} t^{-1}[x_t(\phi, h) - \phi] = \lim_{t\downarrow 0} t^{-1}[\pi_1 T(t)(\phi, h) - \pi_1\{\phi, h\}] = \pi_1 A\{\phi, h\} = \phi',$$

and if we rely on the chain rule for Fréchet differentials (see Dieudonné [1], p. 151), (5.4.40) is proven.

The next proposition provides further information on the infinitesimal generator A.

Proposition 5.4.6.   *Consider the semigroup $T(t)$, $t \in R_+$, defined by (5.4.24), under the assumptions of Fréchet differentiability for G, and inequalities (5.4.19), (5.4.20). Then the following properties are satisfied:*

(1) *there exists $\lambda_0 > 0$, such that for $\lambda > \lambda_0$ the operator $(\lambda I - A)^{-1}$ exists with domain X, and is Lipschitz continuous, i.e., there exists $C_1 > 0$, with the property*

$$|(\lambda I - A)^{-1}\{\phi, h\} - (\lambda I - A)^{-1}\{\psi, k\}| \leq C_1 \lambda^{-1} |\{\phi, h\} - \{\psi, k\}|, \quad (5.4.42)$$

for any $\{\phi, h\}, \{\psi, k\} \in X$;
(2) *for all $\{\phi, h\} \in X$,*

$$\lim_{\lambda \to \infty} \lambda(\lambda I - A)^{-1}\{\phi, h\} = \{\phi, h\}; \quad (5.4.43)$$

(3) *D_A is dense in X;*
(4) *A is closed.*

Proof. To prove (5.4.42), we consider the equation

$$(\lambda I - A)\{\psi, k\} = \{\phi, h\}, \quad (5.4.44)$$

and see whether it has a unique solution $\{\psi, k\} \in X$. From the Corollary, equation (5.4.44) is equivalent to

$$\lambda\psi - \psi' = \phi, \quad \lambda k - G'(\psi)\psi' = h, \quad \psi(0) = k, \quad (5.4.45)$$

which, in turn, in equivalent to

$$\psi(u) = e^{\lambda u}k + \int_u^0 e^{\lambda(u-s)}\phi(s)\,ds, \quad \lambda k - G'(\psi)\psi' = h. \quad (5.4.46)$$

We now define a linear semigroup of contractions $T_1(t)$, $t \in R_+$, in the space L_ρ^p, by letting $(T_1(t)\phi)(u) = \phi(t + u)$, a.e. on $(-\infty, -t)$, and $(T_1(t)\phi)(u) = 0$ for almost all $u \in (-t, 0)$. The infinitesimal generator of $T_1(t)$, $t \in R_+$, is given by $A_1\phi = \phi'$, with D_A, consisting of all $\phi \in L_\rho^p$, such that ϕ is locally absolutely continuous, $\phi' \in L_\rho^p$, and $\phi(0) = 0$. For $\lambda > 0$ we easily obtain

$$((\lambda I - A_1)^{-1}\phi)(u) = \int_u^0 e^{\lambda(u-s)}\phi(s)\,ds,$$

(see, for instance, Yosida [1], or Dunford and Schwartz [1], on the representation of the resolvent of the infinitesimal generator), which leads to the estimates

$$|(\lambda I - A_1)^{-1}\phi| \leq \lambda^{-1}|\phi|, \quad \forall\phi \in L_\rho^p, \quad (5.4.47)$$

$$\lim_{\lambda \to \infty} (\lambda I - A_1)^{-1}\phi = \phi, \quad \forall\phi \in L_\rho^p. \quad (5.4.48)$$

Let us further consider the map $K: E \to E$, given by

$$K(k) = \lambda^{-1}[G'(e^{\lambda u}k + (\lambda I - A_1)^{-1}\phi)(\lambda e^{\lambda u}k + \lambda(\lambda I - A_1)^{-1}\phi - \phi) + h],$$

for all $k \in E$,

where $u \to e^{\lambda u}k$, from R_- into E, belongs to L^p_ρ. Then, for $k, \bar{k} \in E$ one obtains

$$|K(k) - K(\bar{k})| \le \lambda^{-1}C(1 + \lambda)|e^{\lambda u}(k - \bar{k})| \le \lambda^{-1}C(1 + \lambda)\left[\frac{\rho(0)}{p\lambda}\right]^{1/p}|k - \bar{k}|.$$

(5.4.49)

If λ is sufficiently large, then K is a contraction on E, and therefore it has a unique fixed point in E. If k is the fixed point of K, then $\{\psi, k\}$ defined by (5.4.46) is the unique solution for (5.4.44).

The above considerations lead to the conclusion that $(\lambda I - A)^{-1}$ is defined on the whole X, provided λ is sufficiently large (so that K becomes a contraction on E, as seen from (5.4.49)). In order to prove (5.4.42), let us assume $\{\psi, k\} = (\lambda I - A)^{-1}\{\phi, h\}, \{\bar{\psi}, \bar{k}\} = (\lambda I - A)^{-1}\{\bar{\phi}, \bar{h}\}$. Taking into account (5.4.46), (5.4.47) and the definition of K we obtain

$$|\psi - \bar{\psi}| \le |e^{\lambda u}(k - \bar{k})| + |(\lambda I - A_1)^{-1}(\phi - \bar{\phi})|$$

$$\le \left[\frac{\rho(0)}{p\lambda}\right]^{1/p}|k - \bar{k}| + \lambda^{-1}|\phi - \bar{\phi}|,$$

$$|k - \bar{k}| \le \lambda^{-1}|[G'(e^{\lambda u}k + (\lambda I - A_1)^{-1}\phi)(\lambda e^{\lambda u}k + \lambda(\lambda I - A_1)^{-1}\phi - \phi) + h]$$

$$- [G'(e^{\lambda u}\bar{k} + (\lambda I - A_1)^{-1}\bar{\phi})(\lambda e^{\lambda u}\bar{k} + \lambda(\lambda I - A_1)^{-1}\bar{\phi} - \bar{\phi}) + \bar{h}]|$$

$$\le \lambda^{-1}\left[C(1 + \lambda)\left[\frac{\rho(0)}{p\lambda}\right]^{1/p}|k - \bar{k}| + |\phi - \bar{\phi}| + |h - \bar{h}|\right].$$

By combination of the last two inequalities, we obtain inequality (5.4.42), with a convenient $C_1 > 0$. This shows that (1) holds.

We now have to prove (5.4.43). We rely on (5.4.46), and inequality (5.4.47), from which we obtain

$$|\pi_1(\lambda I - A)^{-1}\{\phi, h\}|$$

$$\le |e^{\lambda u}k| + |(\lambda I - A_1)^{-1}\phi|$$

$$\le \left[\frac{\rho(0)}{p\lambda}\right]^{1/p}|k| + \lambda^{-1}|\phi|$$

$$\le \left[\frac{\rho(0)}{p\lambda}\right]^{1/p}\lambda^{-1}|G'(\pi_1(\lambda I - A)^{-1}\{\phi, h\})(\pi_1\lambda(\lambda I - A)^{-1}\{\phi, h\} - \phi) + h| + \lambda^{-1}|\varphi|$$

$$\le \left[\frac{\rho(0)}{p\lambda}\right]^{1/p}\lambda^{-1}C((1 + \lambda)|\pi_1(\lambda I - A)^{-1}\{\phi, h\}| + |\phi| + |h|) + \lambda^{-1}|\phi|,$$

where the existence of a constant $C_2 > 0$, such that

$$|\pi_1(\lambda I - A)^{-1}\{\phi, h\}| \leq C_2$$

for sufficiently large λ, is now obvious (it depends on $\{\phi, h\}$). Using again (5.4.46) and the last inequality obtained above, we have

$$|\pi_1\lambda(\lambda I - A)^{-1}\{\phi, h\} - \phi|$$

$$\leq |\lambda e^{\lambda u} k| + |\lambda(\lambda I - A_1)^{-1}\phi - \phi|$$

$$\leq \left[\frac{\rho(0)}{p\lambda}\right]^{1/p} C(|\pi_1(\lambda I - A)^{-1}\{\phi, h\}| + |\pi_1\lambda(\lambda I - A)^{-1}\{\phi, h\} - \phi| + |h|)$$

$$+ |\lambda(\lambda I - A_1)^{-1}\phi - \phi|$$

$$\leq \left[\frac{\rho(0)}{p\lambda}\right]^{1/p} C(C_2 + |\pi_1\lambda(\lambda I - A)^{-1}\{\phi, h\} - \phi| + |h|)$$

$$+ |\lambda(\lambda I - A_1)^{-1}\phi - \phi|.$$

Then (5.4.48) and the last inequality imply

$$\lim_{\lambda\to\infty} \pi_1\lambda(\lambda I - A)^{-1}\{\phi, h\} = \phi. \qquad (5.4.50)$$

We also need to prove that

$$\lim_{\lambda\to\infty} \pi_2\lambda(\lambda I - A)^{-1}\{\phi, h\} = h, \qquad (5.4.51)$$

in order to conclude that (5.4.43) holds. Indeed, taking into account (5.4.20), (5.4.46) and the estimate obtained above for $\pi_1(\lambda I - A)^{-1}\{\phi, h\}$, we have

$$|\pi_2\lambda(\lambda I - A)^{-1}\{\phi, h\} - h|$$

$$= |G'(\pi_1(\lambda I - A)^{-1}\{\phi, h\})(\pi_1\lambda(\lambda I - A)^{-1}\{\phi, h\} - \phi)|$$

$$\leq C(1 + |\pi_1(\lambda I - A)^{-1}\{\phi, h\}|)|\pi_1\lambda(\lambda I - A)^{-1}\{\phi, h\} - \phi|$$

$$\leq C(1 + C_2)|\pi_1\lambda(\lambda I - A)^{-1}\{\phi, h\} - \phi|,$$

from which (5.4.51) follows. Property (2) in the Proposition 5.4.6 is thus established.

To obtain property (3), we notice that it is a consequence of property (2), while property (4) of Proposition 5.4.6 is obtained as follows: property (1) shows that $(\lambda I - A)^{-1}$ is continuous (hence closed) for sufficiently large λ. This means that $\lambda I - A$ is also closed, for large λ. But this means that A is closed, which establishes property (4).

Proposition 5.4.6 is thus proven.

The last proposition we shall provide in connection with the semigroup $T(t)$, $t \in R_+$, defined by (5.4.24), relates to the case when the Banach space E is reflexive.

Proposition 5.4.7. *Let E be a reflexive Banach space, and $G: L_\rho^p \to L_\rho^p$ be Lipschitz continuous. Then the semigroup $T(t)$, $t \in R_+$, defined by (5.4.24) satisfies for every $\{\phi, h\} \in D_A$,*

$$T(t)\{\phi, h\} \in D_A \quad \text{a.e. on } R_+. \tag{5.4.52}$$

If, moreover, G is Fréchet differentiable on L_ρ^p, and estimates (5.4.19), (5.4.20) hold, then (5.4.52) is satisfied for all $t \in R_+$.

Proof. To prove the first assertion, let $\{\phi, h\} \in D_A$. There exists a constant $C > 0$, such that $|T(t)\{\phi, h\} - \{\phi, h\}| \le Ct$, as soon as t is small enough ($t > 0$). Using Theorem 5.4.2, we have for $0 \le s \le u$

$$|x(\phi, h)(u) - x(\phi, h)(s)| = |G(x_u(\phi, h)) - G(x_s(\phi, h))|$$

$$\le \beta|\pi_1 T(u)\{\phi, h\} - \pi_1 T(s)(\phi, h)|$$

$$= \beta|\pi_1 T(s)T(u - s)\{\phi, h\} - \pi_1 T(s)\{\phi, h\}|$$

$$\le \beta M e^{\omega s}|T(u - s)\{\phi, h\} - \{\phi, h\}|$$

$$\le \beta M e^{\omega s} C(u - s).$$

This implies that $x(\phi, h)(u)$ is Lipschitz continuous in u, and since E is reflexive, $x(\phi, h)(u)$ is a.e. differentiable. But Lipschitz continuity also implies that $\dot{x}(\phi, h)(u)$ is bounded in u on any finite interval of R_+. Therefore, for $u \ge 0$, $x_u(\phi, h)$ is locally absolutely continuous, and $(d/du)x_t(\phi, h)(u)$ exists a.e. on R_-, and $(d/du)x_t(\phi, h) \in L_\rho^p$. For almost all $t > 0$

$$u^{-1}[G(\pi_1 T(t + u)\{\phi, h\}) - G(\pi_1 T(t)\{\phi, h\})]$$

$$= u^{-1}[\pi_2 T(t + u)\{\phi, h\} - \pi_2 T(t)\{\phi, h\}]$$

$$= u^{-1}[x(\phi, h)(t + u) - x(\phi, h)(t)]$$

converges to $(d/dt)x(\phi, h)(t)$. From Theorem 5.4.2, we conclude that $T(t)\{\phi, h\} \in D_A$ for almost all $t \in R_+$.

With regard to the second assertion of Proposition 5.4.7, let $t > 0$ and, using the first part of the proof, let us consider a sequence $t_n \to t$, with the property that $T(t_n)\{\phi, h\} \in D_A$. Then $T(t_n)\{\phi, h\} \to T(t)\{\phi, h\}$ according to property (2) in Theorem 5.4.4, and $AT(t_n)\{\phi, h\} = \{x_{t_n}(\phi, h)', G'(x_{t_n}(\phi, h))x_{t_n}(\phi, h)'\}$, from the Corollary to Theorem 5.4.4. From the continuity of G', we obtain that $G'(x_{t_n}(\phi, h))x_{t_n}(\phi, h)'$ converges to $G'(x_t(\phi, h))x_t(\phi, h)'$. Hence, $AT(t_n)\{\phi, h\}$ converges in X, and the closedness of A yields $T(t)\{\phi, h\} \in D_A$.

Remark 1. The conditions of Proposition 5.4.7 are implied if we consider the Volterra equation (5.4.1) under the assumptions (a), (b) for Lipschitz continuity of G, and under assumptions (a'), (b'), (c'), $p > 1$, for Fréchet differentiability of G.

Remark 2. Under the stronger assumptions of Proposition 5.4.7, which are implied (as seen above) by the assumptions (a), (b), (a′), (b′), (c′), and $p > 1$, we can state that there is a dense set of pairs $\{\phi, h\}$ in X, $\{\phi, h\} \in D_A$, such that

$$(d/dt)\,T(t)\{\phi, h\} = A\,T(t)\{\phi, h\}, \quad t \in R_+, \ T(0)\{\phi, h\} = \{\phi, h\}.$$

This property means that equation (5.4.1) can be regarded as an abstract ordinary differential equation in X, or that the dynamics described by (5.4.1) are, in fact, the dynamics of a convenient nonlinear semigroup. Notice that (5.4.1) is a 'convolution' type equation or, better, an equation involving a time-invariant operator.

5.5   Global existence for a nonlinear integrodifferential equation in Hilbert space

In this section we shall consider integrodifferential equations of the form

$$x'(t) + \int_0^t a(t - s)G(x(s))\,ds = F(t), \quad t \in R_+, \tag{5.5.1}$$

with $a(t)$ a scalar kernel of positive type, $x, f : R_+ \to H$, a Hilbert space, and G a (nonlinear) monotone operator on H.

The real Hilbert space H contains two real reflexive Banach spaces V and W, both with dense domain and continuous imbedding in H. Since H can be identified with its own dual H', we can write the inclusions

$$V \subset H \subset V', \quad W \subset H \subset W'. \tag{5.5.2}$$

The scalar product in H will be denoted by $\langle \cdot, \cdot \rangle$, while the pairing between V' and V (or W' and W), will be denoted by $(\cdot, \cdot)$.

Let $A: V \to V'$ and $M: W \to W'$ be two mappings, single-valued and everywhere defined, such that the following conditions are verified.

(1) A is linear, continuous, monotone, coercive, and $(Au, v) = (Av, u)$ for all $u, v \in V$.

If we let $\phi(x) = \tfrac{1}{2}(Ax, x)$, then $A = \partial\phi$, the Fréchet differential of ϕ. The condition of coerciveness is then $\phi(x)/|x|_V \to \infty$ as $|x|_V \to \infty$.

(2) M is cyclically maximal monotone and coercive.

Under assumption (2), M can be represented as the subdifferential, $M = \partial\psi$, of a functional $\psi: W \to R$, with ψ lower semicontinuous. Moreover, the coerciveness condition can be written as $\psi(x)/|x|_W \to \infty$ as $|x|_W \to \infty$. For details, see Barbu [4].

(3) M maps bounded sets on W into bounded sets of W'.

(4) If, for an arbitrary $T > 0$, the sequence $\{x_n\} \subset AC_2([0, T], H) \cap L^\infty([0, T], W)$ satisfies $x_n \to x$ strongly in $L^2([0, T], H)$ and weakly-star in $L^\infty([0, T], W)$, then $\{Mx_n\}$ contains a subsequence which converges in the weak-star topology of $L^\infty([0, T], W')$ to Mx.

By $AC_2([0, T], H)$ we denote the space of absolutely continuous functions from $[0, T]$ into H, such that their first derivative is in $L^2([0, T], H)$. The norm is $x \to |x(0)| + |x'|_{L^2}$.

We shall define now two auxiliary operators A_H and M_H in the product $H \times H$, as follows:

$$\left.\begin{array}{l} A_H x = Ax, \quad x \in D_{A_H} = \{x \in V; Ax \in H\}, \\ M_H x = Mx, \quad x \in D_{M_H} = \{x \in W; Mx \in H\}. \end{array}\right\} \tag{5.5.3}$$

If we now consider the functionals $\phi_H(x) = \phi(x)$, $x \in V$, $\phi_H(x) = +\infty$ for $x \in H \setminus V$; and $\psi_H(x) = \psi(x)$, $x \in W$, $\psi_H(x) = +\infty$, $x \in H \setminus W$, then we have

$$A_H = \partial \phi_H, \quad M_H = \partial \psi_H. \tag{5.5.4}$$

We need one more assumption on A and M, in order to be able to formulate an existence result.

(5) For any $\varepsilon > 0$, $(I + \varepsilon M_H)^{-1} x \in V$, and $\phi(I + \varepsilon M_H)^{-1} x) \leq \phi(x)$,

$$\forall x \in V.$$

We can now formulate the following existence result for equation (5.5.1), in which

$$G = A + M.$$

Theorem 5.5.1. *Consider equation* (5.5.1), *with G given as above, and assume that conditions* (1)–(5) *hold. Assume further that a, f, and x_0 satisfy the following conditions:*

(i) $a \in C(R_+, R) \cap C^{(2)}((0, \infty), R)$, $a \not\equiv 0$;

(ii) $(-1)^k a^{(k)}(t) \geq 0$, $t > 0$, $k = 0, 1, 2$;

(iii) $f, f' \in L^1_{loc}(R_+, H)$;

(iv) $x_0 \in V \cap W$.

If, in addition, the injection $V \subset H$ is compact, then equation (5.5.1) *has at least one solution such that $x(0) = x_0$, and*

$$u \in C(R_+, H) \cap L^\infty_{loc}(R_+, V \cap W), \tag{5.5.5}$$

$$u' \in L^2_{loc}(R_+, H) \cap L^\infty_{loc}(R_+, (V \cap W)'), \tag{5.5.6}$$

$$u'' \in L^1_{loc}(R_+, (V \cap W)'). \tag{5.5.7}$$

Before we provide the proof of Theorem 5.5.1, we need to state some auxiliary propositions. Let us start with a remark concerning the existence of the derivative.

Remark.   Equation (5.5.1) is considered in the sense of the theory of distributions, i.e., with a distributional derivative. It is useful to notice that for $u: [0, T] \to X$ (a reflexive Banach space), with u, $u' \in L^1([0, T], X)$, where u' is taken in the distributional sense, one can establish the absolute continuity of u, its almost everywhere differentiability, as well as the fact that u' coincides a.e. on $(0, T)$ with the usual (strong) derivative.

Lemma 5.5.2.   *Under conditions* (i), (ii) *for* $a(t)$, *the convolution operator* $x \to a * x$ *is positive on* $L^2([0, T], \tilde{H})$, *where* $T > 0$ *and* $\tilde{H}$ *is an arbitrary real Hilbert space. In other words, for any* $T > 0$ *and* $x \in L^2([0, T], \tilde{H})$,

$$\int_0^T \left\langle \int_0^t a(t - s)x(s) \, ds, x(t) \right\rangle dt \geq 0. \tag{5.5.8}$$

The proof of Lemma 5.5.2 can be found, for instance, in Barbu [4]. The conclusion is actually stronger than (5.5.8) shows (strong positiveness).

Lemma 5.5.3.   *Let* $u: R_+ \to R_+$ *be continuous, and* $k: R_+ \to R_+$ *be locally integrable. If the inequality*

$$u^2(t) \leq C + 2 \int_0^t k(s)u(s) \, ds \tag{5.5.9}$$

holds on R_+, *then*

$$u(t) \leq \sqrt{C + 2 \int_0^t k(s) \, ds}, \quad t \in R_+. \tag{5.5.10}$$

Proof. Let $U(t) = \sup u(s), 0 \leq s \leq t$. Then (5.5.9) yields

$$U^2(t) - 2 \left(\int_0^t k(s) \, ds \right) U(t) - C \leq 0,$$

which obviously implies

$$U(t) \leq \int_0^t k(s) \, ds + \left\{ \left(\int_0^t k(s) \, ds \right)^2 + C \right\}^{1/2}. \tag{5.5.11}$$

From (5.5.11), (5.5.10) follows directly.

Remark.   The inequality (5.5.9) is an integral inequality of Bihari's type, and can be dealt with as such (see, for instance, C. Corduneanu [6]).

Lemma 5.5.4. *Assume that* $A = \partial\phi$, *where* $\phi\colon H \to (-\infty, +\infty]$ *is a lower semi-continuous proper convex functional, and* H *is a real Hilbert space. Then, the functional* ϕ_λ *defined for* $\lambda > 0$ *by*

$$\phi_\lambda(x) = \inf\left\{\frac{|x - y|^2}{2\lambda} + \phi(y)\colon y \in H\right\}, \quad x \in H, \tag{5.5.12}$$

is convex, Fréchet differentiable, $\partial\phi_\lambda = A_\lambda$ *where*

$$A_\lambda = \lambda^{-1}(I - J_\lambda), \quad J_\lambda = (I + \lambda A)^{-1}, \tag{5.5.13}$$

and A_λ *is Lipschitz continuous on* H. *Moreover, for all* $x \in H$.

$$\lim_{\lambda \to 0} \phi_\lambda(x) = \phi(x), \quad \phi(J_\lambda x) \leq \phi_\lambda(x) \leq \phi(x). \tag{5.5.14}$$

The proof of this result, due to M. G. Crandall and A. Pazy, can be found in Barbu [4] (see p. 58).

We shall now consider an integrodifferential inclusion related to the basic equation (5.5.1), namely

$$x'(t) + m\partial\phi(x(t)) + \int_0^t a(t - s)\partial\phi(x(s))\,ds \ni f(t), \quad \text{a.e. on } R_+, \tag{5.5.15}$$

with $m > 0$, and under initial condition $x(0) = x_0$. The reason we consider the inclusion in (5.5.15), instead of the equality, is motivated by the fact that the subdifferential ∂ is, in general, a multivalued operator. As usual, $\phi\colon H \to (-\infty, +\infty]$ is lower semicontinuous *proper* convex functional.

Proposition 5.5.5. *Assume that the following hypotheses hold with regard to the integrodifferential inclusion* (5.5.15):

  (a) *the injection* $V \subset H$ *is continuous and compact;*
  (b) $D_\phi \subset V$, *and* $\phi(x) + |x|^2 \to \infty$ *as* $|x|_V \to \infty$;
  (c) *the scalar kernel,* a, *satisfies conditions* (i) *and* (ii) *of Theorem* 5.5.1;
  (d) $f \in L^2_{loc}(R_+, H)$.

Then there exists a solution $x\colon R_+ \to H$ *of inclusion* (5.5.15), *satisfying the initial condition* $x(0) = x_0 \in D\phi$, *and such that*

$$x \in C(R_+, H) \cap L^\infty_{loc}(R_+, V), \tag{5.5.16}$$

$$x' \in L^2_{loc}(R_+, H), \tag{5.5.17}$$

$$\partial\phi(x) \in L^2_{loc}(R_+, H). \tag{5.5.18}$$

Proof. Let us denote, as above, $A = \partial\phi$. Without loss of generality, we can assume $\phi(x) \geq 0$. The following auxiliary equations will be considered, for every $\lambda > 0$:

$$x'_\lambda(t) + mA_\lambda x_\lambda(t) + \int_0^t a(t-s)A_\lambda x_\lambda(s)\,ds = f(t), \quad t \in R_+. \qquad (5.5.19)$$

Since A_λ is Lipschitz continuous, and a is bounded on R_+, equation (5.5.19) has a unique solution $x_\lambda(t)$ defined on R_+, with values in V, such that $x(0) = x^0 \in D_\phi$. In Section 3.3, we proved such existence results for abstract Volterra equations in the linear case (but only Lipschitz continuity was really used). On the other hand, if we integrate both sides of (5.5.19) from 0 to $t > 0$, we obtain an integral equation like those discussed in Theorem 5.1.1. It is obvious that $x_\lambda(t)$ satisfies (5.5.16) and (5.5.17).

In order to obtain some useful estimates for $x_\lambda(t)$, we shall multiply (scalarly) both sides of (5.5.19) by $A_\lambda x_\lambda(t)$, and integrate the result on $[0,t]$. If we take into account the positiveness of the convolution operator with kernel $a(t)$ (see Lemma 5.5.2), then we obtain

$$\phi_\lambda(x_\lambda(t)) + m \int_0^t |A_\lambda x_\lambda(s)|^2\,ds \le \int_0^t \langle A_\lambda x_\lambda(s), f(s)\rangle\,ds + \phi_\lambda(x_0). \qquad (5.5.20)$$

Let us now fix a $T > 0$, and consider (5.5.20) on the interval $[0, T]$. It yields

$$m \int_0^T |A_\lambda x_\lambda(s)|^2\,ds \le \left(\int_0^T |A_\lambda x_\lambda(s)|^2\,ds\right)^{1/2} \left(\int_0^T |f(s)|^2\,ds\right)^{1/2} + \phi_\lambda(x_0).$$
$$(5.5.21)$$

From (5.5.21) we have

$$\int_0^T |A_\lambda x_\lambda(s)|^2\,ds \le C_1, \quad \forall \lambda > 0, \qquad (5.5.22)$$

with C_1 a positive constant depending on T, f, and m, but independent of λ (because (5.5.14) takes place).

On the other hand, if we multiply both sides of (5.5.19) by $x'_\lambda(t)$, and integrate from 0 to $T > 0$, we obtain

$$\int_0^T |x'_\lambda(t)|^2\,dt + m\phi_\lambda(x_\lambda(t))$$

$$\le m\phi_\lambda(x_0) + \int_0^T \langle f(t), x'_\lambda(t)\rangle\,dt + \int_0^T \left|\left\langle x'_\lambda(t), \int_0^t a(t-s)A_\lambda x_\lambda(s)\,ds \right\rangle\right| dt.$$
$$(5.5.23)$$

Taking (5.5.22) into account, we easily obtain an inequality of the form

$$\int_0^T \left|\int_0^t a(t-s)A_\lambda x_\lambda(s)\,ds\right|^2 dt \le C_2, \quad \forall \lambda > 0. \qquad (5.5.24)$$

Using (5.5.24), and processing (5.5.23) in the same way as (5.5.21), we obtain an inequality of the form

$$\int_0^T |x_\lambda'(s)|^2 \, ds \leq C_3, \quad \forall \lambda > 0. \tag{5.5.25}$$

Consider now the maps $y_\lambda(t) = J_\lambda x_\lambda(t)$, $\lambda > 0$, and notice that Lemma 5.5.4 yields $\phi(y_\lambda(t)) \leq \phi_\lambda(x_\lambda(t)) \leq C_4$, the existence of C_4 independent of $\lambda > 0$ being the consequence of (5.5.20) and (5.5.22). Since J_λ is a contraction,

$$\int_0^T |y_\lambda'(t)|^2 \, dt \leq \int_0^T |x_\lambda'(t)|^2 \, dt \leq C_3, \quad \forall \lambda > 0. \tag{5.5.26}$$

We obtain the compactness of the set $\{y_\lambda; \lambda > 0\} \subset L^2([0, T], H)$, if we take into account hypotheses (a) and (b). Consequently, we can find sequences $\{\lambda_n\}$, $\lambda_n \downarrow 0$, such that $y_{\lambda_n}(t) \to x(t)$ as $n \to \infty$, strongly in $L^2([0, T], H)$.

In the norm of $L^2([0, T], H)$ we can write

$$|x_{\lambda_n} - x| \leq |x_{\lambda_n} - y_{\lambda_n}| + |y_{\lambda_n} - x|,$$

and, from (5.5.13),

$$|y_{\lambda_n} - x_{\lambda_n}| = \lambda_n |A_{\lambda_n} x_{\lambda_n}|.$$

Taking (5.5.22) into account, we conclude from the inequalities above that

$$x_{\lambda_n} \to x \text{ as } n \to \infty, \text{ strongly in } L^2([0, T], H). \tag{5.5.27}$$

Further, from (5.5.22), (5.5.25) we can derive the existence of a subsequence of $\{\lambda_n\}$, which we agree to denote also by $\{\lambda_n\}$, such that

$$x_{\lambda_n} \to x, \quad \text{strongly in } L^2([0, T], H), \tag{5.5.28}$$

$$x_{\lambda_n}' \to x', \quad \text{weakly in } L^2([0, T], H), \tag{5.5.29}$$

$$A_{\lambda_n} x_n \to Ax, \quad \text{weakly in } L^2([0, T], H), \tag{5.5.30}$$

$$\int_0^t a(t - s) A_{\lambda_n} x_{\lambda_n}(s) \, ds \to \int_0^t a(t - s) Ax(s) \, ds, \text{ weakly in } L^2([0, T], H). \tag{5.5.31}$$

The property of A of being demiclosed (i.e. closed in the weak topology) has been used above (see Rockafellar [1]), together with the compactness criterion given by Aubin [1].

If in equation (5.5.19) we now let $\lambda = \lambda_n$, and take the limit as $n \to \infty$, we obtain inclusion (5.5.15), taking (5.5.28)–(5.5.31) into account. Of course, inclusion (5.5.15) is obtained only on $[0, T]$, and it is satisfied a.e. on this interval. But $T > 0$ can be chosen arbitrarily large, because the solution x_λ of (5.5.19) is (unique and) defined on R_+.

From the above considerations we easily derive properties (5.5.16), (5.5.17) and (5.5.18) for the solution $x(t)$ of (5.5.15).

Proposition 5.5.5 is thus proven.

Proof of Theorem 5.5.1. First, notice that for the operators A_H and M_H given by (5.5.3) we have $\bar{D}_{A_H} = \bar{D}_{M_H} = H$. Then, using condition (5) and a property of maximal monotone operators (see, for instance, Brézis [1], Prop. 2.17), we obtain the maximal monotonicity of $A_H + M_H = G$, which is densely defined in H. Therefore, $G = \partial(\phi_H + \psi_H)$, and

$$\overline{D_{A_H} \cap D_{M_H}} = \overline{V \cap W} = H.$$

Further, we can write

$$V \cap W \subset H \subset (V \cap W)',$$

the norm in the intersection being the sum of the norms in the spaces V and W. Consider now the regularized equation

$$x_\varepsilon'(t) + \varepsilon G x_\varepsilon(t) + \int_0^t a(t - s) G x_\varepsilon(s)\,ds = f(t) \qquad (5.5.32)$$

on the semiaxis R_+, where $\varepsilon > 0$ is an arbitrary 'small' parameter. As usual, we consider (5.5.32) under the initial condition $x_\varepsilon(0) = x_0$. Under our assumptions, Proposition 5.5.5 is applicable, and we can assert the existence of a solution $x_\varepsilon: R_+ \to H$, such that

$$x_\varepsilon(t) \in C(R_+, H) \cap L^\infty_{\mathrm{loc}}(R_+, V \cap W),$$

$$x_\varepsilon'(t) \in L^2_{\mathrm{loc}}(R_+, H), \quad G x_\varepsilon(t) \in L^2_{\mathrm{loc}}(R_+, H).$$

We shall now prove that for any $T > 0$, with $a(T) > 0$, the following properties hold:

$$\{x_\varepsilon; \varepsilon > 0\} \text{ is bounded in } L^\infty([0, T], V \cap W), \qquad (5.5.33)$$

$$\{G x_\varepsilon; \varepsilon > 0\} \text{ is bounded in } L^\infty([0, T], (V \cap W)'), \qquad (5.5.34)$$

$$\left\{\int_0^t G x_\varepsilon(s)\,ds, t \in R_+; \varepsilon > 0\right\} \text{ is bounded in } L^\infty([0, T], H), \qquad (5.5.35)$$

$$\{x_\varepsilon'; \varepsilon > 0\} \text{ and } \{\sqrt{\varepsilon}\ G x_\varepsilon; \varepsilon > 0\} \text{ are bounded in}$$

$$L^2([0, T], H) \cap L^\infty([0, T], (V \cap W)'). \qquad (5.5.36)$$

The proof of above properties will be carried out by using the energy function

$$E_\varepsilon(t) = (\phi + \psi)x_\varepsilon(t) + \frac{1}{2}a(t)\left|\int_0^t G x_\varepsilon(s)\,ds\right|^2$$

$$-\frac{1}{2}\int_0^t a'(t - s)\left|\int_s^t G x_\varepsilon(u)\,du\right|^2 ds, \quad t \in [0, T]. \qquad (5.5.37)$$

In order to estimate the derivative $E_\varepsilon'(t)$, we notice that

$$\frac{d}{dt}(\phi + \psi)x_\varepsilon(t) = \langle Gx_\varepsilon(t), x_\varepsilon'(t) \rangle,$$

a.e. on $[0, T]$, which leads to

$$E_\varepsilon'(t) = \frac{1}{2}a'(t)\left|\int_0^t Gx_\varepsilon(s)\,ds\right|^2$$

$$- \frac{1}{2}\int_0^t a''(t - s)\left|\int_s^t Gx_\varepsilon(u)\,du\right|^2 ds + \langle Gx_\varepsilon(t), f(t) - y_\varepsilon(t) \rangle,$$

also a.e. on $[0, T]$, where

$$y_\varepsilon(t) = f(t) - x_\varepsilon'(t) - \int_0^t a(t - s)Gx_\varepsilon(s)\,ds.$$

But our assumptions state that $a'(t) \le 0$, $a''(t) \ge 0$, while $y_\varepsilon(t) = \varepsilon G(x_\varepsilon(t))$. From the formula above for $E_\varepsilon'(t)$ we obtain

$$E_\varepsilon'(t) \le \langle f(t), Gx_\varepsilon(t) \rangle \quad \text{a.e. on } [0, T]. \tag{5.5.38}$$

Integrating (5.5.38) over $[0, t]$, one has

$$E_\varepsilon(t) \le (\phi + \psi)(x_0) + \langle f(t), F_\varepsilon(t) \rangle - \int_0^t \langle f'(s), F_\varepsilon(s) \rangle\,ds, \quad t \in [0, T] \tag{5.5.39}$$

where we have denoted

$$F_\varepsilon(t) = \int_0^t Gx_\varepsilon(s)\,ds, \quad t \in R_+. \tag{5.5.40}$$

Combining (5.5.37) and (5.5.40) we obtain on $[0, T]$ the inequality

$$(\phi + \psi)x_\varepsilon(t) + \frac{1}{2}a(t)|F_\varepsilon(t)|^2$$

$$\le (\phi + \psi)x_0 + \frac{4}{a(t)}|f(t)|^2 + \frac{a(t)}{4}|F_\varepsilon(t)|^2 + \int_0^t |f'(s)|\,|F_\varepsilon(s)|\,ds. \tag{5.5.41}$$

But condition (1) implies (more precisely, the coerciveness) that $\phi(x)$ is bounded below, while condition (3) implies that $\psi(x)$ is bounded below. Therefore, (5.5.41) implies

$$|F_\varepsilon(t)|^2 \le C + \int_0^t |f'(s)|\,|F_\varepsilon(s)|\,ds, \quad t \in [0, T], \tag{5.5.42}$$

with a convenient constant $C(\ge 0)$. Inequality (5.5.42) is of the form (5.5.9), and Lemma 5.5.3 yields

$$|F_\varepsilon(t)| \le \sqrt{C} + 2\int_0^T |f'(s)|\,ds, \quad t \in [0, T],$$

which by assumption (iii) in Theorem 5.3.1, leads to

$$|F_\varepsilon(t)| \leq C_1, \quad t \in [0, T], \quad \forall \varepsilon > 0. \tag{5.5.43}$$

In other words, property (5.5.35) is now established.

Since (5.5.41) also implies $(\phi + \psi)x_\varepsilon(t) \leq C_2$, $t \in [0, T]$, $\forall \varepsilon > 0$, we obtain (5.5.33).

Property (5.5.34) is a consequence of (5.5.33), and of the boundedness of the operators A and M on $V \cap W$.

It remains to prove the validity of (5.5.36). Taking the scalar product of both sides of (5.5.1) with $Gx_\varepsilon(t)$, and integrating over $[0, T]$, we obtain after using condition (iii):

$$\varepsilon \int_0^T |Gx_\varepsilon(t)|^2 \, dt \leq C_3, \tag{5.5.44}$$

with C_3 independent of $\varepsilon > 0$. The estimate (5.5.44) shows that $\{\sqrt{\varepsilon} \ Gx_\varepsilon; \varepsilon > 0\}$ is bounded in $L^2([0, T], H)$. Combined with (5.5.34), this gives (5.5.36) for $\{\sqrt{\varepsilon} \ Gx_\varepsilon; \varepsilon > 0\}$. Of course, it suffices to restrict ε to $(0, 1]$.

Finally, in order to derive the property (5.5.36) for $\{x_\varepsilon'; \varepsilon > 0\}$, we shall use (5.5.34) and (5.5.35), as well as (5.5.44) and (iii). Then equation (5.5.32) yields the property.

Now, fixing $T > 0$ as indicated above, in order to state properties (5.5.33)–(5.5.36), we again use the compactness criterion of Aubin [1] to derive the existence of a sequence $\{\varepsilon_n\}$, $\varepsilon_n \downarrow 0$ as $n \to \infty$, such that

$$x_{\varepsilon_n} \to x \text{ strongly in } C([0, T], H). \tag{5.5.45}$$

Without loss of generality, we may assume that $\{\varepsilon_n\}$ is such that

$$x_{\varepsilon_n} \to x \text{ weakly-star in } L^\infty([0, T], V \cap W), \tag{5.5.46}$$

and $x_{\varepsilon_n}' \to x'$ weakly in $L^2([0, T), H)$, and weakly-star in $L^\infty([0, T], (V \cap W)')$. The fact that A is linear, and hypothesis (4) of Theorem 5.5.1, yield

$$Gx_{\varepsilon_n} \to Ax + Mx = Gx \text{ weakly-star in } L^\infty([0, T], (V \cap W)').$$

Therefore,

$$F_{\varepsilon_n}(t) \to F(t) = \int_0^t Gx(s) \, ds \text{ weakly-star in } L^\infty([0, T], H),$$

and

$$\varepsilon_n Gx_{\varepsilon_n} \to 0 \text{ strongly in } L^\infty([0, T], (V \cap W)').$$

If one takes $\varepsilon = \varepsilon_n$ in (5.5.32), and also $x_{\varepsilon_n}(0) = x_0$, letting $n \to \infty$ one derives that $x(t)$ satisfies (5.5.1) on $[0, T]$, $x(0) = x_0$, and

$$x \in C([0, T], H) \cap L^\infty([0, T], V \cap W),\qquad(5.5.47)$$

$$x' \in L^2([0, T], H) \cap L^\infty([0, T], (V \cap W)'),\qquad(5.5.48)$$

$$F(t) = \int_0^t Gx(s)\,\mathrm{d}s \in L^\infty([0, T], H).\qquad(5.5.49)$$

Moreover, differentiating (5.5.1) on $(0, T)$ leads to

$$x''(t) + a(0)Gx(t) + \int_0^t a'(t - s)Gx(s)\,\mathrm{d}s = f'(t),$$

a.e. on $[0, T]$, in $(V \cap W)'$, which implies

$$x'' \in L^1([0, T], (V \cap W)').\qquad(5.5.50)$$

The solution $x(t)$ has been constructed on $[0, T]$, with $T > 0$ arbitrarily chosen. In order to extend it to $[T, \infty)$, one can proceed as follows. Consider the problem

$$y'(t) + \int_0^t a(t - s)Gy(s)\,\mathrm{d}s = f(t + T) - \int_0^T a(t + T - s)Gx(s)\,\mathrm{d}s\qquad(5.5.51)$$

on the interval $[0, T]$, with the initial condition $y(0) = x(T)$. We have $y(0) \in V \cap W$, as seen from (5.5.47), and $F(T) \in H$ follows from $\langle F_{\varepsilon_n}(T), x \rangle \to \langle F(T), x \rangle$, $\forall x \in V \cap W$. If we denote by $f_1(t)$ the right hand side of (5.5.51), and use properties (iii), (5.5.49), we obtain $f_1, f_1' \in L^1([0, T], H)$. Then, proceeding as above in the proof of Theorem 5.5.1, one constructs the solution $y(t)$ on $[0, T]$, $y: [0, T] \to H$, for equation (5.5.51), such that (5.5.47)–(5.5.49) hold with y instead of x. It remains to extend x by putting $x(t) = y(t - T), t \in [T, 2T]$. The process can be continued, which means that the solution $x(t)$ is defined on R_+, and satisfies (5.5.5)–(5.5.7).

Theorem 5.5.1 is thereby proven.

Remark 1. The hypothesis (4) becomes unnecessary if we assume that the injection $V \subset W$ is compact.

Remark 2. From Theorem 5.5.1 one can derive without difficulty an existence result for the equation with infinite delay

$$x'(t) + \int_{-\infty}^t a(t - s)(A + M)x(s)\,\mathrm{d}s = f(t), \quad t \in R_+,$$

with an initial condition of the form $x(t) = h(t), t \in R_-$.

Let us conclude this section with an example of a partial integrodifferential equation occurring in *viscoelasticity*.

Consider a bounded domain $\Omega \subset R^n$, with smooth boundary Γ, and let us choose $H = L^2(\Omega, R)$, $V = H_0^1(\Omega, R)$, and $W = L^p(\Omega, R)$, with $2 \le p < \infty$. Recall that the space $H_0^1(\Omega, R)$ is the Sobolev space obtained by taking the closure of $C_0^{(\infty)}(\Omega, R)$ with respect to the norm $u \to (\int_\Omega [u^2 + (\nabla u)^2] \, dx)^{1/2}$. We choose $A = \Delta$, and M is defined by

$$(Mu)(x) = \beta(u(x)) \quad \text{a.e. on } \Omega, \quad u \in L^p(\Omega, R), \qquad (5.5.52)$$

where $\beta: R \to R$ is a (nonlinear) continuous function satisfying the following assumptions:

$$\beta \text{ is monotone}, \quad \beta(0) = 0; \qquad (5.5.53)$$

$$|\beta(r)| \le K(|r|^{p-1} + 1), \quad r \in R, K > 0; \qquad (5.5.54)$$

$$r\beta(r) \ge L(|r|^{p-1} - 1), \quad r \in R, L > 0. \qquad (5.5.55)$$

Under conditions (5.5.53)–(5.5.55), the operator M defined by (5.5.52) is acting from W into W'.

It has been shown in Barbu [4] that conditions (1)–(5) are implied for the spaces defined above and operators A and M. The following result holds for the equation

$$u_t(t, x) + \int_0^t a(t - s)[-\Delta u(s, x) + \beta(u(s, x))] \, ds = f(t, x), \qquad (5.5.56)$$

considered for $t \in R_+$, and $x \in \Omega$.

Assume $a(t)$ satisfies (i) *and* (ii) *from Theorem 5.5.1, and let β satisfy* (5.5.53)–(5.5.55). *If f and $u_0(x)$ satisfy*

$$f, f_t \in L^1_{loc}(R_+, L^2(\Omega)), \quad u_0 \in H_0^1(\Omega, R) \cap L^p(\Omega, R),$$

then there exists at least one solution $u(t, x)$ of equation (5.5.56) *such that the following conditions are verified*:

$$u(0, x) = u_0(x), x \in \Omega, \quad \text{and } u(t, x) = 0, \quad t > 0, x \in \Gamma;$$

$$u \in L^\infty_{loc}(R_+, H_0^1(\Omega) \cap L^p(\Omega)) \cap C(R_+, L^2(\Omega));$$

$$u_t \in L^2_{loc}(R_+, L^2(\Omega)) \cap L^\infty_{loc}(R_+, H^{-1}(\Omega) + L^q(\Omega)),$$

where $p^{-1} + q^{-1} = 1$;

$$u_{tt} \in L^1_{loc}(R_+, H^{-1}(\Omega) + L^q(\Omega)).$$

Results similar to the existence theorem for equation (5.5.56), as well as many other results related to viscoelastic materials, can be found in the monograph by Renardy, Hrusa and Nohel [1].

Bibliographical notes

The origin of the results included in the first three sections of this chapter is indicated in each section. The results given in Section 5.4 are due to Webb [1], while those in Section 5.5 are due to Aizicovici [1], [2].

The literature on integral and related equations in Banach/Hilbert spaces is quite extensive. I shall here indicate a few more references, in addition to those included in the sections themselves.

With regard to the theory of integral equations involving bounded operators, the following references are useful: Banaś [1], Buhgeim [1], Deimling [2], Diekmann [1], Engl [1], Gripenberg [6], Rzepecki [1], [2].

The theory of semigroups has numerous followers in the theory of integral and related equations, and represents one of the principal tools of investigation for infinite-dimensional problems. Unfortunately, the method applies only to convolution equations (time-invariant operators). The following references are related to the use of semigroup theory in studying integral equations: Chen and Grimmer [1], [2], Delfour [1], Desch and Schappacher [1], [2], Diekmann and van Gils [1], Fitzgibbon [1], [2], Grimmer and Pritchard [1], Kappel [1], Kappel and Schappacher [1], [2], Staffans [6], Tanabe [1], Vrabie [1].

The type of results given in Section 5.5 is very often encountered in the recent literature, because such equations occur in various applied problems (viscoelastic materials, heat conduction in materials with memory, etc.). The following references are pertinent to the type of equation investigated in Section 5.5: Barbu [1], [2], [3], [5], Barbu, DaPrato and Ianneli [1], Carr and Hannsgen [1], Clément and Nohel [1], [2], Crandall and Nohel [1], DaPrato and Iannelli [1], [2], Heard [1], Hirano [1], [2], London and Nohel [1], Lunardi and Sinestrari [2], MacCamy [2], Pruss [2], Sinestrari [1], Vrabie [1].

Most of the basic problems involving integral or related equations in Banach spaces are still unaddressed, or in a very early stage of development. For instance, in relation to the semigroup theory applied to integral equations in a Banach/Hilbert space, which has received more attention on the part of researchers, the following problem has not yet been considered: consider the abstract Volterra equation $\dot{x}(t) = (Lx)(t) + f(t)$, in which L is a linear Volterra operator acting on a function space $S(R_+, X)$, with X a Banach space. Assuming that L is time-invariant (i.e., $(Lx)_t = Lx_t$) and densely defined, find conditions under which L generates a semigroup of linear bounded operators on X. Such a result would constitute a strong generalization of Hale's classical result regarding autonomous equations with finite delay (see Hale [1]).

More comments and references related to integral equations in Banach spaces will be included at the end of the next chapter, in which applications of such equations are dealt with.

6

Some applications of integral and integrodifferential equations

This chapter contains some applications of integral or integrodifferential equations to various problems occurring in contemporary research. The first section is dedicated to the investigation of an integrodifferential equation occurring in coagulation processes, while the second section deals with the Pontryagin maximum principle for dynamic processes described by means of Volterra integral equations. The last two sections of the chapter deal with stability problems from the theory of nuclear reactors.

The literature in this field (of applications) is considerable. I have attempted to illustrate the use of integral or related equations beyond the level reached in the preceding chapters of this book. References will be made to many contributions relating to the applications of integral equations to various fields of applied science. In particular, in the fields of population dynamics and continuum mechanics there are significant contributions that are available in book form (see references under the names of G. F. Webb, J. M. Cushing, J. A. Nohel, M. Renardy, W. J. Hrusa).

6.1 An integrodifferential equation describing coagulation processes

The following integrodifferential equation is encountered in the mathematical description of coagulation processes, under certain simplifying assumptions:

$$\frac{\partial}{\partial t} f(t, x) = \frac{1}{2} \int_0^x \phi(x - y, y) f(t, x - y) f(t, y) \, dy - f(t, x) \int_0^\infty \phi(x, y) f(t, y) \, dy,$$

$$(6.1.1)$$

with the solution $f = f(t, x)$ being sought in the quadrant

$$t \geq 0, \quad x \geq 0. \qquad (6.1.2)$$

The physical processes usually considered for investigation by means of (6.1.1) are the Brownian coagulation processes in solutions, the gravitational coagulation processes (such as those appearing in the formation of stars), or certain meteorogical processes. If we think about coagulation in colloidal solutions, for instance, then the following interpretation of equation (6.1.1) is usually put forward. The number of particles, per unit of volume, whose masses are in the interval $(x, x + dx)$ at time t, is $f(t, x)\,dx$. The kernel function $\phi(x, y) = \phi(y, x) \geq 0$ is assumed given (known), and represents the coalescence of particles of masses x and y. It is assumed that the number of particles is large enough to allow us to use the density function $f(t, x)$. Conservation of mass implies that $\partial f/\partial t$ is equal to the difference between the production of particles x by coalescence of particles y and $x - y$, $0 \leq y \leq x$, and the disappearance of particles x, caused by their coalescence with particles y, $0 \leq y < \infty$. This is, roughly speaking, the motivation for equation (6.1.1).

Sometimes, a process of breakdown of particles is also taken into consideration, a function $\psi(x, y) \geq 0$, with $\psi(x, y) = 0$ for $y > x$, describing the process. In this case, equation (6.1.1) must be substituted by

$$\frac{\partial}{\partial t} f(t, x) = \frac{1}{2} \int_0^x \phi(x - y, y) f(t, x - y) f(t, y)\,dy - f(t, x) \int_0^\infty \phi(x, y) f(t, y)\,dy$$

$$+ \int_0^\infty \psi(y, x) f(t, y)\,dy - x^{-1} f(t, x) \int_0^x y\psi(x, y)\,dy.$$

$$(6.1.3)$$

With equation (6.1.1) or (6.1.3), an initial condition must be associated. Usually, it takes the form

$$f(0, x) = f_0(x), \quad x \geq 0, \tag{6.1.4}$$

and, taking into account the physical meaning of f, it is assumed

$$f_0(x) \geq 0 \quad \text{for } x \geq 0. \tag{6.1.5}$$

Both equations (6.1.1) and (6.1.3) are integrodifferential equations in $f(t, x)$, with quadratic right hand side. If we integrate with respect to t, either equation becomes an integral equation for the function $f(t, x)$. For instance, from (6.1.1) and (6.1.4) we derive

$$f(t, x) = f_0(x) + \frac{1}{2} \int_0^t \int_0^x \phi(x - y, y) f(s, x - y) f(s, y)\,dy\,ds$$

$$- \int_0^t \int_0^\infty f(s, x) \phi(x, y) f(s, y)\,dy\,ds. \tag{6.1.6}$$

We notice that equation (6.1.6) is not of Volterra type in (t, x) because of the second term which contains the integral from 0 to ∞. We can say it is of Volterra type in t only.

We shall now introduce an operator which takes pairs of functions into single functions, which will be denoted by $[f, g]$, and is formally given by

$$[f, g](x) = \frac{1}{2}\left\{\int_0^x f(y)g(x - y)\phi(x - y, y)\,dy\right.$$

$$\left. - f(x)\int_0^\infty g(y)\phi(x, y)\,dy - g(x)\int_0^\infty f(x)\phi(x, y)\,dy\right\}. \quad (6.1.7)$$

This notation allows us to write equation (6.1.1) in the concise form

$$\frac{\partial}{\partial t}f(t, x) = [f(t, x), f(t, x)], \quad (6.1.8)$$

and the integral equation (6.1.6) in the form

$$f(t, x) = f_0(x) + \int_0^t [f(s, x), f(s, x)]\,ds. \quad (6.1.9)$$

It is easy to see that if $\phi(x, y)$ is continuous and bounded in the quadrant $x \geq 0$, $y \geq 0$, while f, $g \in L^\infty([0, \infty), R) \cap L^1([0, \infty), R)$, (6.1.7) does indeed define an operator on $\{L^\infty([0, \infty), R) \cap L^1([0, \infty), R)\}^2$, with values in the space $L^\infty([0, \infty), R) \cap L^1([0, \infty), R)$.

More precisely, we shall establish some estimates for $[f, g]$, in the norms of L^∞ and L^1. Such estimates will be useful subsequently when proving the existence and uniqueness of a solution to (6.1.1), (6.1.4).

Let us denote by A a positive number such that

$$0 \leq \phi(x, y) \leq A, \quad x \geq 0, y \geq 0. \quad (6.1.10)$$

Then the following estimates hold for the operator $[f, g]$:

$$|[f, g]|_{L^\infty} \leq \frac{3A}{4}\{|f|_{L^\infty}|g|_{L^1} + |f|_{L^1}|g|_{L^\infty}\}, \quad (6.1.11)$$

$$|[f, g]|_{L^1} \leq \frac{3A}{2}|f|_{L^1}|g|_{L^1}. \quad (6.1.12)$$

The proof of these estimates can be carried out without difficulty. For instance, in order to obtain (6.1.11) we proceed as follows:

$$|[f, g]| \leq \frac{A}{2}\left\{\int_0^x |f(y)||g(x - y)|\,dy + |f(x)|\int_0^\infty |g(y)|\,dy + |g(x)|\int_0^\infty |f(y)|\,dy\right\}, \quad (6.1.13)$$

and taking into account

$$\int_0^x |f(y)||g(x - y)|\,dy \leq \min\{|f|_{L^\infty}|g|_{L^1}, |f|_{L^1}|g|_{L^\infty}\},$$

as well as the elementary inequality $2\min(a,b) \le a + b$, we obtain from above (6.1.11).

The inequality (6.1.12) can also be obtained by integrating both sides of (6.1.7) on $[0, \infty)$.

We can now proceed to establish a local existence theorem for (6.1.1), which we prefer to be considered in its integral form (6.1.9).

Let us seek a solution to (6.1.9) in the form of a power series with respect to t, namely,

$$f(t, x) = \sum_{k=0}^{\infty} f_k(x) t^k, \qquad (6.1.14)$$

with $f_0(x)$ as in (6.1.4) for obvious reasons. We proceed formally, then we shall deal with the convergence of the resulting series solution. Substituting (6.1.14) into (6.1.9), and equating the coefficients of the same powers of t, we obtain

$$f_{k+1}(x) = \frac{1}{k+1} \sum_{i+j=k} [f_i, f_j], \quad k \ge 0. \qquad (6.1.15)$$

We assume now that the coefficients $f_k(x)$, $k = 0, 1, 2, \ldots$, in (6.1.14) satisfy

$$f_k \in L^{\infty}([0, \infty), R) \cap L^1([0, \infty), R). \qquad (6.1.16)$$

Let

$$A_k = |f_k|_{L^1}, \quad B_k = |f_k|_{L^{\infty}}, \quad k \ge 0. \qquad (6.1.17)$$

Taking into account (6.1.11) and (6.1.12), we obtain from (6.1.15) for $k \ge 0$:

$$A_{k+1} \le \frac{3A}{2} \frac{1}{k+1} \sum_{i+j=k} A_i A_j, \qquad (6.1.18)$$

$$B_{k+1} \le \frac{3A}{2} \frac{1}{k+1} \sum_{i+j=k} A_i B_j. \qquad (6.1.19)$$

We shall prove now that, in terms of A_0 and B_0 which are given by (6.1.17), the following type of estimate is valid:

$$A_k \le A_0 m^k, \quad B_k \le B_0 m^k, \quad k \ge 1, \qquad (6.1.20)$$

for a suitable positive m. Indeed, if we assume (6.1.20) holds for $k = 1, 2, \ldots, n$, then (6.1.18) allows us to obtain

$$A_{n+1} \le \frac{3A}{2} \frac{1}{n+1} \sum_{i+j=n} A_0 m^i A_0 m^j = \frac{3A}{2} \frac{1}{n+1} A_0^2 m^n (n+1) = A_0 \left(\frac{3AA_0}{2m} \right) m^{n+1}. \qquad (6.1.21)$$

Therefore, if we choose $m > 0$ such that $3AA_0 = 2m$ ($m = 3AA_0/2$), (6.1.21) shows us that (6.1.20) remains true for $k = n + 1$. By induction, the validity of the first estimate in (6.1.20) is proven for $k \ge 1$. Similarly, for the second estimate in

(6.1.20) we obtain its validity for any $k \geq 1$, with the same assumption on m:

$$m = \frac{3AA_0}{2}.$$ (6.1.22)

The estimates (6.1.20) show us that the terms of the series (6.1.14) are dominated, in absolute value, by $A_0(mt)^k$, $k \geq 0$. Hence, the series (6.1.14) converges for $mt < 1$ and $x \geq 0$ and converges uniformly in (t, x) in any strip $0 \leq t \leq \delta < m^{-1}, x \geq 0$.

Therefore, if $f_0(x)$ is assumed to be continuous, and belonging to $L^\infty \cap L^1$, then all $f_k(x)$, $k \geq 1$, are continuous, and the series (6.1.14) converges uniformly for $0 \leq t \leq \delta < m^{-1}$, $x \geq 0$, towards a continuous function $f(t, x)$.

It is useful to point out the validity of the following estimates for $f(t, x)$ and its derivatives in t, regarding t as a parameter, $0 \leq t < m^{-1}$:

$$|f(t, x)|_{L^\infty} \leq \frac{B_0}{1 - mt}, \quad |f(t, x)|_{L^1} \leq \frac{A_0}{1 - mt},$$ (6.1.23)

$$\left| \frac{\partial}{\partial t} f(t, x) \right|_{L^\infty} \leq \frac{B_0 m}{(1 - mt)^2}, \quad \left| \frac{\partial}{\partial t} f(t, x) \right|_{L^1} \leq \frac{A_0 m}{(1 - mt)^2}$$ (6.1.24)

$$\left| \frac{\partial^2}{\partial t^2} f(t, x) \right|_{L^\infty} \leq \frac{2B_0 m^2}{(1 - mt)^3}, \quad \left| \frac{\partial^2}{\partial t^2} f(t, x) \right|_{L^1} \leq \frac{2A_0 m^2}{(1 - mt)^3}.$$ (6.1.25)

As a result of the discussion conducted above with regard to the construction of a local solution (in t) to equation (6.1.1), we can formulate the following result.

Proposition 6.1.1. *Consider the integrodifferential equation (6.1.1), under initial condition (6.1.4), and assume the following conditions hold:*

(1) *the function $\phi(x, y)$ is continuous and symmetric in the quadrant $x \geq 0, y \geq 0$, and verifies (6.1.10);*
(2) *$f_0(x)$ is continuous, bounded and integrable on $x \geq 0$.*

Then there exists a solution $f(t, x)$ on $[0, m^{-1}) \times [0, \infty)$, represented by the series (6.1.14), with $f_k(x)$, $k \geq 1$, constructed as above. This solution is continuous in (t, x), analytic in t for fixed x, and satisfies (6.1.23)–(6.1.25).

Remark. We did not use the condition (6.1.5) in order to derive Proposition 6.1.1. If this condition is assumed, it is logical to see whether $f(t, x) \geq 0$ is satisfied. Otherwise, $f(t, x)$ could not be regarded as a 'density function', and the model provided by equation (6.1.1) becomes meaningless.

Proposition 6.1.2. *Under the assumptions of Proposition 6.1.1 and (6.1.5), the solution $f(t, x)$ constructed above satisfies $f(t, x) \geq 0$, for $x \geq 0, 0 \leq t < m^{-1}$.*

Proof. Let us consider the solution $f(t, x)$ in the 'thin' strip, $x \geq 0, 0 \leq t \leq \tau$, with $0 < \tau < m^{-1}$. For a fixed positive n we define for $k = 0, 1, 2, \ldots, n - 1$,

$$f_{0n}(x) = f_0(x), \quad f_{(k+1)n}(x) = f_{kn}(x) + \tau n^{-1}[f_{kn}(x), f_{kn}(x)]. \qquad (6.1.26)$$

Let

$$T_{kn} = |f_{kn}|_{L^1}, \quad L_{kn} = |f_{kn}|_{L^\infty}. \qquad (6.1.27)$$

Hence, $T_{0n} = A_0, L_{0n} = B_0$. From (6.1.26) we have

$$T_{(k+1)n} \leq T_{kn} + \frac{3AA_0\tau}{2n} T_{kn}^2, \qquad (6.1.28)$$

if we take (6.1.12) into account, and

$$L_{(k+1)n} \leq L_{kn} + \frac{3AA_0\tau}{2n} T_{kn}L_{kn}, \qquad (6.1.29)$$

if we take (6.1.11) into account.

We now want to obtain upper bounds for T_{kn} and L_{kn}, independent of k or n. Let us first consider (6.1.28). If we consider the recursion formula, with $\varepsilon > 0$,

$$\tilde{T}_{(k+1)n} = \tilde{T}_{kn}\left(1 + \frac{\tau\varepsilon}{n}\right) + \frac{3AA_0}{2n} \tilde{T}_{kn}^2, \qquad (6.1.30)$$

for $k = 0, 1, 2, \ldots, n - 1$, with $\tilde{T}_{0n} = T_{0n} = A_0$, then we have, for every $k \geq 1$, $T_{kn} < \tilde{T}_{kn}$. Hence, it will suffice to find an upper estimate for $\tilde{T}_{nn}$, using (6.1.30). Let us rewrite (6.1.30) in a more convenient form, by denoting for $0 \leq k \leq n - 1$

$$a = 1 + \frac{\tau\varepsilon}{n}, \quad \alpha_k = \frac{3AA_0\tau}{2na} \tilde{T}_{kn}. \qquad (6.1.31)$$

Then (6.1.30) becomes

$$\alpha_{k+1} = a(\alpha_k + \alpha_k^2), \qquad (6.1.32)$$

and letting $h(x) = ax(1 + x)$, we can substitute for (6.1.32) the recursion formula

$$\alpha_{k+1} = h(\alpha_k), \quad \alpha_0 = \frac{3AA_0\tau}{2na} A_0. \qquad (6.1.33)$$

It is easy to see that the iterations of $h(x)$ satisfy the inequalities

$$0 < h^{(k)}(x) < a^k x \Big/ \left(1 - \sum_{i=0}^{k-1} a^i x\right), \qquad (6.1.34)$$

for those positive x for which the denominator in (6.1.34) is positive. Since

$$a^k \leq a^n = \left(1 + \frac{\tau\varepsilon}{n}\right)^n < \exp(\tau\varepsilon),$$

and

$$\sum_{i=0}^{k-1} a^i = \frac{a^k - 1}{a - 1}, \quad a \neq 1,$$

from (6.1.33) and (6.1.34) one derives

$$\frac{2na}{3AA_0\tau}\alpha_n = \tilde{T}_{nn} < \frac{3AA_0 \exp(\tau\varepsilon)}{3A - 2\varepsilon[\exp(\tau\varepsilon) - 1]}. \tag{6.1.35}$$

The positiveness of the denominator in (6.1.35) is secured if we assume $\tau, \varepsilon > 0$ are chosen in such a way that

$$\tau < \varepsilon^{-1}\ln\left(1 + \frac{3A}{2\varepsilon}\right). \tag{6.1.36}$$

The right hand side of (6.1.36) tends to ∞ as $\varepsilon \downarrow 0$, which means that we can always choose $\varepsilon > 0$ in order to satisfy (6.1.36), for any $\tau < m^{-1}$.

On the other hand, for such a choice of ε, (6.1.35) provides an upper estimate for $\tilde{T}_{nn}$, and consequently for each T_{kn} given by (6.1.27), which is independent of k or n. In other words, there exists a positive constant C_1 such that

$$T_{kn} < C_1, \quad n \geq 1, 0 \leq k \leq n - 1. \tag{6.1.37}$$

If we now proceed in the same way with (6.1.29) as we did with (6.1.28), and take into account the estimate (6.1.37) for T_{kn}, then a similar result is obtained for the quantities L_{kn}:

$$L_{kn} \leq L_{nn} \leq C_2, \quad n \geq 1, 0 \leq k \leq n - 1, \tag{6.1.38}$$

where $C_2 > 0$ is independent of k or n.

Indeed, (6.1.29) and (6.1.37) yield

$$L_{(k+1)n} \leq \left(1 + \frac{3AA_0C_1\tau}{2n}\right)L_{kn}, \tag{6.1.39}$$

from which we derive

$$L_{nn} \leq \left(1 + \frac{3AA_0C_1\tau}{2n}\right)^n B_0 < B_0 \exp\left(\frac{3AA_0C_1\tau}{2}\right) = C_2. \tag{6.1.40}$$

The estimates (6.1.37) and (6.1.40) show that the functions $f_{kn}(x)$, defined by (6.1.26) are uniformly bounded and uniformly integrable in x, on $[0, \infty)$. From their construction it is obvious that they are continuous, if $f_0(x)$ is continuous. Moreover, since we assume $f_0(x) \geq 0$, we also have $f_{kn}(x) \geq 0$, $k = 0, 1, 2, \ldots, n$. Indeed, from (6.1.7) and (6.1.26)

$$f_{(k+1)n}(x) = \frac{\tau}{2n}\int_0^x f_{kn}(y)f_{kn}(x - y)\phi(x - y, y)\,dy$$

$$+ f_{kn}(x)\left[1 - \frac{\tau}{n}\int_0^\infty f_{kn}(y)\phi(x, y)\,dy\right]. \tag{6.1.41}$$

If we choose n sufficiently large, namely $n > n_0 = \tau(AC_1 + 1)$, then the term in the square brackets in (6.1.41) is positive. By induction, we then obtain from (6.1.41) that all $f_{kn}(x)$ are non-negative, $k = 1, 2, \ldots, n$.

We will now define some functions $F_n(t, x)$, by means of the following procedure:

$$F_n(t, x) = f_{kn}(x), \quad \frac{k\tau}{n} \leq t < \frac{(k + 1)\tau}{n}, \tag{6.1.42}$$

for $k = 0, 1, 2, \ldots, n - 1$. Our goal is to prove now that the (local) solution $f(t, x)$ can be approximated with any degree of accuracy by $F_n(t, x)$, which will imply $f(t, x) \geq 0$ in the strip $x \geq 0, 0 \leq t \leq \tau$.

Let us introduce the quantities λ_k and μ_k by means of the following formulas:

$$\lambda_k = \sup\left\{|f(t, x) - F_n(t, x)|; x \geq 0, \frac{k\tau}{n} \leq t < \frac{(k + 1)\tau}{n}\right\}$$

$$= \sup\{|f(t, x) - f_{kn}(x)|; x \geq 0\}, \tag{6.1.43}$$

and

$$\mu_k = \sup\left\{\int_0^\infty |f(t, x) - F_n(t, x)|\, dx; \frac{k\tau}{n} \leq t < \frac{(k + 1)\tau}{n}\right\}$$

$$= \sup\left\{\int_0^\infty |f(t, x) - f_{kn}(x)|\, dx; \frac{k\tau}{n} \leq t < \frac{(k + 1)\tau}{n}\right\}. \tag{6.1.44}$$

In order to find upper bounds for λ_k and μ_k, we will first establish some recursion formulas. If we assume $k \geq 1$, and take into account the existence of the first two derivatives for $f(t, x)$, then we obtain

$$f(t, x) - f_{kn}(x) = f(t, x) - f\left(t - \frac{\tau}{n}, x\right) + f\left(t - \frac{\tau}{n}, x\right)$$

$$- f_{(k-1)n}(x) - \frac{\tau}{n}[f_{(k-1)n}(x), f_{(k-1)n}(x)]$$

$$= \frac{\tau}{n}\frac{\partial}{\partial t}f\left(t - \frac{\tau}{n}, x\right) + \frac{\theta\tau^2}{n^2}\frac{\partial^2}{\partial t^2}f\left(t - \frac{\tilde{\theta}\tau}{n}, x\right)$$

$$+ f\left(t - \frac{\tau}{n}, x\right) - f_{(k-1)n}(x) - \frac{\tau}{n}[f_{(k-1)n}(x), f_{(k-1)n}(x)]$$

$$= f\left(t - \frac{\tau}{n}, x\right) - f_{(k-1)n}(x)$$

$$+ \frac{\tau}{n}\left[f\left(t - \frac{\tau}{n}, x\right) + f_{(k-1)n}(x), f\left(t - \frac{\tau}{n}, x\right) - f_{(k-1)n}(x)\right]$$

$$+ \frac{\theta\tau^2}{n^2}\frac{\partial^2}{\partial t^2}f\left(t - \frac{\tilde{\theta}\tau}{n}, x\right), \tag{6.1.45}$$

where θ and $\tilde{\theta}$ are such that $0 \le \theta$, $\tilde{\theta} \le 1$. We have taken into consideration the fact that

$$\frac{\partial}{\partial t} f\left(t - \frac{\tau}{n}, x\right) = \left[f\left(t - \frac{\tau}{n}, x\right), f\left(t - \frac{\tau}{n}, x\right) \right],$$

which is nothing other than equation (6.1.1), in the form (6.1.8). From (6.1.45) we obtain the inequality

$$|f(t,x) - f_{kn}(x)| \le \left| f\left(t - \frac{\tau}{n}, x\right) - f_{(k-1)n}(x) \right|$$

$$+ \frac{\tau}{n} \left| \left[f\left(t - \frac{\tau}{n}, x\right) + f_{(k-1)n}(x), f\left(t - \frac{\tau}{n}, x\right) - f_{(k-1)n}(x) \right] \right|$$

$$+ \frac{\theta \tau^2}{n^2} \left| \frac{\partial^2}{\partial t^2} f\left(t - \frac{\tilde{\theta} \tau}{n}, x\right) \right|. \tag{6.1.46}$$

But, taking into account the estimates (6.1.23)–(6.1.25), (6.1.37) and (6.1.40), we can write

$$\left| f\left(t - \frac{\tau}{n}, x\right) + f_{(k-1)n}(x) \right| \le C_3,$$

$$\int_0^\infty \left| f\left(t - \frac{\tau}{n}, x\right) + f_{(k-1)n}(x) \right| dx \le C_4,$$

$$\left| \frac{\partial^2}{\partial t^2} f\left(t - \frac{\tilde{\theta} \tau}{n}, x\right) \right| \le C_5,$$

where C_i, $i = 3, 4, 5$, are some positive constraints, independent of k or n. From (6.1.46), by using the estimates (6.1.11), (6.1.12) and (6.1.23)–(6.1.25), and the notation (6.1.43) and (6.1.44) we obtain the following inequalities:

$$\lambda_k \le \lambda_{k-1} + \frac{3A\tau}{4n}(C_3 \mu_{k-1} + C_4 \lambda_{k-1}) + \frac{C_5 \tau^2}{n^2}, \tag{6.1.47}$$

$$\mu_k \le \mu_{k-1} + \frac{3A\tau}{2n} C_3 \mu_{k-1} + \frac{C_k \tau^2}{n^2}, \tag{6.1.48}$$

where C_6 is determined from

$$\int_0^\infty \left| \frac{\partial^2}{\partial t^2} f(t,x) \right| dx \le C_6, \quad 0 \le t \le \tau < m^{-1},$$

according to (6.1.25). If we rely on the mean value theorem, we can write

$$\lambda_0 \le C_7 \frac{\tau}{n}, \quad \mu_0 \le C_8 \frac{\tau}{n}, \tag{6.1.49}$$

the existence of C_7 and C_8 being guaranteed by (6.1.24).

Since (6.1.48) can be rewritten as

$$\mu_k \leq \left(1 + \frac{3AC_3\tau}{2n}\right)\mu_{k-1} + \frac{C_6\tau^2}{n^2},$$

we obtain, by the procedure used above, the estimates

$$\mu_k \leq \mu_n \leq \frac{C_9}{n}, \tag{6.1.50}$$

and

$$\lambda_k \leq \lambda_n \leq \frac{C_{10}}{n}, \tag{6.1.51}$$

where C_9 and C_{10} are new positive constants resulting from the recursion inequalities (6.1.48) and (6.1.47).

From (6.1.43), (6.1.44), (6.1.50) and (6.1.51) we obtain

$$\lim_{n\to\infty} |f(t,x) - F_n(t,x)| = 0, \quad 0 \leq t \leq \tau$$

as well as

$$\lim_{n\to\infty} \int_0^\infty |f(t,x) - f_n(t,x)|\, dx = 0.$$

This completes the proof of Proposition 6.1.2, because, as noticed above, we can choose τ as close as we want to m^{-1}.

The next step in obtaining an improvement of the existence result is to extend the interval $[0, m^{-1})$, in which Proposition 6.1.1 holds, to the whole positive semiaxis $t \geq 0$.

Theorem 6.1.3. *Under the hypotheses of Propositions 6.1.1 and 6.1.2, there exists a solution $f(t,x)$ of equation (6.1.1), satisfying (6.1.4), which is defined for $x \geq 0$, $t \geq 0$, is continuous, bounded, nonnegative, analytic in t for each fixed $x \geq 0$, and integrable in x for each $t \geq 0$. Moreover, the solution is unique.*

Proof. Under the assumptions of Theorem 6.1.3, the existence of a local solution $f(t,x)$ is assumed in the strip $x \geq 0, 0 \leq t < m^{-1}$. If we fix some $t_1, 0 < t_1 < m^{-1}$, then $f(t_1,x)$ is, according to Propositions 6.1.1 and 6.1.2, continuous, bounded, integrable and nonnegative on $x \geq 0$. Therefore, $f(t_1,x)$ can be taken as initial data, and Proposition 6.1.1 guarantees the existence of a solution defined in the strip $x \geq 0, t_1 \leq t \leq t_2$. Again, $f(t_2,x)$ satisfies the conditions required to be taken as initial functions for (6.1.1) and the process can be continued indefinitely. As a result, we obtain an increasing sequence $\{t_m\}$, such that (6.1.1) has a solution with the stated properties in the strip $x \geq 0, 0 \leq t \leq t_m$. If $t_m \uparrow +\infty$, then Theorem

6.1.3 is proven. The case $t_m \uparrow T$, $T < +\infty$, cannot actually occur. This is the fact that we need to prove, in order to conclude the proof of Theorem 6.1.3.

Notice that, if $T < +\infty$, we must necessarily have $t_{m+1} - t_m \to 0$ as $m \to \infty$. But in Proposition 6.1.1 we obtained for m the value (6.1.22). If by $A_0(t_m)$ we denote the L^1-norm of $f(t_m, x)$, so that $A_0 = A_0(0)$, then by (6.1.22) we can write

$$t_{m+1} - t_m \geq \frac{1}{3AA_0(t_m)}, \quad m \geq 0. \tag{6.1.52}$$

Consequently, if $T < +\infty$, one must have

$$A_0(t_m) \to \infty \text{ as } m \to \infty. \tag{6.1.53}$$

Now let us consider, in accordance with the above notation, the function

$$A_0(t) = |f(t, x)|_{L^1}. \tag{6.1.54}$$

In order to estimate $A_0(t)$, we shall integrate both sides of equation (6.1.9) from 0 to ∞, which is equivalent to (6.1.1) and the initial condition (6.1.4). We obtain

$$A_0(t) = A_0 - \frac{1}{2} \int_0^t \int_0^\infty \int_0^\infty f(t, x) f(t, y) \phi(x, y) \, dx \, dy \, dt, \tag{6.1.55}$$

if we take into account the relationship

$$\int_0^\infty \int_0^x \phi(x - y, y) f(t, x - y) f(t, y) \, dy \, dx = \int_0^\infty \int_0^\infty \phi(x, y) f(t, x) f(t, y) \, dx \, dy.$$

All the integrals appearing in (6.1.55) or in the formula above are convergent (absolutely) because of our assumptions on ϕ, and because $f(t, x)$ is integrable on $[0, \infty)$ for each $t < T$. From (6.1.55) we obtain

$$A_0(t) \leq A_0, \quad 0 \leq t < T, \tag{6.1.56}$$

because the integral in the right hand side of (6.1.55) is nonnegative. But (6.1.56) is incompatible with (6.1.53), which means that the hypothesis $T < +\infty$ leads to a contradiction. Therefore, the first part of Theorem 6.1.3 concerning the global existence of the solution $f(t, x)$ is proven.

We shall now prove the uniqueness of the solution, within the class of functions satisfying the properties specified in the statement for $f(t, x)$. In other words, we assume there exists another solution $g(t, x)$ of (6.1.1), (6.1.4), such that it is continuous for $x \geq 0$, $t \geq 0$. It is integrable in x for each fixed t, and $|g(t, x)|_{L^1}$ remains bounded on any finite interval $0 \leq t \leq T < +\infty$.

If we start with (6.1.10) and the similar relationship for $g(t, x)$, then we obtain

$$f(t, x) - g(t, x) = \int_0^t [f(s, x) + g(s, x), f(s, x) - g(s, x)] \, ds.$$

Taking the absolute value in both sides of the preceding equation, and then integrating in x from 0 to ∞, we have

$$\int_0^\infty |f(t,x) - g(t,x)|\,dt \leq \int_0^t \int_0^\infty |[f(s,x) + g(s,x), f(s,x) - g(s,x)]|\,dx\,ds,$$

and by using (6.1.12) we obtain

$$\int_0^\infty |f(t,x) - g(t,x)|\,dx$$

$$\leq \frac{3A}{2} \int_0^t \left\{ \int_0^\infty |f(s,x) + g(s,x)|\,dx \int_0^\infty |f(s,x) - g(s,x)|\,dx \right\} ds. \qquad (6.1.57)$$

Changing the order of integration is valid under our assumptions. But on the basis of the properties established for f, and the fact that $|g(t,x)|_{L^1}$ is bounded on any finite interval $0 \leq t \leq T < \infty$, we can write

$$\int_0^\infty |f(t,s) + g(t,x)|\,ds \leq K, \quad 0 \leq t \leq T,$$

for some $K = K(T) > 0$. Then (6.1.57) leads to the following inequality for

$$h(t) = \int_0^\infty |f(t,x) - g(t,x)|\,dx, \quad 0 \leq t \leq T:$$

$$h(t) \leq \frac{3AK}{2} \int_0^t h(s)\,ds, \quad 0 \leq t \leq T. \qquad (6.1.58)$$

But (6.1.58) and $h(0) = 0$ imply $h(t) = 0$ on $[0, T]$. Since T is arbitrary, we obtain $|f - g|_{L^1} = 0$ for every $t \geq 0$. If we also keep in mind that f and g are continuous functions, we obtain $f(t,x) \equiv g(t,x)$, $x \geq 0$, $t \geq 0$.

This ends the proof of Theorem 6.1.3.

6.2 Optimal control processes governed by Volterra integral equations

We shall deal in this section with the problem of minimizing a functional of the form $g^0(x(T))$, with respect to all $x(t)$ which satisfy the integral equation

$$x(t) = f(t) + \int_0^t [k(t,s,x(s)) + g(s,x(s),u(s))]\,ds \qquad (6.2.1)$$

on the interval $[0, T]$, $T > 0$, with the control function

$$u : [0, T] \to U \subset R^m \qquad (6.2.2)$$

assumed to be (Lebesgue) measurable.

In addition, we impose some constraints at the end of the interval $[0, T]$, namely

$$g_j(x(T)) \leq 0, \quad j = 1, 2, \ldots, p, \tag{6.2.3}$$

$$g_j(x(T)) = 0, \quad j = p + 1, \ldots, q. \tag{6.2.4}$$

More precise conditions on the functions involved will be given below.

First, notice that a more typical problem of optimal control can be reduced to the above formulated one. Indeed, if we introduce one more variable by the formula

$$y(t) = g^0(x(t)), \tag{6.2.5}$$

then the vector $\xi(t) = \text{col}(x(t), y(t))$ satisfies an equation of the form (6.2.1), while the functional to be minimized is, according to (6.2.5), $y(T)$. Hence, we have a problem of the type mentioned above.

We shall now formulate the conditions under which we will conduct the discussion of the problem.

(1) $g^0: R^n \to R$, and $g_i: R^n \to R$, $i = 1, 2, \ldots, q$, are continuously differentiable;
(2) $f: [0, T] \to R^n$ is continuous;
(3) $k: [0, T] \times [0, T] \times R^n \to R^n$ is continuous, together with its first partial derivatives $\partial k_i / \partial x_j$;
(4) $g: [0, T] \times R^n \times U \to R^n$ is continuous together with its first partial derivatives $\partial g_i / \partial x_j$, with $U \subset R^m$ a closed set;
(5) g and $\partial g / \partial x$ are locally bounded in $[0, T] \times R^n \times U$, which means that, for every compact $K \subset R^n$, there exist constants $M_g, N_g > 0$, such that

$$\left. \begin{array}{l} |g(t, x, u)| \leq M_g, \\[2mm] \left| \dfrac{\partial}{\partial x} g(t, x, u) \right| \leq N_g \end{array} \right\} \quad (t, x, u) \in [0, T] \times K \times U. \tag{6.2.6}$$

We shall now formulate the definition of *admissible* pairs (x, u), where (x, u): $[0, T] \to R^{n+m}$. The pair (x, u) will be called *admissible* with respect to the optimal control problem $\min g^0(x(T))$, under constraints (6.2.1)–(6.2.4), if $x: [0, T] \to R^n$ is continuous, $u: [0, T] \to U \subset R^m$ is Lebesgue measurable, and all the relations (6.2.1)–(6.2.4) hold.

Another condition that will be assumed throughout this section is:

(6) there exists at least one admissible pair (x, u) for our problem.

As usual, we shall refer to x as the admissible trajectory which corresponds to the admissible control u.

Another concept that we shall need in finding necessary conditions (the maximum principle) for the existence of an *optimal pair* (x^0, u^0), i.e., such that $g^0(x^0(T))$ attains the minimum value, is that of *perturbation* of an admissible control.

Let (x^0, u^0) be an admissible pair, and consider a positive integer N, a partition of $[0, T]$, $0 < t_1 \le t_2 \le \cdots \le t_N < T$, such that each t_i is a Lebesgue point for $g(t, x^0(t), u^0(t))$, and some elements $v^i \in U$, $i = 1, 2, \ldots, N$. Let $S = \{(t_i, v^i); i = 1, 2, \ldots, N\}$. Further, let $A_N = \{a = (a_1, a_2, \ldots, a_N) \in R^N, a_i \ge 0, i = 1, 2, \ldots, N\}$, and $J(i) = \{j; 1 \le j < i,$ with $t_j = t_i\}$. The following constants are defined by

$$b_i = \begin{cases} 0 & \text{when } J(i) = \phi, \\ \displaystyle\sum_{j \in J(i)} a_j & \text{when } J(i) \ne \phi. \end{cases}$$

We will need the intervals $I_i = [t_i + b_i, t_i + b_i + a_i)$.

Notice that $I_i \subset [0, T)$, provided $|a|$ is sufficiently small. Also, for sufficiently small $|a|$ one has $I_i \cap I_j = \phi$ for $i, j = 1, 2, \ldots, N$, $i \ne j$.

The *perturbed controls* are defined on $[0, T]$ by

$$u(t; a) = \begin{cases} v^i & \text{when } t \in I_i, \\ u^0(t) & \text{when } t \in C\left(\displaystyle\bigcup_{i=1}^{N} I_i \right). \end{cases} \tag{6.2.7}$$

As noticed above, for sufficiently small $|a|$, $a \in A_N$, the perturbed controls $u(t; a)$ are Lebesgue measurable functions satisfying $u(t; a) \in U$ a.e. for $t \in [0, T]$.

It is of primary interest to show that for each perturbed control $u(t; a)$ defined by (6.2.7), there exists an associated trajectory $x(t; a)$. In order to achieve this goal, we need the following auxiliary result.

Lemma 6.2.1. *There exist a constant $L > 0$, and a neighborhood $W \ni \theta \in R^N$, such that for all $a \in A_N \cap W$ one has*

$$\int_0^T |g(t, x^0(t), u(t; a)) - g(t, x^0(t), u^0(t))| \, dt \le L|a|. \tag{6.2.8}$$

Proof. If we denote by $y(T; a)$ the left hand side of (6.2.8), and take into account (6.2.7), then

$$y(T; a) = \sum_{i=1}^{N} \int_{t_i + b_i}^{t_i + b_i + a_i} |f(t, x^0(t), v^i) - f(t, x^0(t), u^0(t))| \, dt.$$

One sees that $y(T; a)$ is differentiable with respect to a for sufficiently small $|a|$, and therefore we have

$$y(T; a) = \frac{\partial}{\partial a} y(T; 0) + |a| \varepsilon(a),$$

where $\varepsilon(a) \to 0$ as $|a| \to 0$. Consequently, there exists $\delta_1 > 0$, such that for all $a \in A_N$ with $|a| < \delta_1$, one has

$$|y(T; a)| \le \left(\left| \frac{\partial}{\partial a} y(T; 0) \right| + 1 \right) |a| = L|a|.$$

We can now proceed to the construction of a 'trajectory' $x(t; a)$, associated with the control $u(t; a)$. We do not look yet for a trajectory satisfying the constraints (6.2.3), (6.2.4).

Proposition 6.2.2. *Let (x^0, u^0) be an admissible pair, and the set S fixed. Then, there exists a neighborhood V_S of $\theta \in R^N$, such that, for all $a \in V_S \cap A_N$, the integral equation*

$$x(t) = f(t) + \int_0^t [k(t, s, x(s)) + g(s, x(s), u(s; a))]\, ds \qquad (6.2.9)$$

has a solution $x = x(t; a)$, defined on $[0, T]$.

Proof. For fixed $b > 0$, let

$$B_b = \{(t, s, x); 0 \le s \le t \le T, |x - x^0(t)| \le b\}.$$

Since B_b is a compact set, the boundedness of k and $\partial k / \partial x$ is assured on B_b.

Now define the operator T by

$$(Tx)(t) = f(t) + \int_0^t [k(t, s, x(s)) + g(s, x(s), u(s; a))]\, ds, \qquad (6.2.10)$$

where $x \in C([0, T], R^n)$. For every fixed $a \in A_N$, the operator T takes the space $C([0, T], R^n)$ into itself. Obviously T is continuous on $C([0, T], R^n)$.

We shall prove that the usual method of successive approximations, starting with $x^0(t)$, leads to a convergent sequence in $C([0, T], R^n)$, provided $|a|$ is small enough. In other words, we shall prove that the sequence $\{T^m x^0\} \subset C([0, T], R^n)$ converges. Let $x^m = T^m x^0$, $m \ge 1$. Then, taking into account that x^0 satisfies equation (6.2.1) with $u = u^0(t)$, we obtain

$$|x^1(t) - x^0(t)| \le \int_0^t |g(s, x^0(s), u(s; a)) - g(s, x^0(s), u^0(s))|\, ds,$$

which by Lemma 6.2.1 yields

$$|x^1(t) - x^0(t)| \le L|a|, \quad t \in [0, T]. \qquad (6.2.11)$$

Consequently, we have $|x^1(t) - x^0(t)| \le b$ on $[0, T]$, as soon as $L|a| \le b$. This is a restriction on $|a|$, which has to be compared with $|a| < \delta_1$, $\delta_1 > 0$, as in Lemma 6.2.1. Similarly,

$$|x^2(t) - x^1(t)| \le \int_0^t |k(t, s, x^1(s)) - k(t, s, x^0(s))|\, ds$$

$$+ \int_0^t |g(s, x^1(s), u(s; a)) - g(s, x^0(s), u(s; a))|\, ds$$

$$\le (K + N_g) \int_0^t |x^1(s) - x^0(s)|\, ds,$$

where $K > 0$ is chosen such that

$$\left|\frac{\partial k}{\partial x}\right| \leq K \text{ in } B_b,$$

while N_g is as in (6.2.6). From (6.2.11) and the above estimate for $|x^2(t) - x^1(t)|$ we obtain the inequality

$$|x^2(t) - x^1(t)| \leq ML|a|t, \tag{6.2.12}$$

with $M > 0$ such that $K + N_g \leq M$. We see from (6.2.12) that $|x^2(t) - x^1(t)| \leq b$ on $[0, T]$, provided a is such that $MLT|a| \leq b$, or $|a| \leq (b/L)(MT)^{-1}$.

We will now assume

$$|a| \leq \min\left(\delta_1, \frac{b}{L}e^{-MT}\right). \tag{6.2.13}$$

This assumption implies both conditions $|a| \leq b/L$ and $|a| \leq b/L(MT)^{-1}$, encountered above.

It is now a straightforward procedure to obtain the following estimates on $[0, T]$:

$$|x^m(t) - x^{m-1}(t)| \leq L|a|\frac{(Mt)^{m-1}}{(m-1)!}, \tag{6.2.14}$$

$$|x^m(t) - x^0(t)| \leq b. \tag{6.2.15}$$

We start with (6.2.14), which reduces to (6.2.12) for $m = 2$, and to (6.2.11) for $m = 1$. If, by induction, we assume $|x^k(t) - x^0(t)| \leq b$ for $k = 1, 2, \ldots, m$, then the recursion formula $x^{m+1} = Tx^m$ leads to

$$|x^{m+1}(t) - x^m(t)| \leq \int_0^t |k(t, s, x^m(s)) - k(t, s, x^{m-1}(s))| \, ds$$

$$+ \int_0^t |g(s, x^m(s), u(s; a)) - g(s, x^{m-1}(s), u(s; a))| \, ds$$

$$\leq M \int_0^t |x^m(s) - x^{m-1}(s)| \, ds \text{ on } [0, T].$$

Consequently, we have the simple recursion inequality

$$|x^{m+1}(t) - x^m(t)| \leq M \int_0^t |x^m(s) - x^{m-1}(s)| \, ds \text{ on } [0, T]. \tag{6.2.16}$$

Let $m = 2, 3, \ldots$ successively; we obtain the validity of (6.2.14) for every m, noticing at the same time that

$$|x^m(t) - x^0(t)| \leq |x^m(t) - x^{m-1}(t)| + \cdots + |x^1(t) - x^0(t)|$$

$$\leq L|a| \left[\frac{(Mt)^{m-1}}{(m-1)!} + \frac{(Mt)^{m-2}}{(m-2)!} + \cdots + 1 \right]$$

$$< L|a|e^{MT} \leq Le^{MT} \frac{b}{L} e^{-MT} = b.$$

Therefore, the validity of (6.2.14), (6.2.15) is established for any m, which means that $x^m(t) \to x(t) \equiv x(t;a)$ uniformly on $[0, T]$. One has in $[0, T]$

$$x(t; a) = f(t) + \int_0^t [k(t, s, x(s; a)) + g(s, x(s; a), u(s; a))]\, ds,$$

as well as $|x(t; a) - x^0(t)| \leq b$, provided a satisfies (6.2.13).

The uniqueness of this solution can be obtained, within the class of those functions $x(t)$ verifying $|x(t) - x^0(t)| \leq b$, in the standard way.

This ends the proof of Proposition 6.2.2.

Under the assumptions formulated at the beginning of this section, one can obtain even the differentiability of the solution $x(t; a)$, with respect to a. We shall formulate this property in the following.

Proposition 6.2.3. *The function $x(t, a)$ defined in Proposition* 6.2.2 *is differentiable with respect to a, at $a = \theta$, and satisfies the system*

$$\frac{\partial}{\partial a_j} x(t; a)\Big|_{a=\theta} = g(t_j, x^0(t_j), v_j) - g(t_j, x^0(t_j), u^0(t_j))$$

$$+ \int_{t_j}^t \left[\frac{\partial}{\partial x} k(t, s, x^0(s)) + \frac{\partial}{\partial x} g(s, x^0(s), u^0(s)) \right] \left[\frac{\partial}{\partial a_j} x(s; a) \right]_{a=\theta} ds$$

$$(6.2.17)$$

on $[t_j, T]$, $j = 1, 2, \ldots, N$.

Proof. Let us fix j, and consider the linear integral equation

$$y(t) = g(t_j, x^0(t_j), v_j) - g(t_j, x^0(t_j), u^0(t_j))$$

$$+ \int_{t_j}^t \left[\frac{\partial}{\partial x} k(t, s, x^0(s)) + \frac{\partial}{\partial x} g(s, x^0(s), u^0(s)) \right] y(s)\, ds \qquad (6.2.18)$$

on the interval $[t_j, T]$. This integral equation is linear in y, and has a continuous kernel. Therefore, (6.2.18) has a unique solution $y = y^j(t)$ on the interval $[t_j, T]$.

Let $\varepsilon > 0$ be sufficiently small, and consider the vector $a^i = (0, \ldots, 0, \varepsilon, 0, \ldots, 0)$ $\in A_N$, i.e., the vector whose coordinates are all zero, except for the coordinate of rank i, $1 \leq i \leq N$. Now we consider for fixed i the vector

$$\Delta(t;\varepsilon) = \frac{x(t;a^i) - x^0(t)}{\varepsilon} - y^i(t), \quad t_i \le t \le T. \qquad (6.2.19)$$

By elementary transformations we find the following inequality for $\Delta(t;\varepsilon)$:

$$
|\Delta(t;\varepsilon)| \le \left| \varepsilon^{-1} \int_{t_i}^{t} \left[k(t,s,x(s;a^i)) - k(t,s,x^0(s)) \right. \right.
$$

$$
\left. \left. - \varepsilon \frac{\partial}{\partial x} k(t,s,x^0(s)) y^i(s) \right] ds \right|
$$

$$
+ \left| \varepsilon^{-1} \int_{t_i}^{t} \left[g(s,x(s;a^i),u^0(s)) - g(s,x^0(s),u^0(s)) \right. \right.
$$

$$
\left. \left. - \varepsilon \frac{\partial}{\partial x} g(s,x^0(s),u^0(s)) y^i(s) \right] ds \right|
$$

$$
+ \left| \varepsilon^{-1} \int_{t_i}^{t_i+\varepsilon} \left[g(s,x(s;a^i),v_i) - g(s,x^0(s),u^0(s)) \right] ds \right.
$$

$$
\left. - \left[g(t_i,x^0(t_i),v_i) - g(t_i,x^0(t_i),u^0(t_i)) \right] \right|
$$

$$
+ \left| \int_{t_i}^{t_i+\varepsilon} \frac{\partial}{\partial x} g(s,x^0(s),u^0(s)) y^i(s) ds \right| = \sum_{j=1}^{4} \omega_j(\varepsilon).
$$

We shall now estimate each $\omega_j(\varepsilon)$, $j = 1, 2, 3, 4$, separately. For $\omega_1(\varepsilon)$ we have obviously

$$
\omega_1(\varepsilon) \le \left| \varepsilon^{-1} \int_{t_i}^{t} \left[k(t,s,x(s;a^i)) - k(t,s,x^0(s)) \right. \right.
$$

$$
\left. \left. - \frac{\partial}{\partial x} k(t,s,x^0(s)) \{ x(s;a^i) - x^0(s) \} \right] ds \right|
$$

$$
+ \int_{t_i}^{t} \left| \frac{\partial}{\partial x} k(t,s,x^0(s)) \right| \cdot |\Delta(s,\varepsilon)| ds.
$$

From the estimate found for $|x^m(t) - x^0(t)|$ at the end of the proof of Proposition 6.2.2, we obtain

$$|x(t;a) - x^0(t)| \le L|a| \exp(MT),$$

and for sufficiently small $\varepsilon > 0$

$$|x(t;a) - x^0(t)| \le L\varepsilon \exp(MT), \qquad (6.2.20)$$

on $[0, T]$. Taking into account (6.2.20), the above inequality for $\omega_1(\varepsilon)$ yields because of the differentiability of $k(t, s, x)$:

$$\omega_1(\varepsilon) \le \sigma_1(\varepsilon) L \exp(MT) + \int_{t_i}^{t} \left| \frac{\partial}{\partial x} k(t,s,x^0(s)) \right| \cdot |\Delta(s,\varepsilon)| ds, \qquad (6.2.21)$$

where $\sigma_1(\varepsilon) \to 0$ as $\varepsilon \to 0$. A similar estimate holds for $\omega_2(\varepsilon)$:

$$\omega_2(\varepsilon) \le \sigma_2(\varepsilon)L\exp(MT) + \int_{t_i}^t \left| \frac{\partial}{\partial x} g(s, x^0(s), u^0(s)) \right| \cdot |\Delta(s, \varepsilon)|\, ds, \qquad (6.2.22)$$

where $\sigma_2(\varepsilon) \to 0$ as $\varepsilon \to 0$. As far as $\omega_3(\varepsilon)$ and $\omega_4(\varepsilon)$ are concerned, it is obvious that they both tend to zero as $\varepsilon \to 0$. Consequently, $\Delta(t; \varepsilon)$ given by (6.2.19) satisfies the following estimate:

$$|\Delta(t; \varepsilon)| \le [\sigma_1(\varepsilon) + \sigma_2(\varepsilon)]L\exp(MT)$$

$$+ \int_{t_i}^t \left\{ \left| \frac{\partial}{\partial x} k(t, s, x^0(s)) \right| + \left| \frac{\partial}{\partial x} g(s, x^0(s), u^0(s)) \right| \right\} |\Delta(s, \varepsilon)|\, ds$$

$$+ \omega_3(\varepsilon) + \omega_4(\varepsilon), \quad t \in (t_i, T].$$

We can write the above inequality in a more concise form, namely

$$|\Delta(t; \varepsilon)| \le \sigma(\varepsilon) + \int_{t_i}^t \left\{ \left| \frac{\partial}{\partial x} k(\ldots) \right| + \left| \frac{\partial}{\partial x} g(\ldots) \right| \right\} |\Delta(s; \varepsilon)|\, ds, \qquad (6.2.23)$$

where

$$\sigma(\varepsilon) = [\sigma_1(\varepsilon) + \sigma_2(\varepsilon)]L\exp(MT) + \omega_3(\varepsilon) + \omega_4(\varepsilon) \to 0$$

as $\varepsilon \to 0$. The linear integral inequality (6.2.23) yields from Gronwall's lemma

$$|\Delta(t; \varepsilon)| \le \sigma(\varepsilon)\exp[(K + N_g)T], \quad t \in [t_i, T]. \qquad (6.2.24)$$

From (6.2.24) one obtains $\Delta(t; \varepsilon) \to 0$ as $\varepsilon \to 0$ (uniformly in t), which means that

$$\frac{\partial}{\partial a^i} x(t; a)\big|_{a=0} = y^i(t), \qquad (6.2.25)$$

and this is what we wanted to show.

This ends the proof of Proposition 6.2.3.

Remark.   It is possible to find an integral representation of the derivative whose existence is proven in Proposition 6.2.3. As shown above, $y^i(t)$ from (6.2.25) is the solution of the integral equation (6.2.18) on $[t_j, T]$. Let us define

$$z^j(t) = \begin{cases} 0 & \text{for } 0 \le t < t_j, \\ y^j(t) & \text{for } t_j \le t \le T. \end{cases} \qquad (6.2.26)$$

The integral equation (6.2.18) becomes

$$z^j(t) = [g(t_j, x^0(t_j), v^j) - g(t_j, x^0(t_j), u^0(t_j))]\chi_j(t),$$

$$+ \int_0^t \left[\frac{\partial}{\partial x} k(t, s, x^0(s)) + \frac{\partial}{\partial x} g(s, x^0(s), u^0(s)) \right] z^j(s)\, ds, \qquad (6.2.27)$$

where $\chi_j(t)$ is the characteristic function of the interval $[t_j, T]$.

In Section 1.3, we showed that for each (continuous) Volterra kernel, there exists a resolvent kernel. In the present situation, the matrix kernel is

$$\frac{\partial}{\partial x} k(t, s, x^0(s)) + \frac{\partial}{\partial x} g(s, x^0(s), u^0(s)), \tag{6.2.28}$$

and we can claim only its measurability. At the same time, because of our assumption (5), the kernel (6.2.28) is bounded. Therefore, it is in L^∞. The construction of the corresponding resolvent kernel is similar to the construction corresponding to the continuous case. For details, see the discussion of this case in Neustadt [1]. Hence, we can represent the (unique) solution of equation (6.2.27) in the form

$$z^j(t) = [g(t_j, x^0(t_j), v^j) - g(t_j, x^0(t_j), u^0(t_j))]\chi_j(t)$$

$$+ \int_0^t R(t, s)[g(t_j, x^0(t_j), v^j) - g(t_j, x^0(t_j), u^0(t_j))]\chi_j(s)\,ds, \tag{6.2.29}$$

where $R(t, s)$ is the resolvent kernel associated with (6.2.28). It also belongs to L^∞.

Taking into account (6.2.26), the formula (6.2.29) reduces to

$$\frac{\partial}{\partial a_j} x(t; a)\big|_{a=\theta} = y^j(t) = \left[I - \int_{t_j}^t R(t, s)\,ds\right] \cdot [g(t_j, x^0(t_j), v^j) - g(t_j, x^0(t_j), u^0(t_j))].$$

$$\tag{6.2.30}$$

This formula yields the integral representation we mentioned above. As usual, I stands for the unit matrix.

We can now discuss the optimal problem we formulated at the beginning of this section. Two more auxiliary propositions will be established, before we can approach the problem in full generality. These propositions are based on optimization in finite dimensions (mathematical programming), and they rely on the Lagrange method of multipliers.

Lemma 6.2.4. *Let the set S be fixed, and assume (x^0, u^0) is an optimal solution for the problem: minimize $g^0(x(T))$, under the constraints (6.2.1)–(6.2.4). Then, there exist multipliers $c_0, c_1, \ldots, c_q$, such that*

$$c_j \geq 0, \quad j = 0, 1, 2, \ldots, p, \tag{6.2.31}$$

$$c_j g_j(x^0(T)) = 0, \quad j = 1, 2, \ldots, p, \tag{6.2.32}$$

$$0 \leq d(c)\left[I - \int_{t_j}^T R(T, s)\,ds\right][g(t_j, x^0(t_j), v^j) - g(t_j, x^0(t_j), u^0(t_j))], \tag{6.2.33}$$

for $j = 1, 2, \ldots, N$, where

$$d(c) = c_0 \frac{\partial}{\partial x} g^0(x^0(T)) + \sum_{j=1}^q c_j \frac{\partial}{\partial x} g_j(x^0(T)). \tag{6.2.34}$$

Proof. Since (x^0, u^0) is an optimal pair for our problem, it means that the finite-dimensional problem

$$\text{minimize } \{g^0(x(T;a)); a \in V_s \cap A_N, g_j(x(T;a)) \leq 0,$$
$$1 \leq j \leq p, g_j(x(T;a)) = 0, p + 1 \leq j \leq q\},$$

with V_s as indicated in Proposition 6.2.2, has an optimal solution for $a = \theta$. Hence, according to a standard result in mathematical programming (see, for instance, Pshenichnyi [1]), there exist multipliers c_j, $0 \leq j \leq q$, such that

$$\sum_{j=0}^{q} |c_j| = 1, \qquad (6.2.35)$$

with both (6.2.31) and (6.2.32) satisfied, and for each $b \in V_s \cap A_N$

$$0 \leq \sum_{i=1}^{N} b_i \left\{ \frac{\partial}{\partial a_i} \left[c_0 g^0(x(T;a)) + \sum_{j=1}^{q} g_j(x(T;a)) \right] \right\}_{a=\theta}.$$

If one takes $b_i = 0$ except for one index, then we obtain from above the inequality

$$0 \leq \frac{\partial}{\partial a_i} \left[c_0 g^0(x(T;a)) + \sum_{j=1}^{q} g_j(x(T;a)) \right] \Big|_{a=\theta}$$

which leads to (6.2.33), if we take into account (6.2.30), (6.2.34).

Lemma 6.2.5. *If (x^0, u^0) is an optimal solution for the problem formulated in Lemma 6.2.4, then there exist multipliers c_j, $j = 0, 2, \ldots, q$, such that (6.2.35) holds, and the inequality*

$$0 \leq d(c) \left[I - \int_t^T R(T, s) \, ds \right] [g(t, x^0(t), v) - g(t, x^0(t), u^0(t))] \qquad (6.2.36)$$

is satisfied a.e. for $t \in [0, T]$, and all $v \in U$.

Proof. Let S denote an arbitrary set as described above when defining perturbed controls. From Lemma 6.2.4 we obtain the existence of the multipliers c_j, $j = 0, 1, \ldots, q$, such that (6.2.31)–(6.2.33) hold true. Let $M(S)$ be the set of all possible multipliers attached to S. The set $M(S)$ is nonempty and compact in R^{q+1} (see (6.2.35)). We want to show that

$$\bigcap_S M(S) \neq \emptyset.$$

Let us start with a finite number of sets S, say S_i, $i = 1, 2, \ldots, r$, and define

$$S_0 = \bigcup_{i=1}^{r} S_i.$$

It can be easily seen that S_0 has the same form as each S_i (if necessary, we rearrange the elements). Hence, $M(S_0) \neq \emptyset$. Moreover, $M(S_0) \subset M(S_i)$ for $i =$

1, 2, ..., r, and this implies

$$M(S_0) \subseteq \bigcap_{i=1}^{r} M(S_i).$$

Therefore, the set $\{M(S)\}$, with all possible S as described above, has the property of finite intersection. And a collection of compact sets (in this case, in R^{q+1}) with the finite intersection property has a nonempty intersection. Therefore, if the multipliers c_j, $0 \leq j \leq q$, are an element in $\bigcap_S \{M(S)\}$, the conditions (6.2.31), (6.2.32) and (6.2.35) are satisfied.

Let us now notice that the set of the points that are non-Lebesque points for the map $s \rightarrow g(s, x^0(s), u^0(s))$ is a set of (Lebesgue) measure zero. Moreover, for any $(s, v) \in [0, T] \times U$, where s is a Lebesque point for the maps $s \rightarrow g(s, x^0(s), u^0(s))$, it belongs to some set S. Hence, we have

$$0 \leq d(c)\left[I - \int_{t}^{T} R(T, s)\, ds\right]\left[g(t, x^0(t), v) - g(t, x^0(t), u^0(s))\right]$$

for all $v \in U$, and almost all $t \in [0, T]$.

This ends the proof of Lemma 6.2.5.

We are now able to provide a set of necessary conditions for the optimality of a pair (x^0, u^0), with respect to the basic problem considered in this section. The next result is similar in content to the Pontryagin's maximum principle, which corresponds to ordinary differential systems (i.e., when $k(t, s, x) \equiv \theta$, and $f(t) = \text{Const.}$).

Theorem 6.2.6. *Consider the optimal control problem*

$$\text{minimize } g^0(x(T)),$$

under constraints (6.2.1)–(6.2.4), and assume the validity of conditions (1)–(5) formulated above. Then if (x^0, u^0) is an optimal pair, there exist constant multipliers λ_j, $0 \leq j \leq q$, and a function $\psi: [0, T] \rightarrow R^n$, such that:

(i) $\displaystyle\sum_{j=0}^{q} |\lambda_j| = 1, \quad \lambda_j \leq 0, 0 \leq j \leq p, \quad \lambda_j g_j(x^0(T)) = 0, \quad 1 \leq j \leq p;$

(ii) $\displaystyle\psi(t) = -d(\lambda)\left[\frac{\partial}{\partial x}k(T, t, x^0(t)) + \frac{\partial}{\partial x}g(t, x^0(t), u^0(t))\right]$

$$+ \int_{t}^{T} \psi(s)\left[\frac{\partial}{\partial x}k(s, t, x^0(t)) + \frac{\partial}{\partial x}g(t, x^0(t), u^0(t))\right]ds;$$

where

$$-d(\lambda) = \lambda_0 \frac{\partial}{\partial x}g^0(x^0(T)) + \sum_{j=1}^{q} \lambda_j \frac{\partial}{\partial x}g_j(x^0(T));$$

(iii) for almost all $t \in [0, T]$,

$$\left[d(\lambda) - \int_t^T \psi(s)\,ds \right] \cdot g(t, x^0(t), u^0(t)) = \max_{v \in U} \left\{ \left[d(\lambda) - \int_t^T \psi(s)\,ds \right] \cdot g(t, x^0(t), v) \right\}.$$

Proof. Lemma 6.2.5 yields the existence of multipliers c_j, $0 \le j \le q$, such that (6.2.31), (6.2.32) and (6.2.35) hold. Let $\lambda_j = -c_j$, $0 \le j \le q$. Then (i) is obviously · verified. From (6.2.36), which is part of Lemma 6.2.5, we obtain

$$0 \ge d(\lambda) \left[I - \int_t^T R(T, s)\,ds \right] \cdot [g(t, x^0(t), v) - g(t, x^0(t), u^0(t))], \quad (6.2.37)$$

a.e. on $[0, T]$, and for all $v \in U$.

Let us now define

$$\psi(t) = d(\lambda) R(T, t), \quad t \in [0, T]. \qquad (6.2.38)$$

From the definition of the resolvent kernel R, we have on $[0, T]$

$$\psi(t) = -d(\lambda) \left[\frac{\partial}{\partial x} k(T, t, x^0(t)) + \frac{\partial}{\partial x} g(t, x^0(t), u^0(t)) \right]$$

$$+ \int_t^T \psi(s) \left[\frac{\partial}{\partial x} k(s, t, x^0(t)) + \frac{\partial}{\partial x} g(t, x^0(t), u^0(t)) \right] ds. \qquad (6.2.39)$$

Indeed, $R(t, s)$ satisfies the so-called equation of the resolvent kernel

$$R(t, s) = -\left[\frac{\partial}{\partial x} k(t, s, x^0(s)) + \frac{\partial}{\partial x} g(s, x^0(s), u^0(s)) \right]$$

$$+ \int_s^t R(t, u) \left[\frac{\partial}{\partial x} k(u, s, x^0(s)) + \frac{\partial}{\partial x} g(s, x^0(s), u^0(s)) \right] du,$$

which combined with (6.2.38) leads immediately to (6.2.39). Hence, (ii) is proven.

If we take now into account (6.2.37), (6.2.38), we can write

$$\left[d(\lambda) - \int_t^T \psi(s)\,ds \right] \cdot g(t, x^0(t), u^0(t)) \ge \left[d(\lambda) - \int_t^T \psi(s)\,ds \right] \cdot g(t, x^0(t), v),$$

for almost all $t \in [0, T]$, and all $v \in U$. The last inequality establishes (iii).

Theorem 6.2.6 is thus proven.

Remark 1. The proof of the maximum principle for the control processes governed by Volterra integral equations can be extended to the more general case when equation (6.2.1) is substituted by the more general one

$$x(t) = f(t) + \int_0^t [k(t, s, x(s)) + h(t, s, x(s)) g(s, x(s), u(s))]\,ds,$$

where h is smooth enough. See this case in Carlson [1], from which we have adapted to the case $h \equiv 1$.

Remark 2. In the special case $k(t, s, x) \equiv \theta$, $f(t) = $ Const., equation (6.2.1) becomes

$$x(t) = C + \int_0^t g(s, x(s), u(s)) \, ds,$$

which is obviously equivalent to

$$x'(t) = g(t, x(t), u(t)) \text{ a.e. on } [0, T], \tag{6.2.40}$$

under the initial condition $x(0) = C$.

Then Theorem 6.2.6 provides the classical result (Pontryagin) about the necessary conditions for an optimal pair (x^0, u^0). Indeed, let us assume that we want to minimize the functional $g^0(x(T))$, over the set of all x satisfying (6.2.40), and the initial condition $x(0) = x^0 \in R^n$. We assume the trajectory x, corresponding to a given measurable control $u: [0, T] \to U$ is not subject to 'terminal' constraints of the form (6.2.3) or (6.2.4). In other words, we can take $g_j(x) \equiv 0$, $j = 1, 2, \ldots, q$.

If we let

$$\phi(t) = -\frac{\partial}{\partial x} g^0(x^0(T)) - \int_t^T \psi(s) \, ds, \quad t \in [0, T],$$

then ϕ satisfies the equation of the 'adjoint' system (obtained immediately from (6.2.39)), $\phi(T) = -(\partial/\partial x) g^0(x^0(T))$, while the condition (iii) from Theorem 6.2.6 takes the well-known form

$$H(t, x^0(t), u^0(t), \phi(t)) \geq H(t, x^0(t), v, \phi(t))$$

for any $v \in U$ and $t \in [0, T]$, where

$$H(t, x, u, \phi) = \phi \cdot g(t, x, u).$$

We shall conclude this section with a discussion of an elementary optimization problem in which integral operators of Fredholm type are involved. There is no connection between this problem and the one discussed above, regarding control processes described by Volterra integral equations. I decided to include this problem in this section because it is another illustration of the usefulness of integral equations/operators in treating applied topics.

Let us consider the integral operator

$$y(t) = (Kx)(t) = \int_a^b k(t, s) x(s) \, ds, \tag{6.2.41}$$

and assume it is acting from $L^2([a, b], R)$ into itself. The kernel $k(t, s)$ is supposed to be square integrable on $[a, b] \times [a, b]$, i.e., satisfies the condition

$$\int_a^b \int_a^b |k(t,s)|^2 \, dt \, ds < +\infty. \tag{6.2.42}$$

In practice, $k(t,s)$ is usually smoother.

In what follows, we shall denote the left hand side of formula (6.2.41) by $y(t;x)$.

The optimization problem we want to consider can be formulated as follows: let $\bar{y}(t) \in L^2([a,b],R)$ be given, and denote

$$I(x) = \int_a^b [y(t;x) - \bar{y}(t)]^2 \, dt; \tag{6.2.43}$$

find min $I(x)$, with respect to all $x(t) \in L^2([a,b],R)$, subject to the constraint

$$|x(t)| \leq 1 \text{ a.e. on } [a,b]. \tag{6.2.44}$$

In order to determine $x_0(t) \in L^2([a,b],R)$, such that $I(x_0) = \min I(x)$ under constraint (6.2.44), we will proceed as follows. For every ε, $0 \leq \varepsilon \leq 1$, let us consider the control

$$x_\varepsilon(t) = x_0(t) + \varepsilon[x(t) - x_0(t)],$$

where $x(t) \in L^2([a,b],R)$ is a fixed control satisfying (6.2.44). It can easily be seen that $x_\varepsilon(t)$ is admissible, i.e., it is in $L^2([a,b],R)$ and verifies (6.2.44). Then, a necessary condition for $x_0(t)$ to provide the minimum of $I(x)$ is obviously

$$\frac{d}{d\varepsilon} I(x_\varepsilon)|_{\varepsilon=0} \geq 0. \tag{6.2.45}$$

Actually, in (6.2.45) we understand the right derivative at $\varepsilon = 0$. Without any difficulty, we find out that (6.2.45) can be rewritten as

$$\int_a^b [y(t;x_0) - \bar{y}(t)][y(t;x) - y(t;x_0)] \, dt \geq 0. \tag{6.2.46}$$

This inequality must be verified by any $x \in L^2([a,b],R)$, under constraint (6.2.44).

Of course, (6.2.46) can be also written in the form

$$\int_a^b [y(t;x_0) - \bar{y}(t)] \int_a^b k(t,s)[x(s) - x_0(s)] \, ds \, dt \geq 0, \tag{6.2.47}$$

if we have in mind (6.2.41) and the meaning of $y(t;x)$. Since $x(s)$ is arbitrary in $L^2([a,b],R)$, subject to (6.2.44), we can choose it in the following manner: for fixed $\theta \in (a,b)$, we let $x_\theta(s) = x = \text{const.}, |s - \theta| \leq \varepsilon$, and $x_\theta(s) = x_0(s), |s - \theta| > \varepsilon$, with $|x| \leq 1$. For $\theta = a$ or $\theta = b$ one proceeds in an obvious manner. It is easy to see that $x_\theta(s)$ is an admissible control. Substituting this control in (6.2.47), dividing both sides by 2ε, and then letting $\varepsilon \to 0$ we obtain the inequality

$$\int_a^b [y(t;x_0) - \bar{y}(t)]k(t,\theta)[x - x_0(\theta)] \, dt \geq 0. \tag{6.2.48}$$

The last inequality can be written as

$$x \int_a^b [y(t; x_0) - \bar{y}(t)]k(t, \theta)\,dt \geq x_0(\theta) \int_a^b [y(t; x_0) - \bar{y}(t)]k(t, \theta)\,dt, \quad (6.2.49)$$

for every $x \in [-1, +1]$, and for almost all $\theta \in [a, b]$. The only way to satisfy this inequality, under the conditions stated above, is to choose

$$x_0(t) = -\text{sign}\left\{ \int_a^b [y(s; x_0) - \bar{y}(s)]k(s, t)\,ds \right\}, \quad t \in [a, b]. \quad (6.2.50)$$

On the other hand, $y(s; x_0)$ is defined by (6.2.41). If we substitute $y(s; x_0)$ in (6.2.50), we obtain the following 'integral' equation

$$x_0(t) = -\text{sign}\left\{ \int_a^b k(s, t)\left[\int_a^b k(s, u)x_0(u)\,du - \bar{y}(t) \right]ds \right\}. \quad (6.2.51)$$

Therefore, the optimal control $x_0(t)$, for the optimization problem stated above, must satisfy the 'integral' equation (6.2.51).

Remark 1. Equations of the form (6.2.41) do appear in connection with the optimal theory of distributed parameters processes (i.e., dynamic processes described by partial differential equations). In this case $k(t, s)$ is a function defined by means of the Green's function of the corresponding boundary value problem. For the discussion of such problems, see for instance Lions [1].

Remark 2. If instead of (6.2.41) we start with a Fredholm integral equation

$$y(t) = x(t) + \int_a^b k(t, s)y(s)\,ds, \quad (6.2.52)$$

for which 1 is not an eigenvalue, then y can be expressed in terms of x and of the resolvent kernel attached to k.

6.3   Stability of nuclear reactors (A)

In this section we shall deal with stability problems for a class of integrodifferential systems occurring in the theory of nuclear reactors. At the same time, this investigation will provide a good example for the problem of stability (more exactly, asymptotic stability) of nonlinear integrodifferential systems with infinite delay by means of transform (Laplace) theory.

A certain amount of auxiliary material is necessary, and we shall begin with the investigation of stability for a linear integrodifferential system which appears in the process of linearization of the nonlinear system encountered in the dynamics of nuclear reactors.

The first problem we want to discuss is the validity of a variation of constants formula for the linear system

$$\dot{x}(t) = (Ax)(t) + f(t), \quad t > 0, \tag{6.3.1}$$

under the initial conditions

$$x(t) = h(t) \text{ for } t < 0, \quad x(0) = x^0 \in R^n. \tag{6.3.2}$$

The operator A in (6.3.1) is formally given by

$$(Ax)(t) = \sum_{j=0}^{\infty} A_j x(t - t_j) + \int_0^t B(t - s)x(s) \, ds, \tag{6.3.3}$$

where A_j are square matrices of order n and $B: [0, \infty) \to \mathscr{L}(R^n, R^n)$ is locally integrable. It is also assumed that the matrices A_j are such that

$$\sum_{j=0}^{\infty} |A_j| < +\infty, \tag{6.3.4}$$

with $|A_j|$ standing for a matrix norm (the same for all j), and $t_j \geq 0, j = 0, 1, 2, \ldots$.

The first remark we want to make about the operator A is that a more general form, such as

$$(Ax)(t) = \sum_{j=0}^{\infty} A_j x(t - t_j) + \int_{-\infty}^t B(t - s)x(s) \, ds,$$

does not lead to a substantially new case, because the first condition (6.3.2) allows us to split the integral from $-\infty$ to t in two parts, the first one

$$\int_{-\infty}^0 B(t - s)x(s) \, ds = \int_{-\infty}^0 B(t - s)h(s) \, ds$$

being a known function. Of course, we assume here that all the integrals occurring in the above discussion are convergent in some sense.

Hence, we can concentrate on the system (6.3.1), with initial data (6.3.2), under the assumptions that A is given by (6.3.3).

A second remark we want to make about the operator A is the fact that A is a Volterra (nonanticipative) operator. Indeed, as we can see from (6.3.3), in order to determine $(Ax)(t)$ we only use values of $x(s)$ for $s \leq t$, because it has been assumed that $t_j \geq 0, j = 0, 1, 2, \ldots$.

Finally, a third remark about the operator A is the fact that this operator is acting on the space $L^2_{loc}(R_+, R^n)$, provided we agree to extend any $x \in L^2_{loc}(R_+, R^n)$ to the whole real axis such that $x \in L^2((-\infty, t], R^n)$ for any $t \in R_+$. In particular, one can always set $x(t) \equiv 0$ on the negative half-axis. It is also understood that (6.3.4) is verified, while B is locally integrable from R_+ into $\mathscr{L}(R^n, R^n)$.

In order to check the property that $A: L^2_{loc}(R_+, R^n) \to L^2_{loc}(R_+, R^n)$, we will rely on elementary properties of L^2-spaces and convolutions.

First of all, because of condition (6.3.4), and the fact that $x \in L^2((-\infty, t], R^n)$

for $t \in R_+$, the convergence of the series

$$\sum_{j=0}^{\infty} A_j x(t - t_j)$$

in the norm of $L^2(I, R^n)$ can be assured when I is any compact interval of the real axis.

Second, the convolution product

$$\int_0^t B(t - s)x(s)\, ds$$

belongs to $L^2(I, R^n)$, for any compact interval $I \subset R_+$, because $B(t)$ is locally integrable on R_+. See formula (2.3.9) for the estimate of the L^2-norm of the convolution product.

Consequently, the operator (6.3.3) acts from $L^2_{loc}(R_+, R^n)$ into itself, provided (6.3.4) holds, and $B(t)$ is locally integrable on R_+. The inequality

$$|Ax|_{L^2} \le A_0(T)|x|_{L^2}, \tag{6.3.5}$$

with

$$A_0(T) = \sum_{j=0}^{\infty} |A_j| + \int_0^T |B(s)|\, ds, \tag{6.3.6}$$

on any interval belonging to $[0, T]$, $T > 0$, is a consequence of the remarks made above, and of the formula (2.3.9).

The above discussion about the operator A shows that A is a Volterra type operator on $L^2_{loc}(R_+, R^n)$. The estimate (6.3.5) shows that we can apply the conclusions reached in Section 3.3 to the operator A, thus obtaining the validity of a variation of constants formula for the system (6.3.1), under initial conditions (6.3.2).

More precisely, we have in mind the formula (3.3.6), which adapted to our case becomes

$$x(t) = X(t, 0)x^0 + \int_0^t X(t, s)f(s)\, ds + \int_0^t X(t, s)(\tilde{B}h)(s)\, ds, \quad t \in R_+. \tag{6.3.7}$$

The operator $\tilde{B}$ appearing in the second integral in (6.3.7) is given by

$$(\tilde{B}h)(t) = \sum_{j=0}^{\infty} A_j h(t - t_j), \quad t \in R_+, \tag{6.3.8}$$

where h is extended to the whole real axis by letting $h(t) \equiv 0$ for $t > 0$. It is obvious that $\tilde{B}h \in L^2_{loc}(R_+, R^n)$, provided $h \in L^2(R_-, R^n)$.

Because of the general form of the operator A, it is possible to simplify formula (6.3.7) somewhat and obtain better information about the function $X(t, s)$. More precisely, we shall prove that

$$X(t, s) \equiv X(t - s, 0), \quad 0 \le s \le t < +\infty, \tag{6.3.8}$$

and $X(t, 0) \equiv X(t)$ is completely determined as the matrix solution of the equation

$$\dot{X}(t) = (AX)(t), \quad t > 0, \tag{6.3.9}$$

with the initial conditions

$$X(t) = 0 \quad \text{for } t < 0, \ X(0) = I, \tag{6.3.10}$$

where I is the unit matrix of order n.

Notice that the system (6.3.1), with initial conditions (6.3.2), can be reduced, by integration from 0 to $t > 0$, to the functional integral equation

$$x(t) = \int_0^t (Ax)(s) \, ds + \int_0^t [f(s) + (\tilde{B}h)(s)] \, ds, \tag{6.3.11}$$

while the first integral on the right hand side of (6.3.11) can be rewritten as

$$\int_0^t (Ax)(s) \, ds = \int_0^t C(t - s)x(s) \, ds, \tag{6.3.12}$$

where

$$C(t) = \sum_{j=0}^{\infty} A_j H(t - t_j) + \int_0^t B(u) \, du, \tag{6.3.13}$$

and $H(t)$ is the Heaviside function ($H(t) = 1$ for $t \ge 0$, $H(t) = 0$ for $t < 0$). The condition (6.3.4) implies the convergence of the series on the right hand side of (6.3.13) in $L^\infty(R_+, R^n)$. In any case, the function $C(t)$ in (6.3.13) belongs to $L^\infty_{\text{loc}}(R_+, \mathscr{L}(R^n, R^n))$, and hence belongs to $L^2_{\text{loc}}(R_+, \mathscr{L}(R^n, R^n))$.

The functional integral equation (6.3.11) can be rewritten in the form of an integral equation

$$x(t) = \int_0^t C(t - s)x(s) + g(t), \quad t \in R_+, \tag{6.3.14}$$

with $C(t)$ given by (6.3.13) and $g(t)$ given by

$$g(t) = \int_0^t [f(s) + (\tilde{B}h)(s)] \, ds, \quad t \in R_+. \tag{6.3.15}$$

In other words, $x(t)$ satisfies a convolution equation, which is equivalent to the initial value problem (6.3.1), (6.3.2).

For the kernel $C(t - s)$ in (6.3.14) there exists a resolvent kernel $\tilde{C}(t - s)$, belonging to $L^\infty_{\text{loc}}(R_+, \mathscr{L}(R^n, R^n))$, such that for any $g(t)$ one can represent the solution of (6.3.14) in the form

$$x(t) = \int_0^t \tilde{C}(t - s)g(s) \, ds + g(t), \quad t \in R_+. \tag{6.3.16}$$

Since in the construction of $X(t, s)$ in Section 3.3 we had

$$X(t, s) = I + \int_s^t \tilde{C}(t - u)\, du,$$

we easily obtain from the above formula $X(t, s) \equiv X(t - s, 0)$, and the formula (6.3.7) becomes

$$x(t) = X(t)x^0 + \int_0^t X(t - s)f(s)\, ds + \int_0^t X(t - s)(\tilde{B}h)(s)\, ds, \quad (6.3.17)$$

if we agree to denote $X(t, 0)$ by $X(t)$.

In order to prove that $X(t)$ is determined by the equation (6.3.9) and the initial condition (6.3.10), we shall notice that the solution of the homogeneous equation $\dot{x}(t) = (Ax)(t), t > 0$, with initial conditions $x(t) = 0$ for $t > 0$, and $x(0) = x^0 \in R^n$, is given according to (6.3.16) by $x(t) = X(t)x^0$. This leads immediately to the conclusion that $X(t)$ must satisfy (6.3.9), as well as the first condition (6.3.10). The second condition (6.3.10) follows from $X(0) = X(0, 0) = I$.

It is interesting, perhaps, to point out the fact that in order to derive the variation of constants formula (6.3.17) for the system (6.3.1), under initial data (6.3.2), we don't have to rely necessarily on formula (6.3.7) that was established in Section 3.3, for the general case of operators acting from $L^2_{\text{loc}}(R_+, R^n)$ into itself.

Indeed, we can establish the equivalence of the initial value problem (6.3.1), (6.3.2) with the convolution integral equation (6.3.14), in which $C(t)$ is given by (6.3.13) and $g(t)$ by (6.3.15), under somewhat different conditions from those listed above.

Namely, we will assume that condition (6.3.4) is verified, as well as the condition

$$B(t) \in L_{\text{loc}}(R_+, \mathscr{L}(R^n, R^n)). \quad (6.3.18)$$

Instead of the condition $h(t) \in L^2(R_-, R^n)$, we will impose the similar one

$$h(t) \in L(R_-, R^n), \quad (6.3.19)$$

which is the same as $\int_0^\infty |h(-s)|\, ds < +\infty$.

Under the above mentioned conditions, $\tilde{B}h \in L_{\text{loc}}(R_+, R^n)$, if we agree as before to let $h(t) \equiv 0$ for $t \geq 0$. Therefore all data involved in (6.3.14) have a meaning. Of course, with regard to $C(t)$ we have as seen above $C(t) \in L^\infty_{\text{loc}}(R_+, \mathscr{L}(R^n, R^n))$.

Now, we can deal with the integral equation (6.3.14) under the conditions resulting from the above hypotheses on the system (6.3.1) and initial data (6.3.2). What we have obtained is a convolution integral equation with locally L^∞-kernel, and absolutely continuous free term $(g(t))$. Of course, this Volterra equation is solvable (uniquely!) for any $g(t)$, and the solution is expressed by the formula (6.3.16). Then we proceed as we did with (6.3.16), to obtain (6.3.17).

The validity of the variation of constants formula (6.3.17) for the initial value problem (6.3.1), (6.3.2), under the conditions (6.3.14), (6.3.18) and (6.3.19), is thus established.

We shall now investigate some *stability* properties of the fundamental matrix $X(t)$, associated with the system (6.3.1). First, we shall define the so-called *symbol* attached to the operator A given by (6.3.3), assuming that (6.3.3) takes place, and $B(t)$ is such that

$$|B(t)| \in L^1(R_+, R). \qquad (6.3.20)$$

In other words, we will consider the analytic function

$$\mathscr{A}(s) = \sum_{j=0}^{\infty} A_j \exp(-t_j s) + \int_0^{\infty} B(t) \exp(-ts)\, dt, \quad \text{Re}\, s \geq 0. \qquad (6.3.21)$$

Our assumptions (6.3.3) and (6.3.20) lead immediately to the conclusion that both the series and the integral in formula (6.3.21) are absolutely convergent for $\text{Re}\, s \geq 0$. Of course, $\mathscr{A}(s)$ is analytic in the half-plane $\text{Re}\, s > 0$.

For the extensive use of the symbol in the theory of integral equations, see Gohberg and Feldman [1].

We can now formulate a stability result for the system (6.3.1), in terms of the fundamental matrix $X(t)$:

$$\dot{X}(t) = (AX)(t),\, t > 0; \quad X(0) = I; \quad X(t) = 0,\, t < 0. \qquad (6.3.22)$$

Theorem 6.3.1. *Under conditions* (6.3.4), (6.3.20) *for the operator A given by* (6.3.3), *the following properties of $X(t)$ are equivalent*:

(1) $\det[sI - \mathscr{A}(s)] \neq 0$ *for* $\text{Re}\, s \geq 0$;
(2) $|X(t)| \in L^1(R_+, R)$;
(3) $|X(t)| \in L^p(R_+, R)$, *for any p satisfying* $1 \leq p \leq \infty$.

Proof. It is easy to show that (2) $\Rightarrow$ (1). Indeed, if $X(t)$ satisfies (2), then $|\dot{X}(t)| = |(AX)(t)|$ is also in $L^1(R_+, R)$ from (6.3.3) and (6.3.20). Hence, we can take Laplace transforms on both sides of $\dot{X}(t) = (AX)(t)$, which leads to

$$s\tilde{X}(s) - I = \mathscr{A}(s)\tilde{X}(s), \quad \text{Re}\, s \geq 0. \qquad (6.3.23)$$

Consequently,

$$[sI - \mathscr{A}(s)]\tilde{X}(s) = I, \quad \text{Re}\, s \geq 0, \qquad (6.3.24)$$

which shows that (1) must hold.

Notice that the validity of (1) implies, according to (6.3.24),

$$\tilde{X}(s) = [sI - \mathscr{A}(s)]^{-1}, \qquad (6.3.25)$$

for those s for which (6.3.24) holds.

We shall now prove that (6.3.25) is indeed satisfied for $\text{Re}\, s \geq \sigma_0$, where $\sigma_0 > 0$ is conveniently chosen. If we let

$$\alpha_0 = \sum_{j=0}^{\infty} |A_j| + \int_0^{\infty} |B(s)| \, ds, \tag{6.3.26}$$

which makes sense because of our assumptions (6.3.3) and (6.3.20), then from (6.3.22) we derive easily

$$\sup_{0 \le s} |X(s)| \le \sqrt{n} + \alpha_0 \int_0^t \left(\sup_{0 \le u \le s} |X(u)| \right) ds.$$

Applying Gronwall's estimate, we obtain from above

$$|X(t)| \le \sqrt{n} \exp(\alpha_0 t), \quad t \ge 0. \tag{6.3.27}$$

Inequality (6.3.27) shows that the Laplace transform of $X(t)$ does exist at least for $\mathrm{Res} > \alpha_0$. In other words, (6.3.25) is satisfied for (at least) those s for which $\mathrm{Res} > \alpha_0$.

Let us now write (6.3.25) in the form

$$\tilde{X}(s) = \left[\frac{s}{s+1} I - \frac{\mathscr{A}(s)}{s+1} \right]^{-1} \frac{I}{s+1}. \tag{6.3.25'}$$

We know that (6.3.25') is valid for $\mathrm{Res} > \alpha_0$. If we now examine the right hand side of (6.3.25'), we notice that $I(s+1)^{-1}$ is the Laplace transform of the matrix-valued function $I \exp(-t)$. The first factor has some properties that will guarantee that the right hand side of (6.3.25') is the Laplace transform of a convenient matrix function.

First of all, notice that the quantity in the brackets in the right hand side of (6.3.25') can be written in the form

$$I - \frac{I}{s+1} - \frac{\mathscr{A}(s)}{s+1},$$

which obviously makes sense for any s, $\mathrm{Res} \ge 0$. Since $I(s+1)^{-1}$ is the Laplace transform of $I \exp(-t)$, and $\mathscr{A}(s)(s+1)^{-1}$ is the product of two Laplace transforms – hence, itself a Laplace transform – it will suffice to show that

$$\inf_{\mathrm{Res} \ge 0} \det \left[\frac{s}{s+1} I - \frac{\mathscr{A}(s)}{s+1} \right] > 0, \tag{6.3.28}$$

in order to be sure that the inverse appearing in the right hand side of (6.3.25') does exist in the function algebra of matrices representable in the form $M + \tilde{\phi}(s)$, where $M \in \mathscr{L}(R^n, R^n)$ and $\tilde{\phi}(s)$ is the Laplace transform of an n by n matrix whose entries belong to $L^1(R_+, R)$. See C. Corduneanu [5] for details on this. For the reader who is not well acquainted with Laplace transform theory, we observe that $\mathscr{A}(s)(s+1)^{-1}$ is the Laplace transform of the matrix-valued function

$$\sum_{j=0}^{\infty} A_j H(t - t_j) \exp\{-(t - t_j)\} + \int_0^t B(t - s) \exp(-s) \, ds,$$

which obviously belongs to $L^1(R_+, \mathscr{L}(R^n, R^n))$.

Let us now prove that (6.3.28) is verified. Indeed, since $\mathscr{A}(s)$ is bounded in the half plane Res ≥ 0, the matrix in the brackets of (6.3.28) tends to I as $|s| \to \infty$. Consequently, we can write

$$\inf \left| \det \left[\frac{s}{s+1} I - \frac{\mathscr{A}(s)}{s+1} \right] \right| \geq \frac{1}{2} \qquad (6.3.29)$$

for Res $\geq 0, |s| \geq \rho_0 > 0$. But in the closed semidisc Res $\geq 0, |s| \leq \rho_0$, the determinant does not vanish according to (1), and therefore it must possess a positive infimum. If we combine this conclusion with (6.3.29), we immediately obtain (6.3.28).

From (6.3.28) we derive the conclusion

$$\left[\frac{s}{s+1} I - \frac{\mathscr{A}(s)}{s+1} \right]^{-1} = I + \tilde{\phi}(s), \quad \text{Res} \geq 0, \qquad (6.3.30)$$

where $\tilde{\phi}(s)$ is the Laplace transform of a square matrix of order n, whose entries belong to $L^1(R_+, R)$. Taking (6.3.30) into account, formula (6.3.25′) becomes

$$\tilde{X}(s) = \frac{I}{s+1} + \frac{\tilde{\phi}(s)}{s+1}, \quad \text{Res} \geq 0, \qquad (6.3.31)$$

which shows that $\tilde{X}(s)$ is the Laplace transform of a matrix whose entries all belong to $L^1(R_+, R)$.

Hence, (1) $\Rightarrow$ (2) and, since (3) $\Rightarrow$ (2), we need only prove that (2) $\Rightarrow$ (3).

This last step is almost immediate if we notice, as above, that $X(t) \in L^1$ implies $\dot{X}(t) \in L^1$. This is an easy consequence of conditions (6.3.3) and (6.3.20). But $\dot{X}(t) \in L^1$ implies $X(t) \in L^\infty$. If we now combine (2) with $X(t) \in L^\infty$, condition (3) is a direct consequence.

Remark 1.   Theorem 6.3.1 is indeed a stability theorem. From $\dot{X}(t) \in L^1$ one derives the existence of the limit $\lim X(t)$ as $t \to \infty$. But $X(t) \in L^1$, which means that

$$\lim_{t \to \infty} |X(t)| = 0. \qquad (6.3.32)$$

The property (6.3.32) shows that the zero solution of $\dot{x}(t) = (Ax)(t), t > 0$, is asymptotically stable.

Remark 2.   The result established in Theorem 6.3.1 has numerous applications with regard to the asymptotic behavior of solutions of the linear system (6.3.1), under the initial conditions (6.3.2). In order to illustrate this statement, we will use the variation of constants formula (6.3.17).

For instance, if we assume that for the system (6.3.1), (6.3.2), the conditions (6.3.3), (6.3.19), (6.3.20), and $t_j \geq 0, j = 0, 1, 2, \ldots$, are satisfied, then for any $f \in L^p(R_+, R^n)$, the corresponding solution also belongs to L^p. This fact follows from Theorem 6.3.1, the properties of the convolution product given in Section 2.3, and the property $h \in L^1(R_-, R^n) \Rightarrow \tilde{B}h \in L^1(R_+, R^n)$, with $\tilde{B}$ defined by (6.3.8).

Remark 3. If we consider nonlinear systems of the form

$$\left.\begin{array}{l} \dot{x}(t) = (Ax)(t) + b\phi(\sigma(t)), \\ \sigma = \langle c, x \rangle, \end{array}\right\} \tag{6.3.33}$$

where $\phi: R \to R$ is continuous, and b, c are n-vectors, under the initial condition (6.3.2), then using formula (6.3.17) again we obtain the following nonlinear integral equation for σ:

$$\sigma(t) = \left\langle c, X(t)x^0 + \int_0^t X(t-s)(\tilde{B}h)(s)\,ds \right\rangle + \int_0^t \langle c, X(t-s)b \rangle \phi(\sigma(s))\,ds. \tag{6.3.34}$$

Such nonlinear integral equations have been investigated by many authors. See C. Corduneanu [4], Miller [1], Londen [2], [4], Staffans [1], [5].

We shall now start the investigation of stability properties of a class of systems of integrodifferential equations which occur in the *dynamics of nuclear reactors*. This kind of system constitutes a generalization of most types of integro-differential systems that have been investigated by researchers during the last two or three decades.

More precisely, we will consider the systems

$$\left.\begin{array}{l} \dot{x}(t) = (Ax)(t) + (b\rho)(t), \\ \dot{\rho}(t) = \sum_{k=1}^M \beta_k A^{-1}[\rho(t) - \eta_k(t)] - PA^{-1}[1 + \rho(t)]v(t), \\ \dot{\eta}_k(t) = \lambda_k[\rho(t) - \eta_k(t)], \quad k = 1, 2, \ldots, M, \\ v(t) = (c^*x)(t) + (\alpha\rho)(t), \end{array}\right\} \tag{6.3.35}$$

where A, b, c^* and α stand for certain difference-integral operators. More precisely, we assume that these operators are formally given by

$$\left.\begin{array}{l} (Ax)(t) = A_0 x(t) + \sum_{j=1}^\infty A_j x(t - t_j) + \int_0^t B(t-s)x(s)\,ds, \\ (b\xi)(t) = b_0 \xi(t) + \sum_{j=1}^\infty b_j \xi(t - t_j) + \int_0^t \beta(t-s)\xi(s)\,ds, \\ (c^*x)(t) = c_0^* x(t) + \sum_{j=1}^\infty c_j^* x(t - t_j) + \int_0^t d^*(t-s)x(s)\,ds, \\ (\alpha\rho)(t) = \alpha_0(t)\lambda\rho(t) + \sum_{j=1}^\infty \alpha_j \rho(t - t_j) + \int_{-\infty}^t \gamma(t-s)\rho(s)\,ds, \end{array}\right\} \tag{6.3.36}$$

with $t_j > 0$, $j = 1, 2, \ldots$, and such that the following conditions hold:

$$
\left.
\begin{aligned}
&\sum_{j=0}^{\infty} |A_j| < +\infty, \quad \int_0^{\infty} |B(t)|\, \mathrm{d}t < +\infty, \\[2mm]
&\sum_{j=0}^{\infty} |b_j| < +\infty, \quad \int_0^{\infty} |\beta(t)|\, \mathrm{d}t < +\infty, \\[2mm]
&\sum_{j=0}^{\infty} |c_j^*| < +\infty, \quad \int_0^{\infty} |d^*(t)|\, \mathrm{d}t < +\infty, \\[2mm]
&\sum_{j=0}^{\infty} |\alpha_j| < +\infty, \quad \int_0^{\infty} |\gamma(t)|\, \mathrm{d}t < +\infty.
\end{aligned}
\right\}
\tag{6.3.37}
$$

In the system (6.3.35), x is the state n-dimensional vector of the regulating system, ρ is the reactor power, and η_k are the flows of the delayed neutrons. These variables are normalized around some nominal values representing the steady state of the nuclear reactor and of the regulating system. When the reactor power takes the value P, the delayed neutron flows take the values $(\beta_k/\lambda_k \Lambda)P$, and the state vector of the control system takes a stationary value which describes the steady state of the reactor.

The physical meaning of the positive constants Λ, β_k and λ_k is as follows: Λ is the mean life of the instantaneous neutrons, λ_k is the radioactive constant of precursors of delayed neutrons in the group of rank k, and β_k is the fraction of delayed terminal neutrons in the same group.

The linear (scalar) functional v stands for the reactivity feedback, and it corresponds to the case of a circulating fuel nuclear reactor.

One more remark is necessary before proceeding further in investigating the system (6.3.35). The fact that the delays t_j are the same for each operator A, b, c^* or α does not cause any loss of generality. Indeed, if different sequences of delays are considered for the above operators, then the union of these sequences is again countable and, therefore, it can be conveniently denoted by $\{t_j\}$. We have to add then some null coefficients A_j, b_j, c_j^* and α_j in the representation of the operators A, b, c^* and α, such that they take the forms (6.3.36). It is worth pointing out the fact that conditions (6.3.37) keep their validity.

Further assumptions will be made on the system (6.3.35), which obviously constitutes an integrodifferential system with infinite delay. The nonlinear part of (6.3.35) occurs only in the right hand side of the second equation and it is of quadratic type.

In order to determine a unique solution of (6.3.35), it is necessary to prescribe some initial data. Taking into account the form of these equations, we have that

$$
x(t) = h(t), \; \rho(t) = \lambda(t), \quad t < 0, \tag{6.3.38}
$$

and

$$x(0+) = x^0, \rho(0+) = \rho_0, \eta_k(0+) = \eta_k^0, \quad k = 1, 2, \dots, M, \quad (6.3.39)$$

constitute the appropriate kind of initial conditions for (6.3.35).

We usually assume that $|h(t)|, |\lambda(t)| \in L^1(R_-, R)$.

Besides (6.3.35), we shall consider the linear system with a real parameter h:

$$\left.\begin{aligned} \dot{y}(t) &= (Ay)(t) + (b\xi)(t), \\ \dot{\xi}(t) &= -\sum_{k=1}^{M} \beta_k \Lambda^{-1}[\xi(t) - \zeta_k(t)] - P\Lambda^{-1}h[(c^*y)(t) + (\alpha\xi)(t)], \\ \dot{\zeta}_k(t) &= \lambda_k[\xi(t) - \zeta_k(t)], \quad k = 1, 2, \dots, M, \end{aligned}\right\} \quad (6.3.40)$$

on which certain assumptions will be made subsequently.

Associated to (6.3.40) is the linear block (i.e., the linear control system with $\mu(t)$ as input),

$$\left.\begin{aligned} \dot{y}(t) &= (Ay)(t) + (b\xi)(t), \\ \dot{\xi}(t) &= -\sum_{k=1}^{M} \beta_k \Lambda^{-1}[\xi(t) - \zeta_k(t)] - P\Lambda^{-1}h[(c^*y)(t) + (\alpha\xi)(t)] + \mu(t), \\ \dot{\zeta}_k(t) &= \lambda_k[\xi(t) - \zeta_k(t)], \quad k = 1, 2, \dots, M, \\ \psi(t) &= (c^*y)(t) + (\alpha\xi)(t). \end{aligned}\right\} \quad (6.3.41)$$

A unique solution for (6.3.40) or (6.3.41) is determined if initial conditions of the form (6.3.38), (6.3.39) are given:

$$y(t) = l(t), \xi(t) = \phi(t), \quad t < 0, \quad (6.3.42)$$

and

$$y(0+) = y^0, \xi(0+) = \xi_0, \zeta_k(0+) = \zeta_k^0, \quad k = 1, 2, \dots, M. \quad (6.3.43)$$

Let us define now the following matrices:

$$A_{h0} = \begin{vmatrix} A_0 & b_0 & 0 & 0 & \cdots & 0 \\ -P\Lambda^{-1}hc_0^* & -\Lambda^{-1}(Ph\alpha_0 + \sum_1^M \beta_k) & \Lambda^{-1}\beta_1 & \Lambda^{-1}\beta_2 & \cdots & \Lambda^{-1}\beta_M \\ \lambda_1 & -\lambda_1 & 0 & \cdots & 0 \\ \lambda_2 & 0 & -\lambda_2 & \cdots & 0 \\ 0 & \vdots & \vdots & \vdots & \\ \lambda_M & 0 & 0 & \cdots & -\lambda_M \end{vmatrix}$$

$$(6.3.44)$$

$$A_{hj} = \begin{vmatrix} A_j & b_j & 0 \\ -P\Lambda^{-1}hc_j^* & -P\Lambda^{-1}h\alpha_j & 0 \\ 0 & 0 & 0 \end{vmatrix}, \quad j = 1, 2, \dots, \quad (6.3.45)$$

and

$$B_h(t) = \begin{vmatrix} B(t) & \beta(t) & 0 \\ -P\Lambda^{-1}hd^*(t) & -P\Lambda^{-1}h\gamma(t) & 0 \\ 0 & 0 & 0 \end{vmatrix}. \qquad (6.3.46)$$

If we consider the vector $z = \mathrm{col}(y, \xi, \zeta_1, \ldots, \zeta_M)$, then system (6.3.40) becomes

$$\dot{z}(t) = A_{h0}z(t) + \sum_{j=1}^{\infty} A_{hj}z(t - t_j) + \int_0^t B_h(t - s)z(s)\,ds + u(t), \qquad (6.3.47)$$

with

$$u(t) = \mathrm{col}\left(\underset{n}{0, \ldots, 0}, -P\Lambda^{-1}h \int_{-\infty}^0 \gamma(t - s)\phi(s)\,ds, \underset{M}{0, \ldots, 0}\right).$$

The linear block (6.3.41) leads to

$$\dot{z}(t) = A_{h0}z(t) + \sum_{j=1}^{\infty} A_{hj}z(t - t_j) + \int_0^t B_h(t - s)z(s)\,ds + \tilde{\mu}(t), \qquad (6.3.48)$$

with

$$\tilde{\mu}(t) = \mu(t)\,\mathrm{col}(\underset{n}{0, \ldots, 0, 1}, 0, \underset{M}{\ldots, 0}) + u(t).$$

Accordingly, the nonlinear system (6.3.35) can be rewritten in the form

$$\dot{w}(t) = A_{10}w(t) + \sum_{j=1}^{\infty} A_{1j}w(t - t_j) + \int_0^t B_1(t - s)w(s)\,ds + f(t; w), \qquad (6.3.49)$$

where $w = \mathrm{col}(x, \rho, \eta_1, \ldots, \eta_M)$ and $f(t; w)$ is given by

$$f(t; w) = \mathrm{col}\left(\underset{n}{0, \ldots, 0}, -P\Lambda^{-1}\left[\int_{-\infty}^0 \gamma(t - s)\phi(s)\,ds + \rho(t)v(t)\right], \underset{M}{0, \ldots, 0}\right).$$
$$(6.3.50)$$

It is now obvious that systems (6.3.47) and (6.3.48) are linear systems of the form we dealt with above, while (6.3.49) constitutes a nonlinear perturbed system. Taking into account conditions (6.3.37), from the definition of the matrices A_{hj} and B_h we have

$$\sum_{j=0}^{\infty} |A_{hj}| < +\infty, \qquad \int_0^{\infty} |B_h(t)|\,dt < +\infty, \qquad (6.3.51)$$

for any real h.

Therefore, we can use the technique sketched above in this section, in order to investigate the stability properties of the systems (6.3.47), (6.3.48) and (6.3.49).

Let us now consider the system (6.3.47) for $h = h_2$, and the linear block (6.3.48) for $h = h_1$. We shall prove that their solutions coincide on the positive half-axis, provided initial data and input function $\mu(t)$ are chosen in a convenient manner.

More precisely, let us denote by $x(t)$, $\rho(t)$, $\eta_k(t)$, $k = 1, 2, \ldots, M$, a solution of

(6.3.35), with initial data (6.3.38), (6.3.39). We assume that this solution is defined on a certain interval $[0, T]$, $T > 0$.

From the above solution of (6.3.35) or (6.3.49), let us construct the solution $\bar{y}(t), \bar{\xi}(t), \bar{\zeta}_k(t), k = 1, 2, \ldots, M$, of (6.3.40) or (6.3.37), for $t \geq T$, corresponding to the following initial data:

$$\bar{y}(t) = \begin{cases} h(t), & t < 0, \\ x(t), & 0 \leq t \leq T, \end{cases} \qquad \bar{\xi}(t) = \begin{cases} \lambda(t), & t < 0, \\ \rho(t), & 0 \leq t \leq T, \end{cases} \qquad (6.3.52)$$

and $\bar{\zeta}_k(T) = \eta_k(T), k = 1, 2, \ldots, M$. The existence and uniqueness of the solution $\bar{y}(t), \bar{\xi}(t), \bar{\zeta}_k(t), k = 1, 2, \ldots, M$, are guaranteed by the results in Section 3.3. As said above, (6.3.40) is considered for $h = h_2$.

Concerning the linear system (block) (6.3.41) or (6.3.48), we are interested in the solution $y(t), \xi(t), \zeta_k(t), k = 1, 2, \ldots, M$, corresponding to the initial data $h(t)$, $x^0, \lambda(t), \rho_0, \eta_k^0, k = 1, 2, \ldots, M$, and to the value h_1 of the parameter h. Moreover, we assume that the input $\mu(t)$ is given by

$$\mu_T(t) = \begin{cases} -P\Lambda^{-1}[1 + \rho(t) - h_1]v(t), & 0 \leq t \leq T, \\ -P\Lambda^{-1}(h_2 - h_1)\bar{v}(t), & t > T, \end{cases} \qquad (6.3.53)$$

with

$$\bar{v}(t) = (c^*\bar{y})(t) + (\alpha\bar{\xi})(t), \quad t > T. \qquad (6.3.54)$$

Lemma 6.3.2. *With the above mentioned notations, the following equalities hold:*

$$y(t) = x(t), \xi(t) = \rho(t), \zeta_k(t) = \eta_k(t), \quad k = 1, 2, \ldots, M, 0 \leq t \leq T, \qquad (6.3.55)$$

and

$$y(t) = \bar{y}(t), \xi(t) = \bar{\xi}(t), \zeta_k(t) = \bar{\zeta}_k(t), \quad k = 1, 2, \ldots, M, t > T. \qquad (6.3.56)$$

Proof. If we let $u = y - x, v = \xi - \rho, w_k = \zeta_k - \eta_k, k = 1, 2, \ldots, M$, on $[0, T)$ and $u = y - \bar{y}, v = \xi - \bar{\xi}, w_k = \zeta_k - \bar{\zeta}_k, k = 1, 2, \ldots, M$, on $[T, \infty)$, then we find from (6.3.35), (6.3.40), (6.3.41) and (6.3.53):

$$\left. \begin{array}{l} \dot{u}(t) = (Au)(t) + (bv)(t), \\[2ex] \dot{v}(t) = -\displaystyle\sum_{k=1}^{M} \beta_k \Lambda^{-1}[v(t) - w_k(t)], \\[2ex] \dot{w}_k(t) = \lambda_k[v(t) - w_k(t)], \quad k = 1, 2, \ldots, M. \end{array} \right\} \qquad (6.3.57)$$

The equations in (6.3.57) hold on R_+. First, we restrict our considerations to the interval $[0, T]$. According to our assumptions in constructing x, ρ, η_k and y, ξ, ζ_k, we know that $u(t) \equiv 0, v(t) \equiv 0, w_k(t) \equiv 0, k = 1, 2, \ldots, M$, on $[0, T]$.

Indeed, u, v, w_k satisfy a homogeneous system and the corresponding initial data are all zero. Then let us consider (6.3.57) as a system with initial data on $(-\infty, T]$. According to (6.3.52) and the above established equalities on $[0, T]$,

the solution $y(t)$, $\xi(t)$, $\zeta_k(t)$ of (6.3.41), with the control function given by (6.3.53), must coincide with the solution $\bar{y}(t)$, $\bar{\xi}(t)$, $\bar{\zeta}_k(t)$ of (6.3.40), i.e., (6.3.57) hold.
This ends the proof of Lemma 6.3.2.

Remark. One obtains from Lemma 6.3.2 that $\psi(t)$ from (6.3.57) can be represented as

$$\psi(t) = \begin{cases} v(t), & 0 \le t \le T, \\ \bar{v}(t), & t > T. \end{cases} \tag{6.3.58}$$

The next lemma is a direct consequence of the validity of the variations of constants formula (6.3.17) for (6.3.41) and (6.3.59). We denote by $\hat{h}(t)$ the map reducing to $h(t)$ for $t < 0$, such that $h(t) = 0$ for $t \ge 0$. $\hat{\lambda}(t)$ is defined similarly.

Lemma 6.3.3. *Let $y(t)$, $\xi(t)$, $\zeta_k(t)$ be the solution of (6.3.41), corresponding to the control function $\mu_T(t)$ given by (6.3.53), the initial data $h(t)$, x^0, $\lambda(t)$, ρ_0, η_k^0, $k = 1, 2, \ldots, M$, and $h = h_1$. Then the solution is identical with that of the linear system*

$$\begin{aligned} \dot{y}_T(t) &= (Ay_T)(t) + (b\xi_T)(t) + \sum_{j=1}^{\infty} A_j\hat{h}(t - t_j) + \sum_{j=1}^{\infty} b_j\hat{\lambda}(t - t_j), \\ \dot{\xi}_T(t) &= -\sum_{k=1}^{M} \beta_k\Lambda^{-1}[\xi_T(t) - \zeta_{kT}(t)] - P\Lambda^{-1}h_1\left[(c^*y_T)(t) \right. \\ &\quad + \sum_{j=0}^{\infty} \alpha_j\xi_T(t - t_j) + \int_0^t \gamma(t - s)\xi_T(s)\,ds\Big] - P\Lambda^{-1}h_1\Big[(c^*\hat{h})(t) \\ &\quad + \sum_{j=1}^{\infty} \alpha_j\hat{\lambda}(t - t_j) + \int_{-\infty}^0 \gamma(t - s)\lambda(s)\,ds\Big] + \mu_T(t), \\ \dot{\zeta}_{kT}(t) &= \lambda_k[\xi_T(t) - \zeta_{kT}(t)], \quad k = 1, 2, \ldots, M, \end{aligned} \tag{6.3.59}$$

with zero initial data on $(-\infty, 0)$ and the data x^0, ρ_0, η_k^0 at $t = 0$.
Before proceeding further with the investigation, let us point out that denoting

$$\begin{aligned} \psi_T(t) &= (c^*y_T)(t) + \sum_{j=0}^{\infty} \alpha_j\xi_T(t - t_j) + \int_0^t \gamma(t - s)\xi_T(s)\,ds, \quad t > 0, \\ \psi_0(t) &= (c^*\hat{h})(t) + \sum_{j=1}^{\infty} \alpha_j\hat{\lambda}(t - t_j) + \int_{-\infty}^0 \gamma(t - s)\lambda(s)\,ds, \quad t > 0, \end{aligned} \tag{6.3.60}$$

and taking into account the relations

$$y(t) = y_T(t) + \hat{h}(t), \quad \xi(t) = \xi_T(t) + \hat{\lambda}(t), \quad \zeta(t) = \zeta_{kT}(t), \tag{6.3.61}$$

and (6.3.58), we obtain

$$\begin{aligned} \psi(t) &= v(t) = \psi_T(t) + \psi_0(t), \quad 0 \le t \le T, \\ \psi(t) &= \bar{v}(t) = \psi_T(t) + \psi_0(t), \quad t > T. \end{aligned} \tag{6.3.62}$$

Lemma 6.3.4.   *Any solution of system* (6.3.35) *that satisfies the initial constraints* $1 + \rho(0) > 0, 1 + \eta_i(0) > 0, i = 1, 2, \ldots, M,$ *also satisfies*

$$1 + \rho(t) > 0, 1 + \eta_i(t) > 0, \quad i = 1, 2, \ldots, M,$$

for $t > 0$.

The proof is elementary and can be found in Rasvan [1]. It is useful to point out that the equations for $\eta_i(t)$ can be written in the form

$$\frac{d}{dt}[1 + \eta_i(t)] + \lambda_i[1 + \eta_i(t)] = \lambda_i[1 + \rho(t)].$$

The following notation is necessary in order to state the main stability results of this paper. First, we consider the rational function

$$R(s) = s^{-1}\left(1 + \Lambda^{-1}\sum_{k=1}^{M}\frac{\beta_k}{s + \lambda_k}\right)^{-1}, \tag{6.3.63}$$

which represents the transfer function corresponding to the effect of the delayed neutrons. Next, let us consider the transfer function

$$\tilde{k}(s) = \tilde{c}^*(s)[sI - \mathscr{A}(s)]^{-1}\tilde{b}(s), \tag{6.3.64}$$

where $\mathscr{A}(s)$ is the symbol of the operator A given by (6.3.21) and $\tilde{b}(s), \tilde{c}^*(s)$ have similar definitions. Further, let us assume $\gamma_0(s) = \tilde{k}(s) + \tilde{\alpha}(s)$ and

$$\gamma_1(s) = R(s)[1 + P\Lambda^{-1}h_1 R(s)\gamma_0(s)]^{-1}, \tag{6.3.65}$$

$$\gamma_2(s) = P\Lambda^{-1}h_1\gamma_0(s)\gamma_1(s), \quad h_1 > 0. \tag{6.3.66}$$

Theorem 6.3.5.   *Consider the system* (6.3.35), *with* A, b, c^* *and* α *given by* (6.3.36). *The conditions* (6.3.37) *are supposed to hold, while constants* $\beta_k, \lambda_k, k = 1, 2, \ldots,$ M, Λ *and* P *are assumed positive. Assume further that:*

(a) $\det[sI - \mathscr{A}(s)] \neq 0$ *for* $\mathrm{Re}\, s \geq 0$, *i.e., the linear system* $\dot{x}(t) = (Ax)(t)$ *is asymptotically stable;*

(b) *there exist some numbers* $h_1, h_2, \delta_1, \delta_2$, *such that*

$$0 < h_1 < 1 < h_2, \delta_0 \geq 0, \delta_2 \geq \delta_1 \geq 0, \text{ with } \delta_1 + \delta_2 > 0, \text{ and}$$

(1°) *the linear system* (6.3.40) *is asymptotically stable for* $h = h_1$ *and* $h = h_2$;

(2°) *the numbers* $\gamma_0 \in (0, 1 - h_1)$ *and* $\bar{\gamma}_0 \in (0, \sqrt{(h_2 - 1)})$ *are so chosen that for the real function*

$$\Phi(\xi) = \xi - \ln(1 + \xi) - (2h_2)^{-1}\xi^2$$

one has

$$\Phi(\xi) \leq \phi(\sqrt{(h_2 - 1)})\min\{1, \beta_k\lambda_k^{-1}\Lambda^{-1}; k = 1, 2, \ldots, M\},$$

as soon as $\xi \in [-\gamma_0, \bar{\gamma}_0]$;

(3°) *if*

$$H(s) = \delta_0[(h_2 - h_1)^{-1} + h_1^{-1}\gamma_2(s)] + \delta_1\gamma_1(s)$$

$$+ \delta_2 P\Lambda^{-1}(h_2 - h_1)|\gamma_1(s)|^2\gamma_0(s), \qquad (6.3.67)$$

then

$$\operatorname{Re} H(i\omega) > 0, \quad \omega \in R, \qquad (6.3.68)$$

holds when $\delta_0 > 0$, *and*

$$\delta_1\Lambda^{-1}\sum_{k=1}^{M} \lambda_k + [\delta_1 h_1 + \delta_2(h_2 - h_1)]\left(\alpha_0 - \sum_{j=1}^{\infty}|\alpha_j|\right) > 0 \quad (6.3.69)$$

holds when $\delta_0 = 0$.

Under the above assumptions, each solution $x(t)$, $\rho(t)$, $\eta_k(t)$, $k = 1, 2, \ldots, M$, *of the system* (6.3.35), *with*

$$|h(t)|, |\lambda(t)| \in L^1(R_-, R) \cap L^2(R_-, R), \qquad (6.3.70)$$

is defined on the positive half-axis and tends to zero at infinity

$$\lim\left(|x(t)| + |\rho(t)| + \sum_{k=1}^{M}|\eta_k(t)|\right) = 0, \text{ as } t \to +\infty,$$

provided certain initial constraints are imposed.

Proof. First, we consider the linear block (6.3.41), for $h = h_1$ and $\mu(t) = \mu_T(t)$ as given by (6.3.53). It is assumed that $T > 0$ is chosen such that the solution $x(t)$, $\rho(t)$, $\eta_k(t)$, $k = 1, 2, \ldots, M$, is defined on $[0, T]$. Let us associate to (6.3.41) the integral index

$$\chi(T) = \delta_0 \int_0^T \mu_T(t)[(h_2 - h_1)^{-1}\mu_T(t) + P\Lambda^{-1}\psi(t)]\,dt$$

$$+ \int_0^T \xi(t)[\delta_1\mu_T(t) + \delta_2 P\Lambda^{-1}(h_2 - h_1)\psi(t)]\,dt. \qquad (6.3.71)$$

The initial data for (6.3.41) are the same as for (6.3.35). Taking into account (6.3.55) and (6.3.53), one obtains by means of elementary calculations

$$\chi(T) = \delta_0(P\Lambda^{-1})^2(h_2 - h_1)^{-1}\int_0^T [h_1 - 1 - \rho(t)][h_2 - 1 - \rho(t)]v^2(t)\,dt$$

$$- \delta_1 h_2[\Omega(T) - \Omega(0)]$$

$$- \delta_1\Lambda^{-1}\sum_{k=1}^{M}\beta_k\int_0^T [\rho(t) - \eta_k(t)]^2\{h_2[1 + \rho(t)]^{-1}[1 + \eta_k(t)]^{-1} - 1\}\,dt$$

$$- (\delta_2 - \delta_1)(h_2 - h_1)\left\{\Omega_1(T) - \Omega_1(0) + \Lambda^{-1}\sum_{k=1}^{M}\beta_k\int_0^T [\rho(t) - \eta_k(t)]^2\right.$$

$$\left. \times [1 + \rho(t)]^{-1}[1 + \eta_k(t)]^{-1}\,dt\right\}, \qquad (6.3.72)$$

with

$$\Omega(t) = \Phi(\rho(t)) + \Lambda^{-1} \sum_{k=1}^{M} \beta_k \lambda_k^{-1} \Phi(\eta_k(t)),$$

$$\Omega_1(t) = \phi(\rho(t)) + \Lambda^{-1} \sum_{k=1}^{M} \beta_k \lambda_k^{-1} \phi(\eta_k(t)),$$

$$\phi(\xi) = \xi - \ln(1 + \xi),$$

where $\Phi(\xi)$ is the function defined in the statement of the Theorem 6.3.5. Indeed, we get from (6.3.53), (6.3.55) and (6.3.58)

$$\int_0^T \mu_T(t)[(h_2 - h_1)^{-1}\mu_T(t) + P\Lambda^{-1}\psi(t)]\,dt$$

$$= (P\Lambda^{-1})^2(h_2 - h_1)^{-1}\int_0^T [1 + \rho(t) - h_1]^2 v^2(t)\,dt$$

$$- (P\Lambda^{-1})^2 \int_0^T [1 + \rho(t) - h_1]v^2(t)\,dt$$

$$= (P\Lambda^{-1})^2(h_2 - h_1)^{-1}\int_0^T [h_1 - 1 - \rho(t)][h_2 - 1 - \rho(t)]v^2(t)\,dt$$

$$+ (P\Lambda^{-1})^2 \int_0^T [1 + \rho(t) - h_1]v^2(t)\,dt$$

$$- (P\Lambda^{-1})^2 \int_0^T [1 + \rho(t) - h_1]v^2(t)\,dt$$

$$= (P\Lambda^{-1})^2(h_2 - h_1)^{-1}\int_0^T [h_1 - 1 - \rho(t)][h_2 - 1 - \rho(t)]v^2(t)\,dt.$$

We have further

$$\int_0^T \xi(t)\mu_T(t)\,dt = \int_0^T \rho(t)\left\{\dot\rho(t) + \Lambda^{-1}\sum_{k=1}^{M}\beta_k[\rho(t) - \eta_k(t)] + P\Lambda^{-1}h_1 v(t)\right\}dt$$

$$= \int_0^T \left\{\rho(t)\dot\rho(t) + \Lambda^{-1}\sum_{k=1}^{M}\beta_k[\rho(t) - \eta_k(t)]^2\right.$$

$$\left. + \Lambda^{-1}\sum_{k=1}^{M}\beta_k\eta_k(t)[\rho(t) - \eta_k(t)]\right\}dt + P\Lambda^{-1}h_1\int_0^T \rho(t)v(t)\,dt$$

$$= \int_0^T \left\{\rho(t)\dot\rho(t) + \Lambda^{-1}\sum_{k=1}^{M}\beta_k\lambda_k^{-1}\eta_k(t)\dot\eta_k(t)\right\}dt$$

$$+ \Lambda^{-1}\sum_{k=1}^{M}\beta_k\int_0^T [\rho(t) - \eta_k(t)]^2\,dt + P\Lambda^{-1}h_1\int_0^T \rho(t)v(t)\,dt$$

$$= \frac{1}{2}\left[\rho^2(t) + \Lambda^{-1}\sum_{k=1}^{M}\beta_k\lambda_k^{-1}\eta_k^2(t)\right]\Bigg|_0^T$$

$$+ \Lambda^{-1}\sum_{k=1}^{M}\beta_k\int_0^T [\rho(t) - \eta_k(t)]^2\,dt + P\Lambda^{-1}h_1\int_0^T \rho(t)v(t)\,dt$$

and

$$\int_0^T \xi(t)\psi(t)\,dt = \int_0^T \rho(t)v(t)\,dt$$

$$= -P^{-1}\Lambda \int_0^T \left[\dot\rho(t) + \Lambda^{-1} \sum_{k=1}^M \beta_k \lambda_k^{-1} \dot\eta_k(t) + P\Lambda^{-1}v(t) \right] dt$$

$$= -P^{-1}\Lambda \left[\rho(t) + \Lambda^{-1} \sum_{k=1}^M \beta_k \lambda_k^{-1} \eta_k(t) \right]\Bigg|_0^T - \int_0^T v(t)\,dt.$$

In order to evaluate the last integral, we shall use again the second equation of (6.3.35). According to Lemma 6.3.4, we can divide both members of the second equation of (6.3.35) by $1 + \rho(t)$, provided ρ_0 is such that $1 + \rho_0 > 0$. Therefore,

$$[1 + \rho(t)]^{-1}\dot\rho(t) = -\Lambda^{-1} \sum_{k=1}^M \beta_k [1 + \rho(t)]^{-1}[\rho(t) - \eta_k(t)] - P\Lambda^{-1}v(t),$$

which leads to

$$[1 + \rho(t)]^{-1}\dot\rho(t) = -\Lambda^{-1} \sum_{k=1}^M \beta_k \lambda_k^{-1}[1 + \rho(t)]^{-1}\dot\eta_k(t) - P\Lambda^{-1}v(t),$$

if one considers also the third equation of (6.3.35). But

$$[1 + \rho(t)]^{-1}\dot\eta_k(t) = [1 + \eta_k(t)]^{-1}\dot\eta_k(t)$$
$$- [1 + \rho(t)]^{-1}[1 + \eta_k(t)]^{-1}[\rho(t) - \eta_k(t)]\dot\eta_k(t),$$

which allows us to write

$$[1 + \rho(t)]^{-1}\dot\rho(t) + \Lambda^{-1} \sum_{k=1}^M \beta_k \lambda_k^{-1}[1 + \eta_k(t)]^{-1}\dot\eta_k(t)$$

$$= \Lambda^{-1} \sum_{k=1}^M \beta_k \lambda_k^{-1}[1 + \rho(t)]^{-1}[1 + \eta_k(t)]^{-1}[\rho(t) - \eta_k(t)]\dot\eta_k(t) - P\Lambda^{-1}v(t)$$

$$= \Lambda^{-1} \sum_{k=1}^M \beta_k [1 + \rho(t)]^{-1}[1 + \eta_k(t)]^{-1}[\rho(t) - \eta_k(t)]^2 - P\Lambda^{-1}v(t).$$

Hence

$$\left\{ \ln[1 + \rho(t)] + \Lambda^{-1} \sum_{k=1}^M \beta_k \lambda_k^{-1} \ln[1 + \eta_k(t)] \right\}\Bigg|_0^T$$

$$= \Lambda^{-1} \sum_{k=1}^M \beta_k \int_0^T [\rho(t) - \eta_k(t)]^2[1 + \rho(t)]^{-1}[1 + \eta_k(t)]^{-1}\,dt$$

$$- P\Lambda^{-1} \int_0^T v(t)\,dt.$$

Summing up the above considerations, we get the formula (6.3.72) for $\chi(T)$.

We shall now write $\chi(T)$ in another form. Namely

$$\chi(T) = \delta_0 \int_0^\infty \mu_T(t)\{(h_2 - h_1)^{-1}\mu_T(t) + P\Lambda^{-1}[\psi_T(t) + \psi_0(t)]\}\,dt$$

$$+ \int_0^\infty \xi_T(t)\{\delta_1\mu_T(t) + \delta_2(h_2 - h_1)P\Lambda^{-1}[\psi_T(t) + \psi_0(t)]\}\,dt$$

$$- (\delta_2 - \delta_1)(h_2 - h_1)P\Lambda^{-1} \int_T^\infty \xi_T(t)[\psi_T(t) + \psi_0(t)]\,dt. \quad (6.3.73)$$

Indeed, taking into account (6.3.53), (6.3.55) and (6.3.62), we see that the integrand is zero, for $t > T$, in the first integral from (6.3.73). Concerning the second integral in (6.3.73), we have

$$\int_0^\infty \xi_T(t)\{\delta_1\mu_T(t) + \delta_2(h_2 - h_1)P\Lambda^{-1}[\psi_T(t) + \psi_0(t)]\}\,dt$$

$$= \delta_1 \int_T^\infty \xi_T(t)\mu_T(t)\,dt + \delta_2(h_2 - h_1)P\Lambda^{-1} \int_T^\infty \xi_T(t)[\psi_T(t) + \psi_0(t)]\,dt$$

$$= -\delta_1(h_2 - h_1)P\Lambda^{-1} \int_T^\infty \xi_T(t)[\psi_T(t) + \psi_0(t)]\,dt$$

$$+ \delta_2(h_2 - h_1)P\Lambda^{-1} \int_T^\infty \xi_T(t)[\psi_T(t) + \psi_0(t)]\,dt$$

$$= (\delta_2 - \delta_1)(h_2 - h_1)P\Lambda^{-1} \int_T^\infty \xi_T(t)[\psi_T(t) + \psi_0(t)]\,dt,$$

which proves the validity of (6.3.73).

Let us point out that, under our assumptions, the second integral in (6.3.73) is convergent. Indeed, taking (6.3.70) into account, we have $\psi_0(t) \in L^1 \cap L^2$. From condition (b) in the statement of the Theorem 6.3.5, one derives $\mu_T(t)$, $\xi_T(t) \in L^p$, $1 \le p \le \infty$, using again the results in Section 2.3. Consequently, the integrand belongs to L^1.

Before using Parseval's formula to find a new form for $\chi(T)$, we shall express conveniently certain Laplace transforms of some functions we deal with. First, let us compute the Laplace transform $\tilde{\xi}_T(s)$ from the system (6.3.59). Taking the Laplace transform of both sides of each equation and substituting the values of $\tilde{y}_T(s)$ and $\tilde{\zeta}_{kT}(s)$ in the second equation, we have

$$\tilde{\xi}_T(s) = \gamma_1(s)\tilde{\mu}_T(s) + \gamma_1(s)[\rho_0 + N(s)], \quad (6.3.74)$$

with $\gamma_1(s)$ given by (6.3.65) and

$$\left.\begin{array}{l} N(s) = \Lambda^{-1} \sum_{k=1}^M \beta_k\eta_k^0(s + \lambda_k)^{-1} - P\Lambda^{-1}h_1 M(s), \\[2mm] M(s) = \tilde{\psi}_0(s) + \tilde{c}^*(s)[sI - \mathscr{A}(s)]^{-1}m(s), \end{array}\right\} \quad (6.3.75)$$

where

$$m(s) = x^0 + \sum_{j=1}^{\infty} A_j e^{-st_j} \int_{-t_j}^{0} h(t)e^{-st}\,dt + \sum_{j=1}^{\infty} b_j e^{-st_j} \int_{-t_j}^{0} \lambda(t)e^{-st}\,dt. \quad (6.3.76)$$

From the definition of $\psi_T(t)$ (see (6.3.60)), we find

$$\tilde{\psi}_T(s) = \tilde{c}^*(s)\tilde{y}_T(s) + \tilde{\alpha}(s)\tilde{\xi}_T(s)$$

$$= \gamma_0(s)\tilde{\xi}_T(s) + \tilde{c}^*(s)[sI - \mathscr{A}(s)]^{-1}m(s)$$

$$= \gamma_0(s)\gamma_1(s)\tilde{\mu}_T(s) + \gamma_0(s)\gamma_1(s)[\rho_0 + N(s)]$$

$$+ \tilde{c}^*(s)[sI - \mathscr{A}(s)]^{-1}m(s). \quad (6.3.77)$$

The proof has to be continued in accordance with the following feature: the parameter $\delta_0 \neq 0$, or $\delta_0 = 0$.

Case 1 ($\delta_0 \neq 0$). In this case, we shall apply a lemma due to Popov [1], based on Parseval's formula, in order to find another form for $\chi(T)$. We start from (6.3.73) and denote:

$$\left.\begin{aligned}
f_1(t) &= -\xi_T(t), \\
f_2(t) &= \delta_1 \mu_T(t) + \delta_2(h_2 - h_1)P\Lambda^{-1}[\psi_T(t) + \psi_0(t)], \\
f_3(t) &= \delta_0(h_2 - h_1)^{-1}\mu_T(t) + \delta_0 P\Lambda^{-1}[\psi_T(t) + \psi_0(t)], \\
f_4(t) &= -\mu_T(t).
\end{aligned}\right\} \quad (6.3.78)$$

Taking into account (6.3.74) and (6.3.77), we find the following formulas for the Fourier transforms of the functions defined by (6.3.78):

$$\tilde{f}_j(i\omega) = U_j(i\omega)\tilde{f}_4(i\omega) + V_j(i\omega), \quad j = 1, 2, 3. \quad (6.3.79)$$

In (6.3.79), the functions U_j and V_j, $j = 1, 2, 3$ are given by

$$\left.\begin{aligned}
U_1(i\omega) &= \gamma_1(i\omega), \\
-U_2(i\omega) &= \delta_1 + \delta_2(h_2 - h_1)P\Lambda^{-1}\gamma_0(i\omega)\gamma_1(i\omega), \\
-U_3(i\omega) &= \delta_0(h_2 - h_1)^{-1} + \delta_0 h_1^{-1}\gamma_2(i\omega), \\
V_1(i\omega) &= -\gamma_1(i\omega)[\rho_0 + N(i\omega)], \\
V_2(i\omega) &= \delta_2(h_2 - h_1)P\Lambda^{-1}\{\gamma_0(i\omega)\gamma_1(i\omega)[\rho_0 + N(i\omega)] + M(i\omega)\}, \\
V_3(i\omega) &= \delta_0 P\Lambda^{-1}\{\gamma_0(i\omega)\gamma_1(i\omega)[\rho_0 + N(i\omega)] + M(i\omega)\}.
\end{aligned}\right\} \quad (6.3.80)$$

Popov's lemma states that

$$\int_0^\infty f_1(t)f_2(t)\,dt + \int_0^\infty f_3(t)f_4(t)\,dt$$

$$\le \frac{1}{8\pi} \int_{-\infty}^\infty \frac{|W(i\omega)|^2\,d\omega}{\operatorname{Re} H(i\omega)}$$

$$+ \frac{1}{4\pi} \int_{-\infty}^\infty [V_1(i\omega)V_2(-i\omega) + V_1(-i\omega)V_2(i\omega)]\,d\omega, \qquad (6.3.81)$$

with $H(s)$ given by (6.3.67), and

$$W(i\omega) = V_1(i\omega)U_2(-i\omega) + V_2(i\omega)U_1(-i\omega) + V_3(i\omega)$$

$$= [\delta_0 h_1^{-1}\gamma_2(i\omega) + \delta_1\gamma_1(i\omega)$$

$$+ 2\delta_2 P\varLambda^{-1}|\gamma_1(i\omega)|^2 \operatorname{Re}\gamma_0(i\omega)[\rho_0 + N(i\omega)]$$

$$+ P\varLambda^{-1}[\delta_0 + \delta_2(h_2 - h_1)\gamma_1(-i\omega)]M(i\omega). \qquad (6.3.82)$$

From (6.3.67) it follows that

$$\lim \operatorname{Re} H(i\omega) = \delta_0(h_2 - h_1)^{-1} > 0, \quad \text{as } |\omega| \to \infty. \qquad (6.3.83)$$

Consequently, one finds $\varepsilon_0 > 0$, such that

$$\operatorname{Re} H(i\omega) \ge \varepsilon_0, \quad \omega \in R. \qquad (6.3.84)$$

If we take into account (6.3.84), it suffices to show that $W(i\omega)$, $V_1(i\omega)$, $V_2(i\omega) \in L^2(R, \mathscr{C})$, in order to make sure that the integrals in the right hand side of (6.3.81) are convergent. Indeed, $R(i\omega)$ given by (6.3.63) behaves at infinity like $|\omega|^{-1}$, while $\gamma_0(i\omega)$ is bounded on R. The boundedness of $\gamma_0(i\omega)$ is a consequence of conditions (6.3.37), the fact that $\tilde{X}(s) = [sI - \mathscr{A}(s)]^{-1}$, and condition (a) from the theorem. Moreover, $\gamma_1(i\omega)$ and $\gamma_2(i\omega)$ also behave at infinity like $|\omega|^{-1}$. It is an elementary matter to show that: $m(s)$ is bounded for $\operatorname{Re} s = 0$; $M(i\omega)$ is a function in L^2; $N(i\omega)$ also belongs to L^2. Summing up the above discussion, we find that $W(i\omega)$, $V_1(i\omega)$ and $V_2(i\omega)$ are in L^2. Therefore, the right hand side of (6.3.81) is finite.

Substituting in (6.3.81) the functions $f_j(t)$, $j = 1, 2, 3, 4$, by their values given by (6.3.78), and taking into account (6.3.73), we obtain

$$-\chi(T) \le \frac{1}{8\pi} \int_{-\infty}^\infty \frac{|W(i\omega)|^2}{\operatorname{Re} H(i\omega)}\,d\omega + \frac{1}{4\pi} \int_{-\infty}^\infty [V_1(i\omega)V_2(-i\omega) + V_1(-i\omega)V_2(i\omega)]\,d\omega,$$

$$+ (\delta_2 - \delta_1)(h_2 - h_1)P\varLambda^{-1} \int_T^\infty \xi_T(t)[\psi_T(t) + \psi_0(t)]\,dt. \qquad (6.3.85)$$

We now want to transform the last integral occurring in the right hand side of (6.3.85). We have to repeat practically some computations encountered in deriving formula (6.3.72). Taking into account that (6.3.40) is asymptotically stable for $h = h_2$, we have

$$P\varLambda^{-1} \int_T^\infty \xi_T(t)[\psi_T(t) + \psi_0(t)]\,dt$$

$$= P\varLambda^{-1} \int_T^\infty \bar{\xi}(t)\bar{v}(t)\,dt$$

$$= -h_2^{-1} \int_T^\infty \bar{\xi}(t)\left[\dot{\bar{\xi}}(t) + \varLambda^{-1} \sum_{k=1}^M \beta_k(\bar{\xi}(t) - \bar{\zeta}_k(t))\right]dt$$

$$= \frac{1}{2h_2}\left[\rho^2(T) + \varLambda^{-1} \sum_{k=1}^M \beta_k\lambda_k^{-1}\eta_k^2(T)\right] - (\varLambda h_2)^{-1} \sum_{k=1}^M \beta_k \int_T^\infty [\bar{\xi}(t) - \bar{\zeta}_k(t)]^2\,dt.$$

Hence, (6.3.85) now becomes

$$-\chi(T) \le \frac{1}{8\pi} \int_{-\infty}^\infty \frac{|W(i\omega)|^2\,d\omega}{\operatorname{Re} H(i\omega)} + \frac{1}{4\pi} \int_{-\infty}^\infty [V_1(i\omega)V_2(-i\omega) + V_1(-i\omega)V_2(i\omega)]\,d\omega$$

$$+ \frac{1}{2h_2}(\delta_2 - \delta_1)(h_2 - h_1)\left[\rho^2(T) + \varLambda^{-1} \sum_{k=1}^M \beta_k\lambda_k^{-1}\eta_k^2(T)\right]$$

$$- (\varLambda h_2)^{-1}(\delta_2 - \delta_1)(h_2 - h_1) \sum_{k=1}^M \beta_k \int_T^\infty [\bar{\xi}(t) - \bar{\zeta}_k(t)]^2\,dt. \qquad (6.3.86)$$

Let us now compare (6.3.72) and (6.3.86). We obtain, after performing elementary operations and neglecting certain terms to strengthen the inequality:

$$\Omega(T) + \delta_1[\delta_1 h_1 + \delta_2(h_2 - h_1)]^{-1}\varLambda^{-1}$$

$$\times \sum_{k=1}^M \beta_k \int_0^T [\rho(t) - \eta_k(t)]^2\{h_2[1 + \rho(t)]^{-1}[1 + \eta_k(t)]^{-1} - 1\}\,dt$$

$$+ \delta_0(h_2 - h_1)^{-1}[\delta_1 h_1 + \delta_2(h_2 - h_1)]^{-1}(P\varLambda^{-1})^2$$

$$\times \int_0^T [1 + \rho(t) - h_1][h_2 - 1 - \rho(t)]v^2(t)\,dt$$

$$\le \Omega(0) + (\delta_2 - \delta_1)(h_2 - h_1)\{2h_2[\delta_1 h_1 + \delta_2(h_2 - h_1)]\}^{-1}$$

$$\times \left[\rho^2(0) + \varLambda^{-1} \sum_{k=1}^M \beta_k\lambda_k^{-1}\eta_k^2(0)\right]$$

$$+ \{4\pi[\delta_1 h_1 + \delta_2(h_2 - h_1)]\}^{-1}$$

$$\times \left\{\int_{-\infty}^\infty \frac{|W(i\omega)|^2\,d\omega}{2\operatorname{Re} H(i\omega)} + \left|\int_{-\infty}^\infty [V_1(i\omega)V_2(-i\omega) + V_1(i\omega)V_2(i\omega)]\,d\omega\right|\right\}.$$

$$(6.3.87)$$

By using (6.3.87), it is possible to prove that the solution of (6.3.35) exists on the whole positive half-axis, provided the initial data satisfy adequate conditions. Let us consider the function of $M + 1$ real variables

$$\Omega(\rho, \eta_1, \ldots, \eta_M) = \Phi(\rho) + \Lambda^{-1} \sum_{k=1}^{M} \beta_k \lambda_k^{-1} \Phi(\eta_k), \qquad (6.3.88)$$

defined on the set

$$0 < 1 + \rho \le \sqrt{h_2}, \quad 0 < 1 + \eta_k \le \sqrt{h_2}, \quad k = 1, 2, \ldots, M. \qquad (6.3.89)$$

The set (6.3.89) contains the origin, i.e., $\rho = \eta_1 = \cdots = \eta_M = 0$. From the definition of $\Phi(\xi)$ we see that $\Omega(0, 0, \ldots, 0) = 0$. But $\Phi(\xi) \ge 0$ for $-1 < \xi < h_2 - 1$, as can be readily seen from its definition, the equality taking place only for $\xi = 0$. Therefore, the minimum value that $\Omega(\rho, \eta_1, \ldots, \eta_M)$ can take on the boundary of the polyhedron (6.3.89) is

$$l = \Phi(\sqrt{h_2} - 1) \min\{1, \beta_k \lambda_k^{-1} \Lambda^{-1}; k = 1, 2, \ldots, M\}. \qquad (6.3.90)$$

If γ_0 and $\bar{\gamma}_0$ are chosen as mentioned in the statement of the theorem, i.e. such that $\Phi(\xi) \le l$ for $\xi \in [-\gamma_0, \bar{\gamma}_0]$, then we impose the following restrictions on the initial values:

$$-1 < -\gamma_0 < \rho(0) < \bar{\gamma}_0 < \sqrt{h_2} - 1, \quad -1 < -\gamma_0 < \eta_k(0) < \bar{\gamma}_0 < \sqrt{h_2} - 1,$$

$$k = 1, 2, \ldots, M. \qquad (6.3.91)$$

It is assumed that $[-\gamma_0, \bar{\gamma}_0]$ is the maximal interval on which $\Phi(\xi) \le l$.

We are going to show that the inequalities

$$-\gamma_0 < \rho(t) < \bar{\gamma}_0, \quad -\gamma_0 < \eta_k(t) < \bar{\gamma}_0, \quad k = 1, 2, \ldots, M, \qquad (6.3.92)$$

hold on the positive half-axis. Indeed, for continuity reasons, the inequalities (6.3.92) must be satisfied on a certain interval $[0, T_0)$, $T_0 > 0$. Again we assume that this is the maximal interval. Therefore, at least one among the inequalities (6.3.92) becomes an equality for $t = T_0$, if T_0 is finite. But (6.3.91) and (6.3.92) imply

$$[1 + \rho(t)][1 + \eta_k(t)] < h_2, \quad k = 1, 2, \ldots, M, t \in [0, T_0), \qquad (6.3.93)$$

$$[1 + \rho(t) - h_1][1 + \rho(t) - h_2] \le 0, \quad t \in [0, T_0), \qquad (6.3.94)$$

the last one being a consequence of $h_1 \le 1 - \gamma_0$. From (6.3.87), (6.3.93) and (6.3.94) we get

$$\Omega(T_0) \le \Omega(0) + (\delta_2 - \delta_1)(h_2 - h_1)\{2h_2[\delta_1 h_1 + \delta_2(h_2 - h_1)]\}^{-1}$$

$$\times \left[\rho_0^2 + \Lambda^{-1} \sum_{k=1}^{M} \beta_k \lambda_k^{-1} \eta_k^2(0) \right] + \{4\pi[\delta_1 h_1 + \delta_2(h_2 - h_1)]\}^{-1}$$

$$\times \left\{ \int_{-\infty}^{\infty} \frac{|W(i\omega)|^2 \, d\omega}{2 \operatorname{Re} H(i\omega)} + \left| \int_{-\infty}^{\infty} [V_1(i\omega) V_2(-i\omega) + V_1(-i\omega) V_2(i\omega)] \, d\omega \right| \right\}.$$

$$(6.3.95)$$

Taking into account (6.3.65), (6.3.66) and (6.3.80), we see that the right hand side of (6.3.95) can be made arbitrarily small, provided $|\rho(0)|$, $|\eta_k(0)|$, $|x^0|$, $|h|_L$, are chosen small enough. If we choose the initial data such that the right hand side in (6.3.95) is smaller than l given by (6.3.90), we find that $\Omega(T_0) < l$. But this does not agree with the fact that T_0 is maximal, which implies $\Omega(T_0) = l$. Hence inequalities (6.3.92) hold for $t \in R_+$.

From (6.3.87) we derive

$$\int_0^T [1 + \rho(t) - h_1][h_2 - 1 - \rho(t)]v^2(t)\,dt \le \text{const.} \qquad (6.3.96)$$

the constant in the right hand side being independent of T. Hence, the integrand belongs to $L^1(R_+, R)$. Since $\rho(t)$ is bounded and $1 + \rho(t) - h_1 > 1 - \gamma_0 - h_1 > 0$, $h_2 - 1 - \rho(t) > h_2 - 1 - \bar{\gamma}_0 > 0$, we obtain $v(t) \in L^2(R_+, R)$. Furthermore, we find that $[1 + \rho(t)]v(t) \in L^2(R_+, R)$. From the last $M + 1$ equations of the system (6.3.35) we have

$$\lim \rho(t) = \lim \eta_k(t) = 0 \text{ as } t \to \infty, \quad k = 1, 2, \ldots, M. \qquad (6.3.97)$$

Now let us consider the variable $x(t)$ of the system (6.3.35). Using the variation of constants formula (6.3.17) we find

$$x(t) = X(t)x^0 + \int_0^t X(t - s)(b\rho)(s)\,ds + (Yh)(t), \quad t \in R_+, \qquad (6.3.98)$$

where

$$(Yh)(t) = \int_0^t X(t - s)(\tilde{B}h)(s)\,ds, \quad t \in R_+, \qquad (6.3.99)$$

with $\tilde{B}$ given by (6.3.98). Consequently, $x(t)$ is also defined on R_+ and, because $|X(t)| \to 0$, $|(Yh)(t)| \to 0$, $|\rho(t)| \to 0$ as $t \to \infty$, we obtain

$$\lim|x(t)| = 0 \text{ as } t \to \infty. \qquad (6.3.100)$$

Case 2 ($\delta_0 = 0$). If $\delta_0 = 0$, then $\chi(T)$ given by (6.3.73) becomes

$$\chi(T) = -\frac{1}{2}\delta_1\rho^2(0) + \int_0^\infty \xi_T(t)\{\delta_1[\mu_T(t) - \dot{\xi}_T(t)] + \delta_2(h_2 - h_1)P\Lambda^{-1}[\psi_T(t)$$

$$+ \psi_0(t)]\}\,dt - (\delta_2 - \delta_1)(h_2 - h_1)P\Lambda^{-1}\int_T^\infty \bar{\xi}(t)\bar{v}(t)\,dt. \qquad (6.3.101)$$

Indeed, the integral of $\xi_T(t)\dot{\xi}_T(t)$ is $\frac{1}{2}\xi_T^2(t)$ and $\xi_T(0) = \rho(0)$, $\xi_T(\infty) = 0$. The last equality is a consequence of the fact that the stability of the system (6.3.40) takes place for $h = h_2$, and of Lemmas 6.3.2 and 6.3.3.

From (6.3.74) we can find $\tilde{\mu}_T(i\omega)$ and, applying Parseval's formula to the first integral occurring in (6.3.101), we obtain

$$\int_0^\infty \xi_T(t)\{\delta_1[\mu_T(t) - \dot\xi_T(t)] + \delta_2(h_2 - h_1)P\varLambda^{-1}[\psi_T(t) + \psi_0(t)]\}\,dt$$

$$= \frac{1}{2\pi}\text{Re}\int_{-\infty}^\infty \tilde\xi_T(-i\omega)\{\delta_1[(\gamma_1(i\omega))^{-1}\tilde\xi_T(i\omega) - N(i\omega)]$$

$$+ \delta_2(h_2 - h_1)P\varLambda^{-1}[\gamma_0(i\omega)\tilde\xi_T(i\omega) + M(i\omega)]\}\,d\omega. \qquad (6.3.102)$$

If we denote

$$V(i\omega) = -\delta_1 N(i\omega) + \delta_2(h_2 - h_1)P\varLambda^{-1}M(i\omega) \qquad (6.3.103)$$

and

$$G(i\omega) = \delta_1(\gamma_1(i\omega))^{-1} + \delta_2(h_2 - h_1)P\varLambda^{-1}\gamma_0(i\omega), \qquad (6.3.104)$$

then (6.3.101) and (6.3.102) lead to

$$\chi(T) = -\frac{1}{2}\delta_1\rho^2(0) + \frac{1}{2\pi}\int_{-\infty}^\infty \text{Re}\,G(i\omega)|\tilde\xi_T(i\omega)|^2\,d\omega$$

$$+ \frac{1}{2\pi}\int_{-\infty}^\infty \text{Re}\,\tilde\xi_T(-i\omega)V(i\omega)\,d\omega$$

$$- (\delta_2 - \delta_1)(h_2 - h_1)P\varLambda^{-1}\int_T^\infty \bar\xi(t)\bar v(t)\,dt. \qquad (6.3.105)$$

Let us remark further that (6.3.67) and (6.3.104) give (for $\delta_0 = 0$)

$$|\gamma_1(i\omega)|^2\,\text{Re}\,G(i\omega) = \text{Re}\,H(i\omega) > 0, \qquad (6.3.106)$$

if (6.3.68) is also considered. Moreover, one derives from (6.3.104), by elementary manipulations

$$\lim_{|\omega|\to\infty} \text{Re}\,G(i\omega) \geq \delta_1\varLambda^{-1}\sum_{k=1}^M \beta_k + [\delta_1 h_1 + \delta_2(h_2 - h_1)]\left(\alpha_0 - \sum_{j=1}^\infty |\alpha_j|\right) > 0,$$
$$(6.3.107)$$

taking into account (6.3.69) also. From (6.3.106), (6.3.107) we obtain the existence of $\varepsilon_0 > 0$, such that

$$\text{Re}\,G(i\omega) \geq \varepsilon_0, \quad \omega \in R. \qquad (6.3.108)$$

Therefore, if ε is such that $0 < 2\varepsilon < \varepsilon_0$, we derive from (6.3.105)

$$\chi(T) = -\frac{1}{2}\delta_1\rho^2(0) + \frac{1}{2\pi}\int_{-\infty}^\infty [\text{Re}\,G(i\omega) - \varepsilon]|\tilde\xi_T(i\omega)|^2\,d\omega$$

$$+ \frac{1}{2\pi}\int_{-\infty}^\infty \text{Re}\,\tilde\xi_T(-i\omega)V(i\omega)\,d\omega$$

$$- (\delta_2 - \delta_1)(h_2 - h_1)P\varLambda^{-1}\int_T^\infty \bar\xi(t)\bar v(t)\,dt + \varepsilon\int_0^\infty \xi_T^2(t)\,dt. \qquad (6.3.109)$$

Simple transformations in (6.3.109) lead to

$$\chi(T) = -\frac{1}{2}\delta_1\rho^2(0) + \frac{1}{2\pi}\int_{-\infty}^{\infty}\left|\sqrt{(\operatorname{Re}G(i\omega) - \varepsilon)}\tilde{\xi}_T(i\omega) + \frac{1}{2}\frac{V(i\omega)}{\sqrt{(\operatorname{Re}G(i\omega) - \varepsilon)}}\right|^2 d\omega$$

$$-\frac{1}{8\pi}\int_{-\infty}^{\infty}\frac{|V(i\omega)|^2}{\operatorname{Re}G(i\varepsilon) - \varepsilon}d\omega - (\delta_2 - \delta_1)(h_2 - h_1)P\Lambda^{-1}\int_T^{\infty}\bar{\xi}(t)\bar{v}(t)\,dt$$

$$+ \varepsilon\int_0^T\rho^2(t)\,dt + \varepsilon\int_T^{\infty}\bar{\xi}^2(t)\,dt, \tag{6.3.110}$$

if we take into account Lemma 6.3.3.

According to the formula preceding (6.3.86), (6.3.110) can be rewritten in the form

$$\chi(T) = -\frac{1}{2}\delta_1\rho^2(0) + \frac{1}{2\pi}\int_{-\infty}^{\infty}|\cdot|^2\,d\omega - \frac{1}{8\pi}\int_{-\infty}^{\infty}|\cdot|\,d\omega$$

$$- (2h_2)^{-1}(\delta_2 - \delta_1)(h_2 - h_1)\left[\rho^2(T) + \Lambda^{-1}\sum_{k=1}^{M}\beta_k\lambda_k^{-1}\eta_k^2(T)\right]$$

$$+ (\Lambda h_2)^{-1}(\delta_2 - \delta_1)(h_2 - h_1)\sum_{k=1}^{M}\beta_k\int_T^{\infty}|\bar{\xi}(t) - \zeta_k(t)|^2\,dt$$

$$+ \varepsilon\int_0^T\rho^2(t)\,dt + \varepsilon\int_T^{\infty}\bar{\xi}^2(t)\,dt. \tag{6.3.111}$$

Another form for $\chi(T)$ has been obtained above, valid for any δ_0. Now taking $\delta_0 = 0$ in (6.3.72) one has

$$\chi(T) = -\delta_1 h_2[\Omega(T) - \Omega(0)]$$

$$- \delta_1\Lambda^{-1}\sum_{k=1}^{M}\beta_k\int_0^T[\rho(t) - \eta_k(t)]^2\{h_2[1 + \rho(t)]^{-1}[1 + \eta_k(t)]^{-1} - 1\}\,dt$$

$$- (\delta_2 - \delta_1)(h_2 - h_1)\left\{\Omega_1(T) - \Omega_1(0) + \Lambda^{-1}\sum_{k=1}^{M}\beta_k\int_0^T[\rho(t) - \eta_k(t)]^2\right.$$

$$\left.\times [1 + \rho(t)]^{-1}[1 + \eta_k(t)]^{-1}\,dt\right\}. \tag{6.3.112}$$

If we now equate the value of $\chi(T)$ given by (6.3.111) and (6.3.112), we obtain after elementary operations an inequality like (6.3.87):

$$\Omega(T) + \delta_1\Lambda^{-1}[\delta_1 h_1 + \delta_2(h_2 - h_1)]^{-1}\sum_{k=1}^{M}\beta_k\int_0^T[\rho(t) - \eta_k(t)]^2\{h_2[1 + \rho(t)]^{-1}$$

$$\times [1 + \eta_k(t)]^{-1} - 1\}\,dt + \varepsilon[\delta_1 h_1 + \delta_2(h_2 - h_1)]^{-1}\int_0^T\rho^2(t)\,dt$$

$$\leq \Omega(0) + [\delta_1 h_1 + \delta_2(h_2 - h_1)]^{-1}\left\{\frac{1}{8\pi}\int_{-\infty}^{\infty}\frac{|V(i\omega)|^2}{\operatorname{Re}G(i\omega) - \varepsilon}d\omega\right.$$

$$\left.+ \frac{1}{2}\delta_1\rho^2(0) + \frac{1}{2h_2}(\delta_2 - \delta_1)(h_2 - h_1)\left[\rho^2(0) + \Lambda^{-1}\sum_{k=0}^{M}\beta_k\lambda_k^{-1}\eta_k^2(0)\right]\right\}. \tag{6.3.113}$$

The inequality (6.3.113) can be used in order to prove that any solution of (6.3.35), such that inequalities (6.3.91) hold, is defined on R_+ and also satisfies (6.3.92). Moreover, (6.3.113) implies $\rho(t) \in L^2(R_+, R)$ because the right hand side in (6.3.113) does not depend on T, and therefore $\int_0^T \rho^2(t)\, dt$ is bounded above by a fixed number (for each set of initial data).

From the first equation of (6.3.35), and $\rho(t) \in L^2(R_+, R)$, we obtain (6.3.100), taking into account that $(b\rho)(t) \in L^2(R_+, R^n)$ and Theorem 6.3.1. Furthermore, the equations in η_k and $\rho(t) \in L^2(R_+, R)$ imply $\eta_k(t) \to 0$ as $t \to +\infty$. Finally, $\rho(t) \to 0$ as $t \to \infty$ because it is bounded on R_+ and this property implies the boundedness of $\dot{\rho}(t)$ on R_+ (hence, $\rho(t)$ is uniformly continuous on R_+). Consequently, (6.3.97) are also verified in Case 2, and together with (6.3.100) yield the conclusion.

This ends the proof of Theorem 6.3.5.

Remark. In order to carry out the proof of existence of solutions on the half-axis R_+, such that (6.3.92) is verified, a local existence result for (6.3.35) is necessary. Such a result is easily obtainable by means of the contraction mapping theorem. See Section 3.3.

6.4   Stability of nuclear reactors (B)

In this section we shall consider the stability problem of an uncontrolled nuclear reactor, under hypotheses that are somewhat different than those accepted in Section 6.3 for the case of controlled nuclear reactors.

Notice that, if we disregard the variable x in the equation (6.3.35), which means that we do not consider the interaction between the controlling system and the nuclear reactor (which is the controlled part of the larger system), we obtain the system

$$\left.\begin{aligned}
\dot{\rho}(t) &= -\sum_{k=1}^{M} \beta_k \Lambda^{-1}[\rho(t) - \eta_k(t)] - P\Lambda^{-1}[1 + \rho(t)]v(t), \\
\dot{\eta}_k(t) &= \lambda_k[\rho(t) - \eta_k(t)], \quad k = 1, 2, \ldots, M, \\
v(t) &= (\alpha\rho)(t),\}
\end{aligned}\right\} \tag{6.4.1}$$

in which the reactivity $v(t)$ can be given by means of the same operator α as defined in (6.3.36), or it may depend on ρ in a different way from that considered in Section 6.3.

The stability analysis we are going to present here is due to Podowski [1], [2]. A first distinct feature with respect to the material presented in Section 6.3 consists in the fact that the second set of equations in (6.4.1) is replaced by the following set of integral relationships:

$$\eta_k(t) = \lambda_k \int_{-\infty}^{t} \rho(s) \exp\{-\lambda_k(t - s)\}\, ds, \quad k = 1, 2, \ldots, M. \tag{6.4.2}$$

This kind of relationship involves infinite memory. The equivalence of (6.4.2) with the set of corresponding differential equations in (6.4.1) is a matter of discussion. Of course, the properties of $\rho(t)$ are essential in establishing such equivalence. While the equations $\dot{\eta}_k(t) = \lambda_k[\rho(t) - \eta_k(t)]$ have solutions for any locally integrable $\rho(t)$, the formulas (6.4.2) obviously have a meaning only for those $\rho(t)$ whose behavior on the negative half-axis makes the integrals in (6.4.2) convergent. Dealing with (6.4.2) instead of the differential equations $\dot{\eta}_k(t) = \lambda_k[\rho(t) - \eta_k(t)]$, $k = 1, 2, \ldots, M$, does not seem to be very restrictive because the functions $\eta_k(t)$ in (6.4.2) are solutions of the above equations whenever the integrals do converge.   Notice that introducing in the first equation in (6.4.1) the $\eta_k(t)$ given by (6.4.2), we obtain the following integrodifferential equation with infinite delay:

$$\dot{\rho}(t) = -\Lambda^{-1}\left(\sum_{k=1}^{M} \beta_k\right)\rho(t) + \Lambda^{-1}\sum_{k=1}^{M} \beta_k\lambda_k \int_{-\infty}^{t} \rho(s)\exp\{-\lambda_k(t - s)\}\,\mathrm{d}s$$
$$- P\Lambda^{-1}[1 + \rho(t)]v(t). \tag{6.4.3}$$

The last equation in (6.4.1) has to be associated with (6.4.3) and, if we want to eliminate v from these two equations, we obviously obtain an integrodifferential equation with infinite memory for $\rho(t)$ only. The only nonlinearity in the resulting equation is given by the term $[1 + \rho(t)](\alpha\rho)(t)$. Even if α is a linear operator, the product is a nonlinear term for the equation

$$\dot{\rho}(t) = -\Lambda^{-1}\left(\sum_{k=1}^{M} \beta_k\right)\rho(t) + \Lambda^{-1}\sum_{k=1}^{M} \beta_k\lambda_k \int_{-\infty}^{t} \rho(s)\exp\{-\lambda_k(t - s)\}\,\mathrm{d}s$$
$$- P\Lambda^{-1}[1 + \rho(t)](\alpha\rho)(t). \tag{6.4.4}$$

If we can obtain stability results for equation (6.4.4), then we can go back to the system (6.4.1), or even better to the formulas (6.4.2), to derive the same property with respect to the variables $\eta_k(t)$, $k = 1, 2, \ldots, M$. The fact that the stability problem has been reduced to the case of a single (scalar) equation, is certainly encouraging from the point of view of obtaining simpler criteria.

Most of the conditions relating to the right hand side of (6.4.4) will be the same as in Section 6.3. One condition which is completely different from condition (6.3.36) on $(\alpha\rho)(t)$ will be now stated.

Namely, we will assume that $\rho \to \alpha\rho$ is a Volterra type (nonanticipative) operator on the space $\tilde{C}(R, R)$ of real-valued continuous functions on the real axis, such that each $\rho \in \tilde{C}(R, R)$ is bounded on any half-axis $(-\infty, t_0)$, $t_0 \in R$. Of course, it suffices to assume, for instance, that ρ is bounded on the negative half-axis $R_- = (-\infty, 0]$. The natural topology of $\tilde{C}(R, R)$ is given by the family of seminorms

$$|\rho|_t = \sup\{|\rho(s)|; s \leq t\},$$

for any $t \in R$. The space $\tilde{C}(R, R)$, endowed with the topology η generated by the seminorms $|\rho|_t$, $t \in R$, is a Fréchet space (i.e., a linear metric space, complete, and with linearly invariant metric).

The following first basic assumption will be made with regard to the operator α: for any $r > 0$, there exists $L(r) > 0$ such that for any $t \in R$ we have

$$|(\alpha\rho_1)(t) - (\alpha\rho_2)(t)| \leq L(r)|\rho_1 - \rho_2|_t, \qquad (6.4.5)$$

as soon as ρ_1, $\rho_2 \in \tilde{C}(R, R)$ verify the estimates $|\rho_1|_t$, $|\rho_2|_t \leq r$. This condition is, in fact, a generalized *Lipschitz condition* which implies the continuity of α.

Another assumption we make on α, to be kept throughout this section, is the *time invariance* of this operator. In other words, if we denote $\rho(t + h) = \rho_h(t)$, $t \in R$, where $h \in R$ is fixed, then $(\alpha\rho_h)(t) = (\alpha\rho)(t + h)$ for any $h \in R$.

It is useful to notice that for every $\rho \in \tilde{C}(R, R)$, one has $\rho_h \in \tilde{C}(R, R)$ for any $h \in R$.

If we now assume that $\rho = \rho(t)$ is a solution of the equation (6.4.4), such that $\rho \in \tilde{C}(R, R)$, then it is not difficult to see that we also have $\dot{\rho}(t) \in \tilde{C}(R, R)$. This assumption can be checked if we rely on the properties of the operator α, and take into account the special form of equation (6.4.4).

Consequently, it is appropriate to look for the solutions of (6.4.4) in the space of those $\rho: R \to R$, such that both ρ and $\dot{\rho} \in \tilde{C}(R, R)$. This set of functions is obviously a subset of $\tilde{C}(R, R)$. It can be organized as a Fréchet space in its own rights, if we take as seminorms the maps $\rho \to |\rho|_t + |\dot{\rho}|_t$, $t \in R_+$. We shall denote this space by $\tilde{C}^{(1)}(R, R)$, and use it in the investigation of stability of the zero solution of equation (6.4.4).

The initial value problem for the equation (6.4.4) that we are going to discuss can be formulated as follows: find a solution of (6.4.4), defined in an interval $[0, T)$, $T > 0$, such that

$$\rho(t) = \lambda(t), \quad t \leq 0, \qquad (6.4.6)$$

where $\lambda(t)$ is continuous and bounded on the half-axis $t \leq 0$, together with its first derivative $\dot{\lambda}(t)$. In other words, $\lambda(t)$ must be the restriction to the negative half-axis of a function which belongs to $\tilde{C}^{(1)}(R, R)$.

The local existence of a solution $\rho(t)$ to the equation (6.4.4), such that (6.4.6) is satisfied, is a consequence of the more general result established in Section 3.4 (Theorem 3.4.1). The global existence (on R_+) will follow from the considerations to be made subsequently.

Let us point out the significant fact that instead of assigning $\rho(t)$ on $t \leq 0$, as in (6.4.6), one could use any other real number t_0 as starting value. This follows immediately from the time-invariance property of the right hand side of the equation (6.4.4). This will enable us to deal with (6.4.4) only for $t_0 = 0$, even when we discuss stability properties.

Before we get involved in the discussion of stability properties of the equation (6.4.4), we shall make one more remark which will help us to understand better the framework in which we conduct this discussion. Apparently, it is desirable to allow more freedom to the derivative $\dot{\rho}(t)$, not assuming it necessarily continuous. One possible approach is to assume that $\rho(t)$ is absolutely continuous, which implies the existence of $\dot{\rho}(t)$ a.e. on R, and both $\rho(t)$ and $\dot{\rho}(t)$ are in $L^{\infty}((-\infty, t_0], R)$ for any $t_0 \geq 0$. Instead of the space $\tilde{C}^{(1)}(R, R)$ considered above, we have to deal with the space $A\tilde{C}(R, R)$, whose topology can also be defined by means of a family of seminorms:

$$\rho \to |\rho(0)| + |\rho|_t + |\dot{\rho}|_t, \quad t \in R_+,$$

in which $|\dot{\rho}|_t = \text{ess-sup}|\dot{\rho}(s)|$ for $s \leq t$. In the framework of the space $A\tilde{C}(R, R)$, the derivative in (6.4.4) should be understood as existing only a.e. (we would obtain what is usually termed as a solution of that equation in Carathéodory sense). We leave to the reader the task of conducting the stability analysis under these (more general) assumptions.

Let us return now to equation (6.4.4) and investigate its stability properties in the space $\tilde{C}^{(1)}(R, R)$. Unlike the approach adopted in Section 6.3, we shall now use the classical method of Liapunov. In other words, we shall use a convenient auxiliary functional, related to equation (6.4.4), and derive the stability properties from those of the Liapunov functional.

We shall use the notation introduced earlier (Section 3.2), namely $\rho_t(s) = \rho(t + s)$, $-\infty < s \leq 0$. Then, the initial condition (6.4.6) can be rewritten as $\rho_0 = \lambda$, where λ is a continuous and bounded function on the negative half-axis, together with its first derivative.

The definition of *stability* we shall use in this section can be formulated as follows: the solution $\rho = 0$ of (6.4.4) is stable if for every $\varepsilon > 0$ there exists $\delta > 0$, such that $|\rho_0|_0 < \delta$ and $|\dot{\rho}_0|_0 < \delta$ imply $|\rho(t)| < \varepsilon$ for $t \in R_+$.

The *asymptotic stability* of the solution $\rho = 0$ of (6.4.4) means the property of stability defined above, plus the existence of a $\delta_0 > 0$ with the property $|\rho_0|_0 < \delta_0$, $|\dot{\rho}_0|_0 < \delta_0$ imply $\rho(t) \to 0$ as $t \to \infty$.

Let us consider the following subset of the space of initial functions (i.e., of the function space consisting of the restrictions to the negative half-axis of the functions in $\tilde{C}^{(1)}(R, R)$):

$$D(\delta_1, \delta_2, \delta_3) = \{\rho(t): \delta_1 < \rho(t) < \delta_2, |\dot{\rho}(t)| < \delta_3, t \in R_-\}, \quad (6.4.7)$$

where $-1 < \delta_1 < 0 < \delta_2, \delta_3 > 0$. The reason we assume $-1 < \delta_1$ is that $1 + \rho(t)$ is supposed to remain positive. See Lemma 6.3.4 in this respect.

An important task is to find estimates on δ_i, $i = 1, 2, 3$, such that $D(\delta_1, \delta_2, \delta_3)$ defined by (6.4.7) is included in the domain of attraction of the origin.

We shall also define some subsets of the space $\tilde{C}^{(1)}(R, R)$, as follows:

$$C(h_1, h_2, h_3) = \{\rho(t): \rho \in \tilde{C}^{(1)}(R, R), h_1 < \inf \rho, \sup \rho < h_2, \sup|\dot{\rho}(t)| < h_3\}, \quad (6.4.8)$$

where $h_1 < 0$ and $h_2, h_3 > 0$.

We shall use Liapunov functionals of the form

$$V(\rho_t) = F(\rho(t)) + Q(\rho_t), \quad t \geq 0, \tag{6.4.9}$$

defined on sets like $C(h_1, h_2, h_3)$, given by (6.4.8). We recall that ρ_t is the restriction of ρ to the half-axis $(-\infty, t]$, and it is usually meant as $\rho_t(s) = \rho(t + s)$, $s \in R_-$.

If taking the derivative of $V(\rho_t)$ along the solutions of equation (6.4.4) we obtain

$$\dot{V}(\rho_t) = -R(\rho_t), \tag{6.4.10}$$

with R a functional of the same nature as $Q(\rho_t)$, then the following result holds.

Theorem 6.4.1. *Assume that there exists a functional $V(\rho_t)$ of the form (6.4.9), whose derivative with respect to the equation (6.4.4) is given by (6.4.10), such that the following conditions are verified:*

(1) *$F(u)$ is continuously differentiable for $u \in (m_1, m_2)$, $m_1 < 0 < m_2$, $uF'(u) > 0$ for $u \neq 0$, and $F(0) = 0$;*

(2) *$Q(\rho_t)$ is a continuous function on R_+, for any $\rho \in C(m_1, m_2, m_3)$, where $m_3 > 0$ is a fixed number;*

(3) *$R(\rho_t)$ is uniformly continuous in t, $t \in R_+$, for any $\rho \in C(m_1, m_2, m_3)$;*

(4) *$Q(\rho_t)$, $R(\rho_t) \geq 0$ for $t \in R_+$, and any $\rho \in C(m_1, m_2, m_3)$;*

(5) *there exists a continuous function $\psi(\sigma_1, \sigma_2, \sigma_3) \geq 0$ defined for $m_1 < \sigma_1 < 0 < \sigma_2 < m_2, 0 < \sigma_3 < m_3, \psi(0, 0, 0) = 0$, such that from $\sigma_1 < \rho(s) < \sigma_2$ and $|\dot{\rho}(s)| < \sigma_3$ for $s \leq t$, one obtains $Q(\rho_t) \leq \psi(\sigma_1, \sigma_2, \sigma_3)$;*

(6) *$\rho \in C(m_1, m_2, m_3)$ and $R(\rho_t) \to 0$ as $t \to \infty$, imply $\rho(t) \to 0$ as $t \to \infty$.*

Then the solution $\rho = 0$ of equation (6.4.4) is asymptotically stable. Moreover, if μ_1, μ_2 are fixed numbers such that $m_1 < \mu_1 < 0 < \mu_2 < m_2$, and $\delta_1, \delta_2, \delta_3$ are some arbitrary numbers such that $\mu_1 < \delta_1 < 0 < \delta_2 < \mu_2, 0 < \delta_3 < m_3$, and

$$F(\delta_2) + \psi(\delta_1, \delta_2, \delta_3) \leq F(\mu_1) = F(\mu_2), \tag{6.4.11}$$

then $D(\delta_1, \delta_2, \delta_3)$ belongs to the domain of attraction of the origin.

Proof. Let $\varepsilon > 0$ be such that $\varepsilon < \min(-m_1, m_2)$. According to our assumptions we can find $\delta > 0$, $\delta < m_3$, such that

$$\max\{F(-\delta), F(\delta)\} + \psi(-\delta, \delta, \delta) < \min\{F(-\varepsilon), F(\varepsilon)\}. \tag{6.4.12}$$

If we now assume that $\rho(t)$ is a solution of (6.4.4) such that $|\rho_0|_0 < \delta < \varepsilon$, and $|\dot{\rho}_0|_0 < \delta < m_3$, then $0 \leq V(\rho_0) < \min\{F(-\varepsilon), F(\varepsilon)\}$, and $\dot{V}(\rho_0) \leq 0$. We claim that this solution $\rho(t)$ satisfies $|\rho(t)| < \varepsilon$ for $t \in R_+$. Indeed, if $\rho(\bar{t}) = \varepsilon$ for some $\bar{t} > 0$, then there exists a number $t_1 > 0$ with the property $\rho(t_1) = \varepsilon$, and $\rho(t) < \varepsilon$ for $0 \leq t < t_1$. Hence, $F(\rho(t_1)) \leq V(\rho_{t_1}) \leq V(\rho_0) < \min\{F(-\varepsilon), F(\varepsilon)\}$, which is obviously contradictory because $\rho(t_1) = \varepsilon$. This proves that $\rho = 0$ is stable for (6.4.4).

In order to derive the asymptotic stability, let us assume that $\rho(t)$ is a solution of (6.4.4) which begins in $C(\delta_1, \delta_2, \delta_3)$, where $\delta_1, \delta_2, \delta_3$ are such that (6.4.11) holds. There results $\mu_1 < \rho(t) < \mu_2$ for $t \in R_+$. But $\dot{V}(\rho_t) \leq 0$ implies the fact that $V(\rho_t)$ is monotonically decreasing as $t \to \infty$. Its derivative $\dot{V}(\rho_t)$ is uniformly continuous on R_+, which implies $\dot{V}(\rho_t) \to 0$ as $t \to \infty$. From condition (6) in the statement, we must have $\rho(t) \to 0$ as $t \to \infty$.

Theorem 6.4.1 is thereby proven.

Remark.   It is obvious from the proof of Theorem 6.4.1 that this result does not apply only to equation (6.4.4). If we assume the existence of a Liapunov functional $V(\rho_t) = F(\rho(t)) + Q(\rho_t)$, with the properties (1)–(6) listed in the statement of Theorem 6.4.1, but instead of equation (6.4.4) we substitute another functional-differential equation (say $\dot{\rho}(t) = \phi(\rho_t)$), then we obtain the asymptotic stability for the zero solution of the new equation.

As we can see from the statement of Theorem 6.4.1, very general conditions are involved with respect to the functions/functionals occurring in the stability problem for equation (6.4.4). It is interesting to construct some functions like $F(u)$, or functionals like $Q(\rho_t)$ or $R(\rho_t)$, such that the conditions (1)–(6) are satisfied. Only in this case can we expect to obtain stability criteria that are significant for applications.

Let us try the following expressions for $F(u)$ and $Q(\rho_t)$:

$$F(u) = a\left[u - (1 + c)\ln\frac{1 + c + u}{1 + c}\right] - \frac{b}{2}u^2, \qquad (6.4.13)$$

$$Q(\rho_t) = \Lambda^{-1}\sum_{k=1}^{M}\beta_k\lambda_k^{-1}F(\eta_k) + q(\rho_t), \qquad (6.4.14)$$

with η_k given by (6.4.2), and q given by

$$q(\rho_t) = \int_{-\infty}^{t} \exp\{-2\mu(t - s)\}$$

$$\times \left[\frac{aP\Lambda^{-1}(1 + \rho)\rho v}{1 + c + \rho} + b\rho p - dpP^{-1}v + ev^2 + fp^2P^{-2}\right]ds, \qquad (6.4.15)$$

where

$$p = \dot{\rho} + \sum_{k=1}^{M} \Lambda^{-1}\beta_k(\rho - \eta_k) = -P\Lambda^{-1}(1 + \rho)v,$$

and the constants involved are such that

$$\mu > 0, \quad a \geq 0, \quad b < a/(1 + c), \quad c > -1. \qquad (6.4.16)$$

The reader will easily recognize that the function $F(u)$ defined by (6.4.13) is basically the same as the function $\phi(\xi)$ defined in the statement of

Theorem 6.3.5. The meaning of $Q(\rho_t)$ or $q(\rho_t)$ is more difficult to explain, but our aim is to produce a valid illustration of the conditions appearing in Theorem 6.4.1.

If for the function $V(\rho_t) = F(\rho(t)) + Q(\rho_t)$, with F and Q given by (6.4.13), (6.4.14), we estimate the derivative along the trajectories of the equation (6.4.4), then for $R(\rho_t)$ in (6.4.10) we obtain

$$R(\rho_t) = F_1(\rho(t), \eta_1(t), \ldots, \eta_M(t)) + r(\rho_t) + 2\mu q(\rho_t), \qquad (6.4.17)$$

where

$$F_1(\rho, \eta_1, \ldots, \eta_M) = \sum_{k=1}^{M} \varLambda^{-1} \beta_k (\rho - \eta_k)^2 \left[\frac{a(1+c)}{(1+c+\rho)(1+c+\eta_k)} - b \right], \qquad (6.4.18)$$

and

$$r(\rho_t) = v^2 [d\varLambda^{-1}(1+\rho) - e - f\varLambda^{-2}(1+\rho)^2]. \qquad (6.4.19)$$

Let us recall that $v(t) = (\alpha\rho)(t)$, with α a Volterra (causal) operator, which justifies the notation (6.4.19).

In order to apply Theorem 6.4.1 to this particular choice of the functional $V(\rho_t)$, we need to impose further restrictions on the constants involved in the above expressions.

Let us start with condition (1) in Theorem 6.4.1. It is obvious that $F(u)$ given by (6.4.13) is continuously differentiable in a neighborhood of the origin ($u = 0$), $F(0) = 0$ and $uF'(u) > 0$ for $u \neq 0$ in the same neighborhood or a smaller one, because

$$F'(u) = \left[\frac{a}{1+c+u} - b \right] u.$$

Hence, for $b \leq 0$ and $a \geq 0$, with $a + |b| \neq 0$, we have $uF'(u) > 0$ for $u \neq 0$, $u > -(1+c)$. If $0 < b < a(1+c)^{-1}$ and $a > 0$ we have $uF'(u) > 0$ for $u \neq 0$, $u > -(1+c)$. In case $0 < b < a(1+c)^{-1}$ and $a > 0$ one has $uF'(u) > 0$ for $-(1+c) < u < ab^{-1} - (1+c)$, $u \neq 0$. This ends the discussion in relation to condition (1) of Theorem 6.4.1. We notice that (6.4.16) excludes the case $a = b = 0$. One can take, therefore, $m_1 = -(1+c)$, $m_2 = ab^{-1} - (1+c)$.

Let us now concentrate on $F_1(\rho, \eta_1, \ldots, \eta_M)$ given by (6.4.18). First of all, notice from (6.4.2) that the following inequalities hold:

$$\inf_{s \leq t} \rho(s) \leq \eta_k(t) \leq \sup_{s \leq t} \rho(s), \quad k = 1, 2, \ldots, M. \qquad (6.4.20)$$

Consequently, $F_1(\rho, \eta_1, \ldots, \eta_M)$ is defined for all $\rho(t)$, provided $\rho(t) > -(1+c)$. In case $b < 0$ and $a \geq 0$, F_1 is nonnegative for any such $\rho(t)$. If $b \geq 0$ and $a > 0$, then an additional restriction must be imposed on $\rho(t)$, namely

$$\sup_t \rho(t) < \sqrt{(a(1+c)b^{-1})} - (1+c), \qquad (6.4.21)$$

in order to be sure F_1 is nonnegative. The estimate (6.4.21) is easily obtained from (6.4.18) if we want every square bracket appearing in the formula to remain positive.

We shall now impose one more restriction on the positive constants d, e and f:

$$d - e\Lambda - f\Lambda^{-1} > 0. \qquad (6.4.22)$$

Consequently, if $\rho(t)$ is such that

$$\frac{d\Lambda}{2f} - \left(\frac{d^2\Lambda^2}{4f^2} - \frac{e\Lambda^2}{f}\right)^{1/2} \le 1 + \rho(t) \le \frac{d\Lambda}{2f} + \left(\frac{d^2\Lambda^2}{4f^2} - \frac{e\Lambda^2}{f}\right)^{1/2}, \qquad (6.4.23)$$

then the inequality

$$d\Lambda(1 + \rho) - e\Lambda^2 - f(1 + \rho)^2 \ge 0 \qquad (6.4.24)$$

holds. (6.4.23) says that the roots of the quadratic polynomial occurring in the left hand side of (6.4.24) must contain $1 + \rho(t)$ in between. It is obvious, from (6.4.22), that (6.4.24) is verified for sufficiently small $|\rho|$.

One more consequence we can derive from the inequality (6.4.22) is the validity of the following inequalities:

$$\left. \begin{array}{l} A_1 = \dfrac{d\Lambda}{2f} - \left(\dfrac{d^2\Lambda^2}{4f^2} - \dfrac{e\Lambda^2}{f}\right)^{1/2} - 1 < 0, \\[4mm] A_2 = \dfrac{d\Lambda}{2f} - \left(\dfrac{d^2\Lambda^2}{4f^2} - \dfrac{e\Lambda^2}{f}\right)^{1/2} - 1 > 0. \end{array} \right\} \qquad (6.4.25)$$

Indeed, these follow from (6.4.23) if we notice that $\rho = 0$ is admissible in (6.4.24), according to (6.4.22).

From (6.4.24) and the definition of p and $q(\rho_t)$ we easily derive that $q(\rho_t)$ makes sense for all $\rho \in C(h_1, h_2, h_3)$ with $-(1 + c) \le h_1 < 0$ and arbitrary $h_2, h_3 > 0$.

The conditions (2) and (3) of Theorem 6.4.1 are obviously satisfied if we assume $v(t) = (\alpha\rho)(t)$ to be uniformly continuous in t on R_+, for any $\rho \in C(m_1, m_2, m_3)$.

In order to assure the validity of condition (4) in Theorem 6.4.1, we will now assume that *the functional $q(\rho_t)$ given by (6.4.15) is nonnegative*. Then, from what we have shown above in relation to the function $F_1(\rho, \eta_1, \ldots, \eta_M)$, and based on the fact that $r(\rho_t) \ge 0$ under the assumption (6.4.22), one obtains from (6.4.17) the fact that $R(\rho_t)$ is nonnegative. From the same assumption and (6.4.14) we also obtain $Q(\rho_t) \ge 0$, which means that condition (4) is satisfied.

Let us now consider condition (5) of Theorem 6.4.1. We have to construct a function $\psi(\sigma_1, \sigma_2, \sigma_3)$ satisfying the requirement of condition (5). It can easily be checked that choosing

$$\psi(\sigma_1, \sigma_2, \sigma_3) = \frac{1}{2\mu}\left\{\frac{a}{\Lambda}\frac{1 + \sigma_2}{1 + c + \sigma_1}\sigma\psi(\sigma) + [b\sigma + d\psi(\sigma)]\left(\sigma_3 + \frac{2\beta}{\Lambda}\sigma\right)\right.$$

$$\left. + e[\psi(\sigma)]^2 + f\left(\sigma_1 + \frac{2\beta}{\Lambda}\sigma\right)^2\right\},$$

where $\sigma = \max(-\sigma_1, \sigma_2)$, $\beta = \beta_1 + \beta_2 + \cdots + \beta_M$, and $\psi(\sigma)$ is such that

$$|v(t)| \le \psi(|\rho|_t), \quad t \in R_+, \tag{6.4.27}$$

we have $q(\rho_t) \le \psi((\sigma_1, \sigma_2, \sigma_3)$, provided ρ is such that $\sigma_1 < \rho(s) < \sigma_2$ and $|\dot{\rho}(s)| < \sigma_3$ for $s \le t$. If we take into account (6.4.5), the function ψ can be chosen as $\psi(\sigma) = L\sigma, \sigma \ge 0$.

We have only to notice, in relation to condition (6) of Theorem 6.4.1, that, because of our assumptions, $R(\rho_t) \to 0$ as $t \to \infty$ implies $r(\rho_t) \to 0$ as $t \to \infty$. Taking (6.4.19) into account, we see that the last condition is verified by the fact that $v(t) = (\alpha\rho)(t) \to 0$ as $t \to \infty$ has as consequence $\rho(t) \to 0$ as $t \to \infty$.

The discussion conducted above in relation to the choice $V(\rho_t) = F(\rho(t)) + Q(\rho_t)$ as a candidate for a Liapunov functional, where $F(u)$ and $Q(\rho_t)$ are given by (6.4.13)–(6.4.15), can be summarized as follows.

Theorem 6.4.2. *Assume there exist numbers m_1, m_2, m_3 and μ, such that $-1 < m_1 < 0 < m_2, m_3 > 0, \mu > 0$, with the following properties:*

(1) *the functional $q(\rho_t)$, $t \in R_+$, given by (6.4.15), with d, e, f satisfying (6.4.22), is nonnegative for any $\rho \in C(m_1, m_2, m_3)$;*

(2) *$v(t) = (\alpha\rho)(t)$ is uniformly continuous on R_+ for every $\rho \in C(m_1, m_2, m_3)$;*

(3) *$(\alpha\rho)(t) \to 0$ as $t \to \infty$ implies $\rho(t) \to 0$ as $t \to \infty$, for any $\rho \in C(m_1, m_2, m_3)$.*

Then, the solution $\rho = 0$ of the equation (6.4.4) is asymptotically stable in $\tilde{C}^{(1)}(R, R)$. Moreover, a set $D(\delta_1, \delta_2, \delta_3)$ of admissible initial values can be estimated by choosing the numbers μ_1, μ_2 such that

$$\max\{m_1, A_1, -(1 + c)\} < \mu_1 < 0, \tag{6.4.27}$$

and

$$0 < \mu_2 < \min\{m_2, A_2, ab^{-1} - (1 + c), \sqrt{(ab^{-1}(1 + c))} - (1 + c)\}, \tag{6.4.28}$$

with $F(\mu_1) = F(\mu_2)$, A_1, A_2 being given by (6.4.25), and then solving in $\delta_1, \delta_2, \delta_3$ the inequality

$$\left[1 + \sum_{k=1}^{M} (\beta_k/\Lambda\lambda_k)\right] F(\delta_2)$$

$$+ \frac{1}{2\mu}\left\{\frac{a}{\Lambda}\frac{(1 + \delta_2)\delta_2\psi(\delta_2)}{1 + c + \mu_1} + \left(\frac{2\beta}{\Lambda}\delta_2 + \delta_3\right)[b\delta_2 + d\psi(\delta_2)] + e[\psi(\delta_2)]^2\right.$$

$$\left. + f\Lambda\left(\frac{2\beta}{\Lambda}\delta_2 + \delta_3\right)^2\right\} \le F(\mu_2), \tag{6.4.29}$$

under the constraint $F(\delta_1) = F(\delta_2)$.

Inequality (6.4.29) is a special case of inequality (6.4.11) in Theorem 6.4.1, corresponding to the special choice of the function $F(u)$, and of the functional $Q(\rho_t)$ as shown above.

With regard to the functional $q(\rho_t)$, given by (6.4.15), the positiveness condition is the only restrictive requirement. Besides restriction (6.4.22) imposed on the parameters, some extra conditions must be imposed. Nevertheless, the restriction is not very strong, as the following considerations illustrate.

If we particularize the parameters, taking $a = 1$, and $b = c = d = e = f = 0$ (in which case (6.4.22) does not hold anymore), and we obtain $F(u) = u - \ln(1 + u)$ and

$$q(\rho_t) = \Lambda^{-1} \int_{-\infty}^{t} \exp\{-2\mu(t - s)\} \rho(s)(\alpha\rho)(s)\,ds,$$

then the positiveness of $q(\rho_t)$ can obviously be regarded as a generalized positiveness condition for the operator α in $v(t) = (\alpha\rho)(t)$. This is quite a natural type of restriction.

The results given above in Theorems 6.4.1 and 6.4.2 can be further generalized, as shown in Podowski's papers [1], [2]. The same method of Liapunov functionals can be adapted to cover the case of essentially nonlinear feedback relationships, such as $v(t) = (\alpha\rho)(t) + (\tilde{\alpha}\rho)(t)$, with α given by (6.3.36), and $\tilde{\alpha}$ a nonlinear Volterra type operator which can be represented by means of Volterra series (see, for instance, Rugh [1], Schetzen [1], Sandberg [6], [9]).

It is interesting to point out that Theorem 6.4.2, if we choose α as defined in (6.3.36), leads to a frequency domain type stability criterion, as shown in Podowski [2]. It is interesting to compare Podowski's result with the findings of Section 6.3.

Bibliographical notes

The material in Section 6.1 is adapted from Melzak [1], this paper representing his Ph. D. thesis written under N. Levinson at MIT. Among the immediate followers are Galkin [1], and Galkin and Dubovskii [1]. More references can be found in these papers. The transport equations are encountered in many areas of physical sciences and there is a varied literature on this subject. Without any intention of being complete, I shall mention here a series of useful references: Cergignani [1], [2], Chandrasekar [1], Colton and Kress [1], C. Corduneanu [4] (a chapter is dedicated to Krein's theory of Wiener–Hopf equations occurring in radiative transfer), Davis [1], Gohberg and Feldman [1], Gohberg and Kaashoek [1], Greenberg, van der Mee and Protopopescu [1], Palczewski [1], Saaty [1].

Section 6.2 contains the proof of the maximum principle for systems governed by Volterra integral equations. The approach presented in this section is due to Carlson [1], and leads immediately to the classical Pontryagin's principle for control systems governed by ordinary differential equations. The following references are pertinent to this topic: Angell [1], Bakke [1], Carlson [2], Chen and Grimmer [3], Conti [1] (infinite dimensional case control problems), Delfour and Karrachou [1], Medhin [1], Vinokurov [1], Warga [1]. Of course, Neustadt [1] is an excellent reference if we keep in mind the fact that the author was the first to try to build-up a control theory for abstract Volterra equations.

334   Some applications of integral and integrodifferential equations

Sections 6.3 and 6.4 are devoted to the stability problem of nuclear reactors. The results included in Section 6.3 are basically those obtained by C. Corduneanu [8], and they constitute generalizations of the results obtained on this subject by Halanay and Rasvan [1]. The first stability results involving integral equations occurring in the dynamics of nuclear reactors were obtained by V. M. Popov, J. J. Levin and J. A. Nohel, by using different methods. The theory has developed during the 1960s, 1970s and 1980s, and the following references will give some guidance on its present status: Akcasu et al. [1], C. Corduneanu [4], Gorjacenko [1], Kappel and DiPasquantonio [1], Luca [2], Nohel and Shea [1], Podowski [1], [2] (whose results constitute the content of Section 6.4), Popov [1], [2], Smets [1], Williams [1]. Stability results for the nuclear reactors described by integrodifferential equations such as those considered in Sections 6.3 and 6.4 are potentially contained in various books on the stability of feedback systems: Desoer and Vidyasagar [1], Distefano [2], Kudrewicz [1], Narendra and Taylor [1].

The number of applications of integral or integrodifferential equations is considerable and I shall not attempt to provide a complete list of references on these matters. Nevertheless, I think it is appropriate to give some indications of the research work conducted on the applications of integral and related equations to population dynamics and viscoelasticity.

As mentioned in the introduction, Volterra himself has considered applications to population dynamics and hereditary mechanics. All his contributions can be found in his 'Opere Matematiche' published in 1954–1955. In Volterra and d'Ancona [1] one finds more about populations, from the mathematical point of view. Another relevant source is Kostitzin [1]. Recently, two monographs have been published on this subject by Cushing [3], and Webb [3]. Further references are provided in these books. From the journal literature I will mention the papers by Thieme (primarily [1]).

The mathematical theory of viscoelastic materials is also one of the users of the theory of integral equations. More precisely, these equations are similar to those investigated in Section 5.5 (integro-partial differential equations). For recent investigations see Hrusa [1], [2], Dafermos and Nohel [1], [2], Leugering [1] (a control problem on viscoelastic liquids is investigated), Leitman and Mizel [1], Nohel [1], [2], as well as the monograph by Renardy, Hrusa and Nohel [1].

Other areas of applied research in which integral or related equations are in current use are: autoaeroelasticity (see the reference Belotserkovskii et al. [1], a monograph entirely dedicated to this field, in which more adequate references are included); electromagnetic theory (see Bloom [1]); dynamics of a rigid body (see Kharlamova and Mozalevskaya [1]); linear programming in infinite-dimensional spaces (see Anderson and Nash [1], where the role of integral constraints or inequalities is emphasized); creep theory (see Arutjunian and Kolmanovskii [1]); network theory and radiophysics (see Ramm [1], in which several other areas are illustrated).

References

N. H. Abel
1. Solution de quelques problèmes à l'aide d'integrales definies. *Mag. Naturvidenskaberne*, Aargang I, Bind 2, Christiania, 1823 (Oeuvres Completes, T.1, B.2 Christiania, 1881, 11–27).
2. Auflösung einer mechanischen Aufgabe. *Crelles J.* (1826), 153–7.

R. d'Adhémar
1. L'équation de Fredholm et les problèmes de Dirichlet et de Neumann. Gauthier-Villars, Paris, 1909.

N. V. Ahmed
1. Nonlinear integral equations on reflexive Banach spaces with applications to stochastic integral equations and abstract evolutions equations, *J. Int. Equations* 1 (1979), 1–5.

S. Aizicovici
1. On a class of functional differential equations, *Rend. Mat.* (3) 8 (1975), 685–706.
2. On a nonlinear integrodifferential equation, *Math. An. Appl.* 63 (1978), 385–95.
3. Un résultat d'existence pour une équation non linéaire du type Volterra, *C. R. hebd Acad. Sci. Paris*, Sér. I Math. 301 (1985), 829–32.

L. Akcasu, G. S. Lellouche, L. M. Shotkin
1. *Mathematical Methods in Nuclear Reactor Dynamics*, Academic Press, New York, 1971.

N. J. Akhiezer
1. Integral operators with Carleman kernels (Russian), *Usp. Mat. Nauk* 2 (1947), No. 5, 93–132.

N. K. Al'bov
1. On a criterion for solvability of Fredholm equations, *Math. USSR, Sbornik* 55 (1986), 113–19.

B. Alfawicka
1. Monotone solutions of Volterra integral equations, *J. Math. An. Appl.* 123 (1987), 39–56.

C. D. Aliprantis, O. Burkinshaw, M. Duhoux
1. Compactness properties of abstract kernel operators, *Pacific J. Math.* 100 (1982), 1–22.

H. Amann
1. Uber die Existenz und iterative Berechnung einer Lösung der Hammersteinschen gleichung, *Aequations Math.* 1 (1968), 242–65.

2. Ein Existenz-und Eindeutigkeits-satz fur die Hammersteinsche gleichung in Banach-räumen, *Math. Zts.* **111** (1969), 175–190.
3. Hammersteinische gleichungen mit kompakten kernen, *Math. Annalen* **186** (1970), 334–40.
4. Existence theorems for equations of Hammerstein type, *Applicable Analysis* **1** (1972), 385–97.
5. On the unique solvability of semi-linear operator equations in Hilbert spaces, *J. Math. Pures. Appl.* **61** (1982), 149–75.

E. J. Anderson, P. Nash
1. *Linear Programming in Infinite-Dimensional Spaces*, John Wiley, New York, 1987.

R. S. Anderssen, F. R. De Hoog, M. A. Lukas
1. *The Application and Numerical Solution of Integral Equations*, Sijthoff & Noordhoff, Leyden, 1980.

T. S. Angell
1. On the optimal control of systems governed by nonlinear Volterra equations, *J. Optim. Th. Appl.* **19** (1976), 29–45.

T. S. Angell, W. E. Olmstead
1. Singular perturbation analysis of an integrodifferential equation modelling filament stretching, *Z. Angew. Math. Phys.* **36** (1985), 487–90.
2. Singularly perturbed Volterra integral equations, *SIAM J. Appl. Math.* **47** (1987), 1–14.

P. M. Anselone
1. (ed.), *Nonlinear Integral Equations*, University of Wisconsin Press, Madison, 1964.
2. *Collectively Compact Operator Approximation Theory, and Applications to Integral Equations*, Prentice-Hall, 1971.

P. M. Anselone, I. H. Sloan
1. Integral equations on the half line, *J. Int. Equations* **9** (1985), Suppl., 3–23.

J. Appell
1. Implicit functions, nonlinear integral equations, and the measure of noncompactness of the superposition operator, *J. Math. An. Appl.* **83** (1981), 251–63.

N. Aronszajn, P. Szeptycki
1. On general integral transformations, *Math. Annalen* **163** (1966), 127–54.

Z. Arstein
1. Continuous dependence of solutions of Volterra integral equations, *SIAM Journal Math. Analysis* **6** (1975), 446–56.
2. Continuous dependence on parameters: On the best possible results, *J. Diff. Equations* **19** (1975), 214–25.

N. H. Arutjunian, V. B. Kolmanovskii
1. *Creep Theory of Nonhomogeneous Bodies* (Russian), Moscow, Nauka, 1983.

I. S. Astapov, S. M. Belotserkovskii, B. O. Kachanov, Yu. A. Kochetkov
1. Systems of integrodifferential equations describing nonstationary motion of bodies in a continuous medium, *Differential Equations* (Transl.) **18** (1982), 1164–71.

K. E. Atkinson
1. *A Survey of Numerical Methods for the Solution of Freedholm Integral Equations of the Second Kind*, SIAM, 1976.

J. P. Aubin
1. Un théoreme de compacité, *C. R. hebd Acad. Sci. Paris*, **256** (1963), 5042–44.

A. Augustynowicz, M. Kwapisz
1. On the existence of continuous solutions of operator equations in Banach spaces, *Casopis pest. mat.* **111** (1986), 267–79.

C. Avramescu
1. Sur l'admissibilité par rapport à un opérateur intégral linéaire. *An. St. Univ. 'Al. I. Cuza' Iasi*, Ia, **27** (1972), 55–64.

N. V. Azbelev
1. Some tendencies towards generalizations of differential equations, *Differential Equations* (Transl.) **21** (1984), 871–82.

N. V. Azbelev, L. M. Berezanskij, P. M. Simonov
1. Theorem on the stability of linear equations with delay (Russian), *Functional Diff. Equations Appl.*, Perm (1986), 3–6.

N. V. Azbelev, Z. B. Caljuk
1. On integral inequalities, I (Russian), *Mat. Sbornik* **56** (1962), 325–42.

N. V. Azbelev, V. P. Maksimov
1. Equations with delayed arguments, *Differential Equations* (Transl.) **18** (1983), 1419–41.

A. F. Bachurskaja, Z. B. Caljuk
1. On the existence of solutions for Volterra systems (Russian), *Yzv. Vyssh. Uc. Zaved., Mat.* (1985), 60–2.

T. H. Baker
1. *The Numerical Treatment of Integral Equations*, Clarendon Press, Oxford, 1977.

V. L. Bakke
1. A maximum principle for an optimal control problem with integral constraints, *J. Opt. Theory Appl.* **13** (1974), 32–55.

S. Banach
1. Sur les opérations dans les ensembles abstraits et leurs application aux équations intégrales, *Fund. Math.* **3** (1922), 133–81.

J. Banaś
1. An existence theorem for nonlinear Volterra integral equations with deviating argument, *Rend. Circ. Mat. Palermo* Series II, **35** (1986), 82–9.

G. Bantas
1. Théoremes d'existence et d'unicité dans la théore des equations integrofonctionnelles de Volterra, *Rend. Acad. Naz. Lincei* **52** (8) (1972), 856–60.
2. On the asymptotic behavior in the theory of Volterra integro-functional equations, *Period. Math. Hungarica* **5** (1974), 323–32.

V. Barbu
1. Integrodifferential equations in Hilbert spaces, *An. St. Univ. 'Al. I. Cuza', Iasi*, Ia, **19** (1973), 365–83.
2. Nonlinear Volterra equations in Hilbert space, *SIAM J. Math. An.* **6** (1975), 728–41.
3. Nonlinear Volterra integrodifferential equations in Hilbert space, Conferenze Seminario Matematico Università Bari, 1976.
4. *Nonlinear Semigroups and Differential Equations in Banach spaces*. Editura Academiei (Bucharest) and Noordhoff International Publishing, Leyden, 1976.
5. On a nonlinear Volterra integral equation in a Hilbert space, *SIAM J. Math. An.* **8** (1977), 345–55.

V. Barbu, G. Daprato, M. Iannelli
1. Stability results for some integrodifferential equations of hyperbolic type in Hilbert space, *J. Int. Equations* **2** (1980), 93–101.

V. Barbu, T. Precupanu
1. *Convexity and Optimization in Banach Spaces*, Editura Academiei, Bucharest and D. Reidel Publ. Co., Dordrecht, 1986.

L. Bass, A. J. Bracken, K. Holmaker, B. R. F. Jefferies
1. Integrodifferential equations for the self-organization of liver zones by competitive exclusion of cell-types, *J. Austral. Math. Soc.* B **29** (1987), 156–94.

P. R. Beesack
1. Systems of multi-dimensional Volterra integral equations and inequalities, *Nonlinear Analysis-TMA* **9** (1985), 1451–86.

S. M. Belotserkovskii
1. A mathematical model in linear nonstationary aeroautoelasticity (Russian), *DAN SSSR* **207** (1972), 557–9.

S. M. Belotserkovskii, Yu. A. Kochetkov, A. A. Krasovskii, V. V. Novitskii
1. *Introduction to Aeroautoelasticity* (Russian), Nauka, Moscow, 1980.

R. Bellman, G. M. Wing
1. *An Introduction to Invariant Imbedding*, John Wiley, 1975.

C. Berge
1. *Espaces Topologiques: Fonctions Multivoques*, Dunod, Paris, 1959.

M. Berger
1. *Nonlinearity and Functional Analysis*, Academic Press, New York, 1977.

A. T. Bharucha-Reid
1. *Random Integral Equations*, Academic Press, New York, 1972.

F. Bloom
1. *Ill-Posed Problems for Integrodifferential Equations in Mechanics and Electromagnetic Theory*, SIAM, Philadelphia, 1981.

M. Bocher
1. *An Introduction to the Study of Integral Equations*, Cambridge Tracts in Mathematics and Mathematical Physics, No. 10, Cambridge, 1909.

P. du Bois-Reymond
1. Bemerkimgen uber $\Delta z = z_{xx} + z_{yy} = 0$, *J. Reine Angew. Math.* **103** (1888), 204–29.

L. Boltzmann
1. Zur Theorie der elastischen Nachwirkung. *Sitz. Akad. Wiss. Wien.* **70** (1874), 275–306.

S. T. Boyd, L. O. Chua
1. Fading memory and the problem of approximating nonlinear operators with Volterra series, *IEEE Trans. CAS* **32** (1985), 1150–61.

F. Brauer
1. On a nonlinear integral equation for population growth problems, *SIAM J. Math. Analysis* **6** (1975), 312–17.

H. Brézis
1. *Opérateurs Maximaux Monotones et Semigroups de Contractions dans les Espaces de Hilbert*, North-Holland, Amsterdam, 1973.

H. Brézis, F. E. Browder
1. Existence theorems for nonlinear integral equations of Hammerstein type, *Bull. AMS* **81** (1975), 73–8.
2. Maximal monotone operators in nonreflexive Banach spaces and nonlinear integral equations of Hammerstein type, *Bull. AMS* **81** (1975), 82–8.
3. Nonlinear integral equations and systems of Hammerstein type, *Adv. Math.* **18** (1975), 115–47.

J. K. Brooks, N. Dinculeanu
1. Weak compactness in spaces of Bochner integrable functions and applications, *Adv. Math.* **24** (1977), 172–88.

F. E. Browder
1. *Problemes non-lineaires*, Presses de l'Université de Montréal, 1966.
2. *Nonlinear Operators and Nonlinear Equations of Evolution in Banach Spaces.* Proceedings of the Symposium in Mathematics, Vol. 18, Part 2, American Mathematical Society, 1976.

E. H. Brunner, P. J. van der Houwen
1. *The Numerical Solution of Volterra Equations*, North-Holland, Amsterdam, 1986.

S. A. Brykalov
1. Boundary value problems for functional-differential inclusions, *Soviet Math. Dokl.* **26** (1982), 410–14.
2. Some criteria for the existence of solutions of nonlinear boundary value problems, *Soviet Math. Dokl.* **32** (1985), 575–8.

H. Buckner
1. *Die praktische Behandlung von Integral-Gleichungen.* Springer-Verlag, Berlin, 1932.

A. L. Buhgeim
1. Volterra operator equations in a scale of Banach spaces, *Soviet Math. Dokl.* **242** (1978), 1084–7.
2. *Volterra Equations and Inverse Problems* (Russian), Nauka, Novosibirsk, 1983.

A. V. Bukhvalov
1. Application of methods of the theory of order-bounded operators to the theory of operators in L^p-spaces, *Russian Math Surveys* **38**, No. 6 (1983), 43–98.

A. I. Bulgakov
1. On the existence of a generalized solution to the functional-integral inclusions, *Differential Equations* (Transl.) **15** (1979), 359–63.

A. I. Bulgakov, L. N. Lyapin
1. Some properties of the set of solutions for the Volterra–Hammerstein integral inclusion, *Differential Equations* (Transl.) **14** (1978), 1043–8.
2. An integral inclusion with a functional operator, *Differential Equations* (Transl.) **15** (1979), 621–6.

T. A. Burton
1. *Volterra Integral and Differential Equations*, Academic Press, New York, 1983.
2. *Stability and Periodic Solutions for Ordinary and Functional Differential Equations*, Academic Press, New York, 1985.

P. J. Bushell
1. On a class of Volterra and Fredholm nonlinear integral equations, *Math. Proc. Camb. Phil. Soc.* **79** (1976), 329–35.

Ya. V. Bykov, D. Ruzikulov
1. *Periodic Solutions of Differential and Integrodifferential Equations and their Asymptotics* (Russian) Frunze, Ilim, 1986.

R. Caccioppoli
1. Un teorema generale sull'esistenza di elementi uniti in una trasformazione funzionale, *Rend. R. Accad. Naz. Lincei* **11** (6) (1930), 794–9.

G. V. Caffarelli, E. G. Virga
1. Sull' unicita della soluzione del problema dinamico della viscoelasticita lineare, *Atti Accad. Naz. Lincei, Rend.* **81** (8) (1987), 379–87.

Z. B. Caljuk
1. Volterra functional inequalities (Russian), *Yzv. Vyssh. Ucheb. Zaved.* No. 3 (1969), 86–95.
2. Volterra integral equations (Russian), *Itogi Nauki, Mat. Analiz* **15** (1977), 131–98.

J. R. Cannon
1. *The One-Dimensional Heat Equation* (Encyclopedia of Mathematics and its Applications, Vol. 23) Addison-Wesley Publ. Co., Menlo Park, CA, 1984. Reissued by Cambridge University Press, 1984.

T. Carleman
1. *Sur les Equations Intégrales Singulières a Noyau Réel et Symetrique*, Uppsala, 1923.

D. A. Carlson
1. An elementary proof of the maximum principle of optimal control problems governed by a Volterra integral equation, *J. Opt. Theory Appl.* **54** (1987), 43–61.
2. Necessary conditions for a generalized Bolza problem with states satisfying a Volterra integral constraint. Southern Illinois University, Carbondale (Preprint), 1987.

R. W. Carr, K. B. Hannsgen
1. Resolvent formulas for a Volterra equation in Hilbert spaces, *SIAM J. Math. An. Appl.* **13** (1982), 459–83.

H. Cassago, Jr, C. Corduneanu
1. The ultimate behavior for certain nonlinear integrodifferential equations, *J. Int. Equations* **9** (1985), 113–24.

M. Cecchi, M. Marini, P. L. Zezza
1. A compactness theorem for integral operators and applications, *Lecture Notes Math.* **799** (1980), 119–25.

C. Cergignani
1. *Theory and Applications of Boltzmann Equation*, Elsevier, New York, 1975. Reissued by Springer Verlag, 1988.
2. (ed.), *Kinetic Theories and the Boltzmann Equation*, Lecture Notes in Mathematics No. 1048, Springer Verlag, Berlin, 1984.

L. G. Chambers
1. *Integral Equations, A Short Course*, Int. Textbook Co., London, 1976.

S. Chandrasekar
1. *Radiative Transfer*, Oxford University Press, 1950 (Dover Publ. Co., New York, 1960).

G. Chen, R. Grimmer
1. Semigroups and integral equations, *J. Int. Equations* **2** (1980), 133–54.
2. Integral equations as evolution equations, *J. Diff. Equations* **45** (1982), 53–74.
3. Asymptotic expansions of a penalty method for computing a regulator problem governed by Volterra equations, Preprint, 1982.

H.-Y. Chen
1. Solutions for certain nonlinear Volterra integral equations, *J. Math. An. Appl.* **69** (1979), 475–88.

S. Cinquini
1. Sulle equationi funzionali del typo di Volterra, *Rend. R. Accad. Naz. Lincei* **17** (1933), 616–21.

Ph. Clément, G. DaPrato
1. Existence and regularity results for an integral equation with infinite delay in a Banach space, *Ann. Sc. Norm. Sup. Pisa*, 1986.

Ph. Clément, R. C. MacCamy, J. A. Nohel
1. Asymptotic properties of solutions of nonlinear abstract Volterra equations, *J. Int. Equations* **3** (1981), 185–216.

Ph. Clément, J. A. Nohel
1. Abstract linear and nonlinear Volterra equations preserving positivity, *SIAM J. Math. An.* **10** (1979), 365–88.

2. Abstract linear and nonlinear Volterra equations with completely positive kernels, *SIAM J. Math. An.* **12** (1981), 514–35.

J. A. Cochran

1. *The Analysis of Linear Integral Equations*, McGraw-Hill, New York, 1972.

C. V. Coffman, J. J. Schäffer

1. Linear differential equations with delays: Admissibility and conditional exponential stability, *J. Diff. Equations* **9** (1971), 521–35.

D. Colton

1. Integral operators and reflexion principles for parabolic equations in one space variable, *J. Diff. Equations* **15** (1974), 551–9.

D. Colton, R. Kress

1. *Integral Equation Methods in Scattering Theory*, John Wiley, New York, 1983.

R. Conti

1. *Infinite Dimensional Linear Autonomous Controllability*, School of Mathematics, University of Minnesota, 1982.

A. Corduneanu

1. The stability on the space C_g for Volterra integral equation, *Bul. Inst. Pol. Iasi*, Series I, **27** (1981), 23–9.

2. Nonlinear integral inequalities in n independent variables, *An. St. Univ. Al. I. Cuza, Iasi*, Ia **31** (1985), 281–8.

3. Integral inequalities in two independent variables, *Revue Roum. Math. Pures Appl.* **32** (1987), 331–41.

C. Corduneanu

1. Une application du théoreme du point fixe a la théorie des équations differentielles, *An. St. Univ. Al. I. Cuza Iasi*, **4** (1958), 43–7.

2. Sur certaines équations functionnelles de Volterra, *Funk. Ekv.* **9** (1966), 119–27.

3. *Almost Periodic Functions*, Interscience Publishers, John Wiley, New York, 1968.

4. *Integral Equations and Stability of Feedback Systems*, Academic Press, New York, 1973.

5. Recent contributions to the theory of differential systems with infinite delay, Inst. Math. Appl., Univ. Catholique de Louvain, Vander, Louvain, 1976.

6. *Principles of Differential and Integral Equations*, Chelsea Publishing Co., New York, 1977.

7. Bounded solutions for certain systems of differential or functional equations, Seminari del Instituto di Matematica Applicata, Universita di Firenze, Gennaio, 1978.

8. Frequency domain criteria for nuclear reactor stability, *Libertas Mathematica* **1** (1981), 91–116.

9. Integrodifferential equations with almost periodic solutions. In *Volterra and Functional Differential Equations*, Marcel Dekker, New York, 1982, pp. 233–44.

10. Ultimate behavior of solutions to some nonlinear integrodifferential equations, *Libertas Mathematica* **4** (1984), 61–72.

11. Bielecki's method in the theory of integral equations, *Ann. Univ. 'Mariae Curie-Sklodowska'*, Section A, **38** (1984), 23–40.

12. An existence theorem for functional equations of Volterra type, *Libertas Mathematica* **6** (1986), 117–24.

13. Periodic solutions of certain integrodifferential equations, *Libertas Mathematica* **7** (1987), 149–54.

14. A singular perturbation approach to abstract Volterra equations, *Nonlinear Analysis and Applications*, Marcel Dekker, New York, 1987, pp. 133–138.

15. Sur une équation fonctionnelle liée à la théorie de la stabilité, *C. R. hebd. Acad. Sci. Paris* **256** (1963), 56–8.

C. Corduneanu, V. Lakshmikantham
1. Equations with unbounded delay: A survey, *Nonlinear Analysis-TMA* **4** (1980), 831–77.

C. Corduneanu, N. Luca
1. The stability of some feedback systems with delay, *J. Math. An. Appl.* **51** (1975), 377–93.

C. Corduneanu, H. Poorkarimi
1. Qualitative problems for some hyperbolic equations, *Trends in Theory and Practice of Nonlinear Analysis* (ed. V. Lakshmikantham), North Holland, 1985, pp. 107–14.

M. G. Crandall, S. O. Londen, J. A. Nohel
1. An abstract nonlinear Volterra integrodifferential equation, *J. Math. An. Appl.* **64** (1978), 701–35.

M. G. Crandall, J. A. Nohel
1. An abstract functional differential equation and a related nonlinear Volterra equation, *Israel J. Math.* **29** (1978), 313–28.

P. Creegan, R. Lui
1. Some remarks about wave speed and travelling wave solutions of a nonlinear integral operator, *J. Math. Biol.* **20** (1984), 59–68.

Th. L. Cromer
1. Asymptotically periodic solutions to Volterra integral equations in epidemic models, *J. Math. An. Appl.* **110** (1985), 483–94.
2. A periodicity threshold theorem for a Volterra integral equation, *Applicable Analysis* **21** (1986), 1–8.

J. M. Cushing
1. Admissible operators and solutions of perturbed operator equations, *Funk. Ekv.* **19** (1976), 79–84.
2. Strongly admissible operators and Banach space solutions of nonlinear equations, *Funk. Ekv.* **20** (1977), 237–45.
3. *Integrodifferential Equations and Delay Models in Population Dynamics*, Lecture Notes in Biomathematics, No. 20, Springer-Verlag, Berlin, 1977.

C. M. Dafermos, J. A. Nohel
1. Energy methods for nonlinear hyperbolic Volterra integrodifferential equations, *Comm. Partial Diff. Equations* **4** (1979), 219–78.
2. A nonlinear hyperbolic Volterra equation in viscoelasticity, *Am. J. Math.*, Suppl. (1981), 87–116.

G. DaPrato, M. Iannelli
1. Linear abstract integrodifferential equations of hyperbolic type in Hilbert spaces, *Rend. Sem. Mat. Univ. Padova* **62** (1980), 191–206.
2. Linear integrodifferential equations in Banach spaces, *Rend. Sem. Mat. Univ. Padova* **62** (1980), 207–19.
3. Distribution resolvents for Volterra equations in a Banach space, Preprint.

G. DaPrato, A. Lunardi
1. Periodic solutions for linear integrodifferential equations with infinite delay in Banach spaces, Preprint.
2. Solvability on the real line of a class of linear Volterra integrodifferential equations of parabolic type, *Annali Mat. Pura Appl.* **150** (4) (1988), 67–118.

C. Daskaloyannis
1. The generalization of the Fredholm alternative for bounded kernels, *J. Math. An. Appl.* **118** (1986), 482–6.

2. On the approximation of the linear equations with bounded kernels, *J. Math. An. Appl.* **133** (1988), 272–81.

H. T. Davis
1. *Introduction to Nonlinear Differential and Integral Equations.* Dover Publ., New York, 1962.

A. L. Dawidowicz, K. Loskot
1. Existence and uniqueness of solution of some integrodifferential equations, *Annales Pol. Math.* **67** (1986), 79–87.

K. Deimling
1. A Carathéodory theory for systems of integral equations, *An. Mat. Pura. Appl.* **86** (1970), 217–60.
2. *Nonlinear Functional Analysis*, Springer-Verlag, Berlin, 1985.

M. C. Delfour
1. The largest class of hereditary systems defining a C_0-semigroup on the product space, *Can. J. Math.* **32** (1980), 969–78.

M. C. Delfour, J. Karrachou
1. State space theory of linear time invariant systems with delays in state, control, and observation variables I, II, *J. Math. An. Appl.* **125** (1987), 361–99, 400–50.

L. M. Delves, J. L. Mohamed
1. *Computational Methods for Integral Equations*, Cambridge University Press, 1985.

L. M. Delves, J. Walsh
1. *Numerical Solution of Integral Equation*, Oxford University Press, London, 1974.

W. Desch, R. Grimmer
1. Initial boundary value problem for integrodifferential equations, *J. Int. Equations* **10** (1985), 73–97.
2. Propagation of singularities for integrodifferential equations, *J. Diff. Equations* **65** (1986), 411–26.

W. Desch, R. Grimmer, W. Schappacher
1. Some considerations for linear integrodifferential equations, *J. Math. An. Appl.* **104** (1984), 219–34.

W. Desch, R. K. Miller
1. Exponential stabilization of Volterra integrodifferential equations in Hilbert space, *J. Diff. Equations* **70** (1987), 366–89.

W. Desch, W. Schappacher
1. On relatively bounded perturbations of linear C_0-semigroups, *Ann. Sc. Norm. Sup. Pisa., Cl. Sci.* **11** (4), (1984), 219–34.
2. A semigroup approach to integrodifferential equations in Banach spaces, *J. Int. Equations* **10** (1985), 99–110.

C. A. Desoer, M. Vidyasagar
1. *Feedback Systems: Input–Output Properties*, Academic Press, New York, 1975.

G. DiBlasio
1. Nonautonomous integrodifferential equations in L^p-spaces, *J. Int. Equations* **10** (1985), 111–21.

O. Diekmann
1. Volterra integral equations and semigroups of operators. Preprint, 1980.

O. Diekmann, S. A. van Gils
1. A variation of constants formula for nonlinear Volterra integral equations of convolution type, *Nonlinear Differential Equations* (Proceedings International Conference Trento, 1980), Academic Press, 1981.

J. Dieudonné
1. *Foundations of Modern Analysis*, Academic Press, New York, 1969.
2. *Treatise on Analysis*, Vol. 7, Academic Press, New York, 1988.

J. J. Dijkstra
1. A sufficient condition for the compactness of integral operators, *Indag. Math.* **46** (1984), 387–90.

N. Distefano
1. A Volterra integral equation in the stability of some linear hereditary phenomena, *J. Math. An. Appl.* **23** (1968), 265–83.
2. *Nonlinear Processes in Engineering*, Academic Press, New York, 1974.

P. G. Dodds, A. R. Schep
1. Compact integral operators on Banach functions spaces, *Math. Zeitschr.* **180** (1982), 249–55.

C. L. Dolph
1. Nonlinear integral equations of the Hammerstein type, *Trans. Am. Math. Soc.* **66** (1949), 289–307.

G. M. Dotseth
1. Admissibility results on subspaces of $C(R, R^n)$ and $LL^p(R, R^n)$, *Math. Syst. Theory* **9** (1975), 10–17.

R. Duduchava
1. *Integral Equations with Fixed Singularities*, Teubner-Texte zur Mathematik, Leipzig, 1979.

N. Dunford, J. T. Schwartz
1. *Linear Operators*, Vol. 1, Interscience Publ., Wiley, New York, 1958.

R. E. Edwards
1. *Functional Analysis: Theory and Applications*, Holt, Rinehart and Winston, New York, 1965.
2. *Fourier Series (A Modern Introduction)*, Vol. 1, 2nd edn, Springer-Verlag, Berlin, 1979.

D. Elstner, A. Pietsch
1. Eigenvalues of integral operators, III, *Math. Nachr.* **132** (1987), 191–205.

H. W. Engl
1. A successive-approximation method for solving equations of the second kind with arbitrary spectral radius, *J. Int. Equations* **8** (1985), 239–47.

H. Engler
1. A version of the chain rule and integrodifferential equations in Hilbert spaces, *SIAM J. Math. An.* **13** (1982), 801–10.

H. Fattorini
1. *The Cauchy Problem*, Encyclopedia Math. Appl. 18, Addison-Wesley Publ. Co., Reading, MA, 1983.

L. G. Fedorenko
1. Stability of functional differential equations, *Differential Equations* (Transl.) **21** (1986), 1031–7.

S. Fenyo, H. W. Stolle
1. *Theorie und Praxis der linearen Integralgleichungen*, Vols. 1–4, Birkhauser, 1982–1984.

G. Fichera
1. *Existence Theorems in Elasticity; Boundary Value Problems of Elasticity with Unilateral Constraints*, Handbuch der Physik, Band 6a/2, Springer-Verlag, Berlin, 1962.

B. Fiedler
1. Global Hopf bifurcation for Volterra integral equations, *SIAM J. Math. An.* **17** (1986), 911–32.

W. E. Fitzgibbon
1. Semilinear integrodifferential equations in Banach space, *Nonlinear Analysis-TMA* **4** (1980), 745–60.
2. Convergence theorems for semilinear Volterra equations with infinite delay, *J. Int. Equations* **8** (1985), 162–74.

I. Fredholm
1. Sur une nouvelle méthode pour la résolution du probleme de Dirichlet. *Kong. Vetensk.-Akd. Froh. Stockholm* (1900), 39–46.
2. Sur une classe d'équations functionnelles. *Acta Math.* **27** (1903), 365–90.

A. Friedman
1. *Foundations of Modern Analysis*, Dover Publications, New York, 1982.

C. N. Friedman
1. Existence and approximation of solutions of a class of functional equations including ordinary differential equations, *J. Math. An. Appl.* **112** (1985), 446–54.

V. A. Galkin
1. Existence and uniqueness of a solution of the coagulation equation, *Differential Equations* (Transl.) **13** (1977), 1014–21.

V. A. Galkin, P. B. Dubovskii
1. Solution of the coagulation equation with unbounded kernels, *Differential Equations* (Transl.) **22** (1986), 373–8.

Yu. L. Gaponenko
1. A condition for the solvability of quasilinear boundary value problems, *Differential Equations* (Transl.) **19** (1984), 1234–8.

D. Gilbarg, N. S. Trudinger
1. *Elliptic Partial Differential Equations of Second Order*, Springer-Verlag, Berlin, 1983.

S. Gillot
1. Équations du type Hammerstein sur les varietétés riemanniennes compactes. *C. R. hebd. Acad. Sci., Paris* **294** (1982), 443–5.

K. Glashoff, J. Sprekels
1. An application of Glicksberg's theorem to set-valued integral equations arising in the theory of thermostats, *SIAM J. Math. An.* **12** (1981), 477–86.
2. The regulation of temperature by thermostat and set-valued integral equations, *J. Int. Equations* **4** (1982), 95–112.

I. C. Gohberg, I. A. Feldman
1. *Convolution Equations and Projection Methods for their Solution*, Transl. of Mat. Monographs No. 41, AMS, Providence, 1974.

I. C. Gohberg, S. Goldberg
1. *Basic Operator Theory*, Birkhauser, Boston-Basel, 1981.

I.C. Gohberg, M. A. Kaashoek
1. *Constructive Methods of Wiener-Hopf Factorization*, Birkhauser, Basel, 1986 (Papers by the authors and collab.).

I. C. Gohberg, M. G. Krein
1. *Theory of Volterra Operators in Hilbert Spaces and Their Applications* (Russian), Nauka, Moscow, 1967.

M. A. Golberg
1. (ed.) *Solution Methods for Integral Equations*, Plenum Press, New York, 1979.

H. E. Gollwitzer
1. Admissibility and integral operators, *Math. Syst. Theory* **7** (1973), 219–31.

H. E. Gollwitzer, R. A. Hager
1. The nonexistence of maximum solutions of Volterra integral equations, *Proc. AMS* **26** (1970), 301–4.

V. D. Gorjacenko
1. *Stability Methods in the Dynamics of Nuclear Reactors* (Russian), Atomizdat, Moscow, 1971.

E. Goursat
1. *A Course in Mathematical Analysis*, Vol. 3, Part 2, Dover Publ., New York, 1964.

D. Graffi
1. Sopra una equazione funzionale e la sua applicazione a un problema di fisica ereditaria. *Annali Mat. Pura. Appl.* 9 (1931), 143–79.

C. D. Green
1. *Integral Equations Method*, Barnes and Nobles, New York, 1969.

W. Greenberg, C. van der Mee, V. Protopopescu
1. *Boundary Value Problems in Abstract Kinetic Theory*, Birkhauser Verlag, Basel-Boston, 1987.

N. E. Gretsky, J. J. Uhl, Jr
1. Carleman and Korotkov operators on Banach spaces, *Acta Sci. Math.* 43 (1981), 207–18.

R. Grimmer
1. Resolvent operators for integral equations in a Banach space, *Trans. AMS* 273 (1982), 333–49.

R. Grimmer, F. Kappel
1. Series expansions for resolvent of Volterra integrodifferential equations in Banach space, *SIAM J. Math. An.* 15 (1984), 595–604.

R. C. Grimmer, R. K. Miller
1. Existence, uniqueness, and continuity for integral equations in Banach spaces, *J. Math. An. Appl.* 57 (1977), 429–47.
2. Well posedness of Volterra integral equations in Hilbert space, *J. Int. Equations* 1 (1979), 201–16.

R. Grimmer, A. J. Pritchard
1. Analytic resolvent operators for integral equations in Banach spaces, *J. Diff. Equations* 50 (1983), 234–59.

R. C. Grimmer, W. Schappacher
1. Weak solutions of integrodifferential equations and resolvent operators, *J. Int. Equations* 6 (1984), 205–29.

G. Gripenberg
1. On some integral and integrodifferential equations in a Hilbert space, *Ann. Mat. Pura. Appl.* 118 (1978), 181–94.
2. Stability problems for some nonlinear Volterra equations, *J. Int. Equations* 2 (1980), 247–58.
3. On some positive definite forms and Volterra integral operators, *Applicable Analysis* 11 (1981), 211–22.
4. On the convergence of solutions of Volterra equations to almost periodic functions, *Quart. Appl. Math.* 39 (1981), 363–73.
5. Decay estimates for resolvents of Volterra equations, *J. Math. An. Appl.* 85 (1982), 473–87.
6. Asymptotic behavior of resolvents of abstract Volterra equations, *J. Math. An. Appl.* 122 (1987), 427–38.
7. On periodic solutions of a thermostat equation, *SIAM J. Math. An.* 18 (1987), 694–702.

G. Gripenberg, S. O. Londen, O. Staffans
1. *Nonlinear Volterra and Integral Equations*, Cambridge University Press, 1990.

C. W. Groetsch
1. *Generalized Inverses of Linear Operators*, Marcel Dekker, New York, 1977.
2. *The Theory of Tikhonov Regularization for Fredholm Equations of the First Kind*, Pitman, Boston-Melbourne-London, 1984.

C. W. Groetsch, J. Guacaneme
1. Arcangeli's method for Fredholm equations of the first kind. *Proc. AMS* **99** (1987), 256–60.

R. B. Guenther, J. W. Lee
1. *Partial Differential Equations of Mathematical Physics and Integral Equations*, Prentice-Hall, Engewood Cliffs, NJ, 1988.

Guo, Dajun
1. The number of nontrivial solutions of Hammerstein integral equations and their applications, *Kexue Tongbao* **27** (1982), 694–98.
2. Multiple positive solutions of nonlinear integral equations and applications, *Appl. An.* **23** (1986), 77–84.

H. Haario, E. Somersalo
1. A regularization method for integral equations of the first kind. In the volume *Theory and Applications of Inverse Problems*, Longman Scientific and Technical, London (1988), pp. 16–26.

A. Halanay
1. *Differential Equations: Stability, Oscillations, Time-Lag*, Academic Press, 1966.

A. Halanay, V. Rasvan
1. Frequency domain criteria for nuclear reactor stability I, II. *Revue Roum. Sci. Techn., Electr. Energ.* **19** (1974), 367–78; **20** (1975), 233–50.

J. K. Hale
1. *Theory of Functional Differential Equations*, Springer-Verlag, Berlin, 1977.

J. K. Hale, J. Kato
1. Phase space for retarded equations with infinite delay, *Funk. Ekv.* **21** (1978), 11–41.

P. R. Halmos, V. S. Sunder
1. *Bounded Integral Operators on L^2-Spaces*, Springer-Verlag, Berlin, 1978.

G. Hamel
1. *Integralgleichungen*, Julius Springer-Verlag, Berlin, 1937.

G. Hammerlin, K. H. Hoffman
1. (eds.) *Constructive Methods for the Practical Treatment of Integral Equations*, Birkhauser, Basel, 1985.

A. Hammerstein
1. Nichtlineare Integralgleichungen nebst Anwendungen, *Acta Math.* **54** (1930), 117–76.

K. B. Hannsgen, R. L. Wheeler
1. Complete monotonicity and resolvents of Volterra integrodifferential equations, *SIAM J. Math. An.* **13** (1982), 962–9.
2. A singular limit problem for an integrodifferential equation, *J. Int. Equations* **5** (1983), 199–209.

M. L. Heard
1. An abstract parabolic Volterra integrodifferential equation, *SIAM J. Math. An.* **13** (1982), 81–105.

M. L. Heard, S. M. Rankin, III
1. A semilinear parabolic Volterra integrodifferential equations, *J. Diff. Equations* **71** (1988), 201–33.

H. P. Heinig
1. Weighted norm inequalities for classes of operators, *Indiana Univ. Math. J.* **33** (1984), 573–82.

References

E. Hellinger, O. Toeplitz
1. *Integralgleichungen und Gleichungen mit unendlichen Unbekannten*, Chelsea, New York, 1953.
T. L. Herdman
1. Existence and continuation properties of solutions of a nonlinear Volterra integral equation. In *Dynamical Systems*, Vol. 2, Academic Press, New York, 1976, pp. 307–10.
2. Behavior of maximally defined solutions of a nonlinear Volterra equation. *Proc. AMS* **67** (1977), 297–302.
T. L. Herdman, S. M. Rankin, III, H. W. Stech
1. (eds.) *Integral and Functional-Differential Equations*, Marcel Dekker, New York, 1981.
G. Herglotz
1. Uber die Integralgleichungen der Elektronen theorie. *Math. Annalen* **65** (1908), 87–106.
L. R. Herrmann
1. On a general theory of Viscoelasticity, *J. Franklin Inst.* **280** (1965), 244–55.
H. B. Heywood, M. Fréchet
1. *L'Equation de Fredholm et ses applications à la physique mathématique*, Herman et Fils, Paris, 1912.
D. Hilbert
1. *Grundzuge einer allgemeinen Theorie der linearen Integralgleichungen*, Teubner, Leigzig, 1912.
E. Hille, R. S. Phillips
1. Functional Analysis and Semigroups. *Am. Math. Soc. Coll. Publ.*, No. 31 (1957).
N. Hirano
1. Asymptotic behavior of solutions of nonlinear Volterra equations, *J. Diff. Equations* **47** (1983), 163–79.
2. Abstract nonlinear Volterra equations with positive kernels, *SIAM J. Math. An.* **17** (1986), 403–14.
H. Hochstadt
1. *Integral Equations*, John Wiley, New York, 1973.
G. Hoheisel
1. *Integral Equations*, Ungar, New York, 1968.
C. S. Hönig
1. *Volterra-Stieltjes Integral Equations*, North Holland/American Elsevier, Amsterdam-New York, 1975.
W. J. Hrusa
1. A nonlinear functional-differential equation in Banach space with applications to materials with fading memory, *Arch. Rat. Mech. Analysis* **84** (1984), 99–137.
2. Global existence of classical solutions to the equation of motion for materials with fading memory, *Phys. Math. Nonlinear Partial Diff. Equations*, Lecture Notes in Pure and Applied Mathematics, 102, Marcel Dekker, 1985, pp. 97–110.
D. S. Hulbert, S. Reich
1. Asymptotic behavior of solutions to nonlinear Volterra integral equations, *J. Math. An. Appl.* **104** (1984), 155–72.
M. Iannelli
1. Mathematical problems in the description of the age structured populations, Univ. degli Studi di Trento, 1984.
M. Imanaliev
1. Asymptotical methods in the theory of singularly perturbed nonlinear integral equations of Volterra type (Russian), *Math. Balkanica* **3** (1973), 145–9.

M. I. Imanaliev, B. V. Khvedelidze, T. G. Gegeliya, A. A. Babaev, A. I. Botashev
1. Integral equations (survey). *Diff. Equations* (Transl.) **18** (1983), 1442–58.
B. Imomnazarov
1. Approximate solution of integro-operator Volterra equations of the first kind, *USSR Comp. Maths. Math. Phys.* **25** (1985), 199–202.
S. V. Israilov
1. The general functional problem for ordinary differential equations (Russian), *Methods of Nonlinear Mechanics and their Applications*, Kiev, 1982.
V. V. Ivanov
1. *The Theory of Approximate Methods and their Application to the Numerical Solution of Singular Integral Equations*, Noordhoff Int. Publ., Leyden, 1976.
M. Janet
1. *Equations Intégrales et Applications a certains problems de la Physique Mathématique*. Mémorial des Sciences Mathématiques, Fasc. C I et C II, Gauthier-Villars, Paris, 1941.
M. A. Jawson, G. T. Symm
1. *Integral Equation Methods in Potential Theory and Elastostatics*, Academic Press, New York, 1977.
A. J. Jerri
1. *Introduction to Integral Equations with Applications*, Marcel Dekker, New York, 1985.
G. S. Jordan, R. L. Wheeler
1. On the asymptotic behavior of perturbed Volterra integral equations, *SIAM J. Math. An.* **5** (1974), 273–7.
2. Linear integral equations with asymptotically periodic solutions, *J. Math. An. Appl.* **52** (1975), 454–64.·
3. Structure of resolvents of Volterra integral and integrodifferential systems, *SIAM J. Math. An.* **11** (1980), 119–32.
4. Weighted L^1 remainder theorems for resolvents of Volterra equations, *SIAM J. Math. An.* **11** (1980), 885–900.
K. Jörgens
1. *Lineare Integraloperatoren*, B. G. Teubner, Stuttgart, 1970.
S. I. Kabanikhin
1. On the solvability of inverse problems for differential equations, *Soviet Math. Dokl.* **30** (1984), 162–4.
R. I. Kačurovskii
1. Tikhonov's fixed point principle and equations with operators weakly closed on a kernel, *Soviet Math. Dokl.* **9** (1968), 1411–14.
H. H. Kagiwada, R. Kalaba
1. *Integral Equations via Imbedding Methods*, Addison-Wesley, Reading, Mass., 1974.
N. J. Kalton
1. Representation of operators between function spaces, *Indiana Univ. Math. J.* **33** (1984), 639–65.
L. V. Kantorovich, G. P. Akilov
1. *Functional Analysis* (second edition), Pergamon Press, Oxford, 1982.
L. V. Kantorovich, V. I. Krylov
1. *Approximate Methods of Higher Analysis*, Interscience Publishers, New York, 1958.
S. Kantrovitz
1. Volterra systems of operators, *J. Math. An. Appl.* **133** (1988), 135–50.
P. Kanwal
1. *Linear Integral Equations (Theory and Technique)*, Academic Press, New York, 1971.

F. Kappel
1. Semigroups and delay equations. In *Semigroup Theory and Applications* (H. Brézis, M. G. Crandall, F. Kappel, eds.), Pitman, 1986, pp. 136–76.

F. Kappel, F. DiPasquantonio
1. Stability criteria for kinetic reactor equations, *Arch. Rat. Mech. An.* **58** (1975), 317–38.

F. Kappel, K. Kunisch
1. Invariance results for delay and Volterra equations in fractional order Sobolev spaces. *Trans. AMS* **304** (1985), 1–51.

F. Kappel, W. Schappacher
1. Nonlinear functional differential equations and abstract integral equations, *Proc. R. Soc. Edin.* **84A** (1979), 71–91.
2. (eds) *Infinite Dimensional Systems*, Springer-Verlag, Berlin, 1984.

G. Karakostas
1. Causal operators and topological dynamics, *Ann. Mat. Pura Appl.* **131** (4) (1982), 1–27.
2. Asymptotic behavior of a certain functional equation via limiting equations, *Czech. Math. J.* **36** (111) (1986), 259–67.

G. Karakostas, Y. G. Sficas, V. A. Staikos
1. On the basic theory of initial value problems for delay differential equations, *Boll. UMI* **1-B** (6) (1982), 1179–98.

N. Kato
1. On the existence and asymptotic behavior of solutions of nonlinear heat flow with memory, *Proc. Japan Acad.* **63** A (1987), 250–3.

W. G. Kelley
1. A Kneser theorem for Volterra integral equations, *Proc. AMS* **40** (1973), 183–90.

E. I. Kharlamova, G. V. Mozalevskaya
1. *Integro-Differential Equations of the Dynamics of a Rigid Body* (Russian), Naukova Dumka, Kiev, 1986.

T. R. Kiffe
1. On nonlinear Volterra equations of nonconvolution type, *J. Diff. Equations* **22** (1976), 349–67.
2. A discontinuous Volterra equation, *J. Int. Equations* **1** (1979), 193–200.

A. Kneser
1. *Die Integralgleichungen und Ihre Anwendungen in der Mathematische Physik* (second edition), Vieweg und Sohn, Braunschweig, 1922.

V. B. Kolmanovskii, V. R. Nosov
1. *Stability of Functional Differential Equations*, Academic Press, New York, 1986.

J. J. Kolodner
1. Equations of Hammerstein type in Hilbert spaces, *J. Math. Mech.* **13** (1964), 701–50.

V. B. Korotkov
1. *Integral Operators* (Russian), Nauka, Novosibirsk, 1983.
2. On systems of integral equations (Russian). *Sib. Math. Zhurnal* **27** (3) (1986), 121–33.

V. A. Kostitzin
1. *Applications des équations intégrales*, Gauthier-Villars, Paris, 1935.

G. Kowalewski
1. *Integralgleichungen*, W. de Gruyter, Berlin, 1930.

S. N. Krachkovskii
1. On Lalesco's condition, *Izv. Akad. Nauk. Latv. SSSR*, **4** (1947), 101–8.

M. A. Krasnoselskii
1. *Topological Methods in the Theory of Nonlinear Integral Equations*, Pergamon Press, Oxford, 1964.

M. A. Krasnoselskii, A. V. Pokrovskii
1. *Systems with Hysteresis* (Russian), Nauka, Moscow, 1983.

M. A. Krasnoselskii, Ja. B. Rutickii
1. *Convex Functions and Orlicz Spaces* (Russian) Gos. Izdat. Fiz. Mat. Lit., Moscow, 1958.

M. A. Krasnoselskii, P. P. Zabreyko, E. I. Pustylnik, P. E. Sobolevskii
1. *Integral Operators in Spaces of Summable Functions*, Noordhoff International Publ., Leyden, 1976.

S. G. Krein
1. *Linear Equations in Banach Spaces*, Birkhäuser, Basel, 1982.

N. Ya. Krupnik
1. *Banach Algebras with Symbol and Singular Integral Operators*, Birkhäuser, Basel, 1987.

J. Kudrewicz
1. *Frequency Methods in the Theory of Nonlinear Dynamical Systems* (Polish), Wydawn. Naukowo-Techn., Warszawa, 1970.

K. Kuratowski
1. *Topology*, Vols. 1, 2, PWN Warszawa, 1966, 1968.

N. S. Kurpel, B. A. Shuvar
1. *Bilateral Operator Inequalities and Their Applications* (Russian), Naukova Dumka, Kiev, 1980.

I. Labuda
1. Le domaine maximal d'extension d'un opératéur intégral. *C. R. hebd Acad. Sci. Paris*, Series I, **301** (1985), 303–6.

V. Lakshmikantham
1. Some problems in integrodifferential equations of Volterra type, *J. Int. Equations* **10** (1985), 137–46.

V. Lakshmikantham, S. Leela
1. *Differential and Integral Inequalities*, Vol. 1, Academic Press, New York, 1969.

T. Lalesco
1. *Introduction à la Théorie des Équations Intégrales*, Gauthier-Villars, Paris, 1912.

C. E. Langenhop
1. Periodic and almost periodic solutions of Volterra integral-differential equations with infinite memory, *J. Diff. Equations* **58** (1985), 391–403.

M. M. Lavrent'ev, V. G. Romanov, S. P. Shishatskii
1. *Ill-posed Problems of Mathematical Physics and Analysis*. Transl. of Math. Monographs, AMS, Vol. 64, 1986.

M. J. Leitman, V. J. Mizel
1. Hereditary laws and nonlinear integral equations on the line, *Adv. Math.* **22** (1976) 220–66.

B. J. Leon, D. J. Schaefer
1. Volterra series and Picard iteration for nonlinear circuits and systems, *IEEE Trans. Circ. Syst.* **25** (1978), 789–93.

G. Leugering
1. Time optimal boundary controllability of a simple linear viscoelastic liquid, *Math. Methods Appl. Sci.* **9** (1987), 413–30.

T. Levi-Civita
1. Sull'inversione degli integrali definiti nel campo reale, *Atti Acad. Sci. Torino, Cl. Fis. Mat. Natur.* **31** (1895), 25–51.

J. J. Levin
1. A bound on the solutions of a Volterra equation, *J. Diff. Equations* **14** (1973), 106–20.

2. Resolvents and bounds for linear and nonlinear Volterra equations. *Trans. AMS* **228** (1977), 207–22.
3. Nonlinearly perturbed Volterra equations, *Tôhoku Math. J.* **32** (1980), 317–35.

M. Lewin
1. On the existence of a weak solution to an equation of Volterra-Skorohod type, *Libertas Mathematica* **7** (1987), 107–23.

J. Lindenstrauss, L. Tzafriri
1. *Classical Banach Spaces*, Vol. 2, *Function Spaces*, Springer-Verlag, Berlin, 1979.

P. Linz
1. *Analytical and Numerical Methods for Volterra Equations*, SIAM Studies in Applied Mathematics, Philadelphia, 1985.

J. L. Lions
1. *Optimal Control of Systems Governed by Partial Differential Equations*, Springer-Verlag, Berlin, 1971.

J. Liouville
1. Sur quelques questions de géometrie et de mechanique, et sur un nouveau genre de calcul pour resoudre ces questions, *J. Ecole Polyt.* **13** (1832), 1–69.
2. Sur le calcul des différentielles à indices quelconques, *J. École Polyt.* **13** (1832), 71–162.
3. Sur le développement des fonctions ou parties de fonctions en séries dont les divers termes sont assujettis a satisfaire a une meme équation différentielle du second ordre contenant un parametre variable. *J. Math. Pures. Appl.* **2** (1837), 16–35.

Z. Lipecki
1. Riesz type representation theorems for positive operators, *Math. Nachr.* **131** (1987), 351–6.

S. O. Londen
1. On an integral equation in a Hilbert space, *SIAM J. Math. An.* **8** (1977), 950–70.
2. On boundedness results of a Volterra equation, *An. St. Univ. Al. I. Cuza Iasi*, Ia **23** (1977), 329–32.
3. An existence result on a Volterra equation in a Banach space, *Trans. AMS* **235** (1978), 285–304.
4. A Volterra equation with L^2-solutions, *SIAM J. Math. An.* **18** (1987), 168–71.

S. O. Londen, J. A. Nohel
1. Nonlinear Volterra integrodifferential equations occurring in heat flow, *J. Int. Equations* **6** (1984), 11–50.

S. O. Londen, O. J. Staffans
1. A note on Volterra equations in a Hilbert space. *Proc. AMS* **70** (1978), 57–62.
2. (eds), *Volterra Equations: Proceedings of the Otaniemi Symposium on Integral Equations*, 1978, Springer-Verlag, Berlin, 1979.

W. V. Lovitt
1. *Linear Integral Equations*, Dover, New York, 1950.

N. Luca
1. The behavior of solutions of a class of nonlinear integral equations of the Volterra type, *Rend. Accad. Naz. Lincei* **62** (1977), 9–61.
2. The stability of the solutions of a class of integrodifferential systems with infinite delay, *J. Math. An. Appl.* **67** (1979), 323–39.

R. Lucchetti, F. Patrone
1. On Nemytskii's operator and its application to the lower semicontinuity of integral functionals, *Indiana Univ. Math. J.* **29** (1980), 703–13.

A. Lunardi
1. Laplace transform methods in integrodifferential equations, *J. Int. Equations* **10** (1985), 185–211.

2. Interpolation spaces between domains of elliptic operators and spaces of continuous functions with applications to nonlinear parabolic equations, *Math. Nachr.* **121** (1985), 295–318.

A. Lunardi, E. Sinestrari
1. Fully nonlinear integrodifferential equations in general Banach spaces, *Math. Zts.* **190** (1985), 225–48.
2. Existence in the large and stability for nonlinear Volterra equations, *J. Int. Equations* **10** (1985), 213–39.

W. A. J. Luxemburg, A. C. Zaanen
1. Compactness of integral operators in Banach function spaces, *Math. Annalen* **149** (1962/3), 150–80.
2. *Riesz Spaces*, Vol. 2, North Holland Publ. Co., Amsterdam, 1983.

L. N. Lyapin
1. Volterra equations in Banach spaces, *Differential Equations* (Transl.) **19** (1983), 801–8.

R. C. MacCamy
1. An integrodifferential equation with applications in heat flow, *Quart. Appl. Math.* **35** (1977/8), 1–19.
2. A model for one-dimensional nonlinear viscoelasticity, *Quart. Appl. Math.* **35** (1977/8), 21–33.

R. C. MacCamy, J. S. W. Wong
1. Stability theorems for some functional equations, *Trans. AMS* **164** (1972), 1–37.

N. A. Magnickii
1. The existence of multiparameter families of solutions of a Volterra integral equation of the first kind, *Soviet Math. Dokl.* **18** (1977), 772–4.
2. Linear integral operations of Volterra, of first and third kind, *Z. Vycsl. Mat. Mat. Fiz.* **19** (1979), 970–88.

G. M. Magomedov, S. N. Dzhalalova
1. On some classes of nonlinear singular integrodifferential equations, *Soviet Math. Dokl.* **27** (1983), 673–5.

O. D. Maksimova
1. On some properties of integral operators on ideal spaces (Russian), *Sib. Mat. Zhurnal* **26** (1985), 127–31.

Ya. D. Mamedov, S. A. Ashirov
1. *Nonlinear Volterra Equations* (Russian), Ashkhabad, 1977.

Ya. D. Mamedov, V. M. Musaev
1. Investigation of the solutions of a system of nonlinear Volterra–Fredholm operator equations, *Soviet Math. Dokl.* **32** (1985), 587–90.

M. Marcus, V. Mizel
1. Limiting equations for problems involving long range memory. Memoirs of the American Mathematical Society, no. 278, **43** (1983), 60 pp.

P. Marocco
1. A study of asymptotic behavior and stability of the solutions of Volterra equations using topological degree, *J. Diff. Equations* **43** (1982), 235–48.

A. A. Martinjuk, R. Gutowski
1. *Integral Inequalities and Stability of Motion* (Russian), Naukowa Dumka, Kiev, 1979.

J. L. Massera, J. J. Schäffer
1. Linear differential equations and function spaces, I, *Ann. Math.* **67** (2) (1958), 517–73.
2. *Linear Differential Equations and Function Spaces*, Academic Press, New York, 1966.

Al. McNabb, G. Weir
1. Comparison theorems for causal functional differential equations, *Proc. AMS* **104** (1988), 449–52.

N. G. Medhin
1. Optimal processes governed by integral equations, *J. Math. An. Appl.* **120** (1986), 1–12.

Z. A. Melzak
1. A scalar transport equation, *Trans. AMS* **85** (1957), 547–60.

S. G. Mikhlin
1. *Integral Equations*, Pergamon Press, Oxford, 1957.

R. K. Miller
1. *Nonlinear Volterra Integral Equations*, W. A. Benjamin, Menlo Park, CA, 1971.
2. A system of Volterra integral equations arising in the theory of superfluidity, *An. St. Univ. 'Al. I. Cuza' Iasi*, Ia **19** (1973), 349–64.
3. Linear Volterra integrodifferential equations as semigroups, *Funk. Ekv.* **17** (1974), 39–55.
4. A system of renewal equations, *SIAM J. Appl. Math.* **29** (1975), 20–34.
5. Volterra integral equations in a Banach space, *Funk. Ekv.* **18** (1975), 163–94.

R. K. Miller, G. Sell
1. *Volterra Integral Equations and Topological Dynamics*, Memoirs of the American Mathematical Society, No. 102, 1970.

R. K. Miller, R. L. Wheeler
1. Asymptotic behavior for linear Volterra integral equations in Hilbert space, *J. Diff. Equations* **23** (1977), 270–84.
2. A remark on hyperbolic integrodifferential equations, *J. Diff. Equations* **24** (1977), 51–6.
3. Well-posedness and stability of linear Volterra integrodifferential equations in abstract spaces, *Funk. Ekv.* **21** (1978), 279–305.

M. Milman
1. Stability results for integral operators, I, *Revue Roum. Math. Pures. Appl.* **22** (1977), 325–33.
2. Some new function spaces and their tensor products. Notas de Matematica No. 20, Universidad de los Andes, Merida, Venezuela, 1978.
3. A note on $L(p, q)$ spaces and Orlicz spaces with mixed norms, *Proc. AMS* **83** (1981), 743–6.

A. B. Mingarelli
1. *Volterra–Stieltges Integral Equations and Generalized Ordinary Differential Equations*, Lecture Notes in Mathematics No. 989, Springer-Verlag, Berlin, 1983.

J. L. Mohamed, J. Walsh (eds.)
1. *Numerical Algorithms*, Clarendon Press, Oxford, 1986.

J. J. Moreau, P. D. Panagiotopoulos, G. Strang (eds.)
1. *Topics in Nonsmooth Mechanics*, Birkhäuser Verlag, Basel, 1988.

J. Moser
1. On nonoscillating networks, *Quart. Appl. Math.* **25** (1967), 1–9.

S. Mossaheb
1. The relationship between various L^p-stabilities of time-varying feedback systems, *Int. J. Control* **38** (1983), 1199–212.
2. Feedback stability of certain non-linear systems, *Int. J. Control* **42** (1985), 1141–4.

A. Myller
1. Sur le mouvement d'une chaine pesante sur une courbe fixe. *Nouvelles Ann. Math.*, Series 4, **9** (1909), 317–26.

A. D. Myshkis
1. New proof of existence of the generalized solution for the general integral equation of the first kind (Russian), *Issled. Integro-Diff. Urav.* **16** (1983), 34–6.

T. Naito
1. A modified form of the variation of constants formula for equations with infinite delay, *Tôkoku Math. J.* **36** (1984), 33–40.

K. S. Narendra, J. H. Taylor
1. *Frequency Domain Criteria for Absolute Stability*, Academic Press, New York, 1973.

M. Z. Nashed
1. (ed.) *Generalized Inverses and Applications*, Academic Press, New York, 1976.

M. Z. Nashed, J. S. W. Wong
1. Some variants of a fixed point theorem of Krasnoselskii and applications to nonlinear integral equations, *J. Math. Mech.* **18** (1969), 767–77.

J. von Neumann
1. *Characterisierung des Spectrums eines Integraloperators*, Act. Sci. Ind., Gauthier-Villars, Paris, 1935.

L. Neustadt
1. *Optimization (A Theory of Necessary Conditions)*, Princeton University Press, 1976.

J. A. Nohel
1. *A Nonlinear Hyperbolic Volterra Equation*, Lecture Notes in Mathematics No. 737, Springer-Verlag, Berlin, 1979, pp. 220–35.
2. *Nonlinear Conservation Laws with Memory*, Computational Methods in Science and Engineering, Vol. 5, North Holland, New York, 1982, pp. 269–80.

J. A. Nohel, D. F. Shea
1. Frequency domain methods for Volterra equations, *Adv. Math.* **3** (1976), 278–304.

T. M. Novitskii
1. On the representation of kernels of integral operators by means of bilinear forms, *Sib. Mat. Zhurnal* **25** (1984), 114–18.

G. O. Okikiolu
1. *Aspects of the Theory of Bounded Integral Operators on L^p-Spaces*, Academic Press, 1971.

W. Okrasinski
1. On a certain nonlinear equation, *Coll. Math.* **63** (1980), 345–9.

W. Orlicz, St. Szufla
1. On some classes of nonlinear Volterra integral equations in Banach spaces. *Bull. Acad. Polon. Sci., Ser. Math. Astr. Phys.* **30** (1982), 239–50.
2. On the structure of L_ϕ-solution sets of integral equations in Banach spaces, *Studia Math* **77** (1984), 465–77.

B. G. Pachpatte
1. A note on some fundamental integrodifferential inequalities, *Tamkang. Math. J.* **13** (1982), 63–7.
2. Existence theorems for nonlinear integrodifferential systems with upper and lower solutions, *An. St. Univ. 'Al. I. Cuza' Iasi*, Ia **30** (1984), No. 3, 31–38.

D. J. Paddon, H. Holstein
1. *Multigrid Methods for Integral and Differential Equations*, Clarendon Press, Oxford, 1985.

A. Palczewski
1. *The Cauchy Problem for the Boltzmann Equation: A Survey of Recent Results*, Lecture Notes in Mathematics No. 1048, Springer-Verlag, Berlin, 1984, pp. 202–6.

S. G. Pandit
1. On Stieltjes–Volterra integral equations, *Bull. Austral. Math. Soc.* **18** (1978), 321–34.

L. Pandolfi
1. On $C_G(I, E^n)$ spaces of continuous functions. *An. St. Univ. 'Al. I. Cuza' Iasi*, Ia **23** (1977), 21–4.

D. Pascali
1. On variational treatment of nonlinear Hammerstein equations, *Libertas Mathematica* **5** (1985), 47–54.

D. Pascali, S. Sburlan
1. *Nonlinear Mappings of Monotone Type*, Ed. Acad. Bucharest, Sijthoff and Noordhoff, 1978.

N. H. Pavel
1. *Nonlinear Evolution Operators and Semigroups*, Lecture Notes in Mathematics No. 1260, Springer-Verlag, Berlin, 1987.

E. A. Pavlov
1. On convolution integral operators (Russian), *Mat. Zametki* **38** (1985), 74–8.

A. Pazy
1. *Semigroups of Linear Operators and Applications to Partial Differential Equations*, Springer-Verlag, New York/Berlin, 1983.

G. H. Peichl
1. A kind of 'History Space' for retarded functional differential equations and representation of solutions, *Funk. Ekv.* **25** (1982), 245–56.

I. G. Petrovskii
1. *Lectures on the Theory of Integral Equations*, Graylock Press, Rochester, New York, 1957.

W. V. Petryshyn
1. Multiple positive fixed points of multivalued condensing mappings with some applications, *J. Math. An. Appl.* **124** (1987), 237–53.

E. Picard
1. Sur quelques applications de l'équation functionnelle de M. Fredholm. *Rend. Circ. Mat. Palermo* **22** (1906), 241–59.

A. Pietsch
1. Eigenvalues of integral operators I, II, *Math. Annalen*, **247** (1980), 169–78; **262** (1983), 343–76.

J. Plemelj
1. *Problems in the Sense of Riemnnan and Klein*, John Wiley (Interscience Publ.), New York, 1964 (Part 2: Integral Equations).

M. Z. Podowski
1. Nonlinear stability analysis for a class of differential-integral systems arising from nuclear reactor dynamics, *IEEE Trans. Aut. Control* AC-31 (1986), 98–107.
2. A study of nuclear reactor models with nonlinear reactivity feedbacks: Stability criteria and power overshoot evaluation, *IEEE Trans. Aut. Control* AC-31 (1986), 108–15.

W. Pogorzelski
1. *Integral Equations and Their Applications*, MacMillan, New York, 1966.

H. Poincaré
1. Les ondes hertziennes et l'équation de Fredholm, *C. R. hebd Acad. Sci. Paris* **148** (1909), 449–53.
2. Remarques diverses sur l'équation de Fredholm, *Acta Math.* **33** (1910), 57–86.

A. V. Ponosov
1. On the Nemytskii conjecture, *Soviet Math. Dokl.* **34** (1987), 231–33.

V. M. Popov
1. A new criterion for the stability of systems containing nuclear reactors, *Rev. d'Electrotecnique Energetique*, Serie A, **8** (1963), 113–30.
2. *Hyperstability of Control Systems*, Springer-Verlag, Berlin, 1973.

J. Prüss
1. *Lineare Volterra Gleichungen in Banachräumen*, Paderborn, 1984.

2. On linear Volterra equations of parabolic type in Banach spaces, *Trans. AMS* **301** (1987), 691–721.
3. Periodic solutions of the thermostat problem, Preprint.

D. Przeworska-Rolewicz

1. Right inverse and Volterra operators, *J. Int. Equations* **2** (1980), 45–56.

D. Przeworska-Rolewicz, S. Rolewicz

1. The only continuous Volterra right inverses in $C_c[0, 1]$ of the operator d/dt are $\int_a^t$, *Coll. Math.* **51** (1987), 281–5.

B. N. Pshenichnyi

1. *Necessary Conditions for an Extremum*, Marcel Dekker, New York, 1971.

V. F. Puljaev

1. ω-periodic solutions of linear Volterra integral equations (Russian), *Kuban. Gos. Univ. Naucnye Trudy* **180** (1974), 132–4.

M. Rama Mohana Rao, S. Sivasundaram

1. Asymptotic stability for equations with unbounded delay, *J. Math. An. Appl.* **131** (1988), 97–105.

M. Rama Mohana Rao, P. Srinivas

1. Positivity and boundedness of solutions of Volterra integro-differential equations, *Libertas Mathematica* **3** (1983), 71–81.

A. G. Ramm

1. Stability of equation systems, *Differential Equations* (Transl.) **14** (1974), 1188–93.
2. *Theory and Applications of Some New Classes of Integral Equations*, Springer-Verlag, New York, 1980.
3. Existence, uniqueness and stability of solutions to some nonlinear oscillations problems, *Applicable Analysis* **11** (1981), 223–32.

V. Rasvan

1. *Absolute Stability of Remote Control Systems with Delay* (Romanian), Academiei Publ. House, Bucharest, 1975.

M. Reed, B. Simon

1. *Methods of Modern Mathematical Physics*, Vols. 2 and 4, Academic Press, New York, 1975, 1978.

S. Reich

1. A fixed point theorem for Fréchet spaces, *J. Math. An. Appl.* **78** (1980), 33–5.
2. Admissible pairs and integral equations, *J. Math. An. Appl.* **121** (1987), 79–90.

H. J. Reinhardt

1. *Analysis of Approximations Methods for Differential and Integral Equations*, Springer-Verlag, 1985.

R. Reissig, G. Sansone, R. Conti

1. *Nichtlineare Differentialgleichungen Höherer Ordnung*, Noordhoff Int. Publ., Leyden, 1974.

M. Renardy, W. J. Hrusa, J. A. Nohel

1. *Mathematical Problems in Viscoelasticity*, Longman Science Technology, New York, 1987.

P. Renno

1. On the Cauchy problem in linear viscoelasticity. *Rend. Accad. Naz. Lincei* **75** (8) (1983), 195–204.

H. J. J. te Riele

1. (ed.) *Colloquium: Numerical Treatment of Integral Equations*, Math. Centrum, Amsterdam, 1979.

F. Riesz, B. Sz.-Nagy

1. *Lécons d'Analyse Fonctionnelle*, Akad. Kiado, Budapest, 1952.

R. T. Rockafellar
1. Convex functions, monotone operators and variational inequalities. *Proceedings NATO Institute, Venice*, 1968, Oderisi Gubio.

V. G. Romanov
1. *Inverse Problems of Mathematical Physics*, VNU Science Press, Utrecht, The Netherlands, 1987.

E. Rouché
1. Mémoire sur le calcul inverse des intégrales définies. *C. R. hebd Acad. Sci., Paris* **151** (1860), 126–8.

J. Le Roux
1. Sur les intégrales des équations linéaires. *An. Ec. Norm. Sup. Paris* **12** (1895), 227–316.

J. L. Rubio de Francia, F. J. Ruiz, J. L. Torrea
1. Calderon-Zygmund theory for operator-valued kernels, *Adv. Math.* **62** (1986), 7–48.

W. J. Rugh
1. *Nonlinear System Theory: The Volterra/Wiener Approach*, Johns Hopkins University Press, Baltimore, MD, 1981.

D. L. Russell
1. A Floquet decomposition for Volterra equations with periodic kernel and a transfer approach to linear recursion equations, *J. Diff. Equations* **68** (1987), 41–71.

A. F. Ruston
1. *Fredholm Theory in Banach Spaces*, Cambridge University Press, 1986.

B. Rzepecki
1. On some classes of Volterra differential-integral equations, *An. St. Univ. 'Al. I. Cuza', Iasi*, Ia **29** (1983), 61–8.
2. On a quasilinear Volterra integral equation, *Demonstr. Math.* **17** (1984), 1003–9.

Th. Saaty
1. *Modern Nonlinear Equations*, Dover Publ., New York, 1981.

C. Sadosky
1. *Interpolation of Operators and Singular Integrals*, Marcel Dekker, New York, 1979.

Y. Sakawa
1. Solution of an optimal control problem in a distributed parameter system, *IEEE Trans. CAS* **9** (1964), 420–42.

P. A. Samuelson
1. *Les fondements de l'analyse économique, Vol. 2: Stabilité des systemes et théorie dynamique*, Gauthier-Villars, Paris, 1971.

I. W. Sandberg
1. Criteria for the response of nonlinear systems to be *L*-asymptotically periodic, *The Bell Syst. Techn. J.* **60** (1981), 2359–71.
2. Expansions for nonlinear systems, *The Bell Syst. Techn. J.* **61** (1982), 159–99.
3. Volterra expansions for time-varying nonlinear systems, *The Bell Syst. Techn. J.* **61** (1982), 201–25.
4. On Volterra expansions for time-varying nonlinear systems, *IEEE Trans. CAS* **30** (1983), 61–7.
5. Volterra-like expansions for solutions of nonlinear integral equations and nonlinear differential equations, *IEEE Trans. CAS* **30** (1983), 68–77.
6. The mathematical foundations of associated expansions for mildly nonlinear systems, *IEEE Trans. CAS* **30** (1983), 441–55.
7. Multilinear maps and uniform boundedness, *IEEE Trans. CAS* **32** (1985), 332–6.
8. On causality and linear maps, *IEEE Trans. CAS* **32** (1985), 392–3.

9. Nonlinear input-output maps and approximate representation, *AT&T Techn J.* **64** (1985), 1967–78.

I. W. Sandberg, J. B. Allen
1. Almost periodic response determination for models of the basilar membrane, *AT&T Techn. J.* **64** (1985), 1775–86.

T. Sato
1. Sur l'équation intégrale non-lineaire de Volterra, *Compositio. Math.* **11** (1953), 271–90.

K. Sawano
1. Some considerations on the fundamental theorems for functional differential equations with infinite delay, *Funk. Ekv.* **25** (1982), 97–104.

W. Schachermayer
1. Integral operators on L^p-spaces I, II, Addendum, *Indiana Univ. Math. J.* **30** (1981), 123–140, 261–6; **31** (1982), 73–81.

J. J. Schäffer
1. Linear differential equations with delays: Admissibility and conditional stability II, *J. Diff. Equations* **10** (1971), 471–84.

A. R. Schep
1. Kernel operators, *Indag. Math.* **41** (1979), 39–53.
2. Compactness properties of an operator which imply that it is an integral operator, *Trans. AMS* **265** (1981), 111–19.
3. Compactness properties of Carleman and Hille-Tamarkin operators, *Can. J. Math.* **37** (1985), 921–33.

M. Schetzen
1. *The Volterra and Wiener Theories of Nonlinear Systems,* John Wiley, New York, 1980.

O. Schlömilch
1. *Analytische Studien,* Greifswald, 1848.

W. Schmeidler
1. *Integralgleichungen mit Anwendungen in Physik und Technik,* Akademisches Verlagsgesellschaft, Leipzig, 1950.

E. Schmidt
1. Zur Theorie der lineareu und nichtlinearen Integralgleichungen I, II, III, *Math. Annalen,* **63** (1907), 433–79; **64** (1907), 161–74; **65** (1908), 370–99.

K. Schumacher
1. On the resolvent of linear nonautonomous partial functional differential equations, *J. Diff. Equations* **59** (1985), 355–87.

St. Schwabik, M. Tvrdy, O. Vejvoda
1. *Differential and Integral Equations,* D. Reidel Publ. Co., Dordrecht, 1979.

L. E. Shaikhet
1. On the optimal control of integral-functional equations, *Prikl. Mat. Mekh. USSR* **49** (1985), 704–12.

L. F. Shampine
1. Solving Volterra integral equations with ODE codes, *IMA J. Num. An.* **8** (1988), 37–41.

E. Sinestrari
1. On the abstract Cauchy problem for parabolic type in spaces of continuous functions, *J. Math. An. Appl.* **107** (1985), 16–66.

I. H. Sloan, A. Spence
1. Integral equations on the half-line: A modified finite-section approximation, *Math. Comp.* **47** (1986), 589–95.

H. B. Smets
1. *Problems in Nuclear Power Reactor Stability*, Presses Université Bruxelles, Bruxelles, 1961.

F. Smithies
1. *Integral Equations*, Cambridge University Press, 1958.

S. L. Sobolev
1. *Partial Differential Equations of Mathematical Physics*, Pergamon Press, Oxford, 1964.

N. Sonine
1. Sur la géneralisation d'une formule d'Abel, *Acta Math.* 4 (1884), 171–6.

A. R. Sourour
1. Pseudo-integral operators, *Trans. AMS* 253 (1979), 339–63.

A. Srazidinov
1. On the asymptotics of solutions and their regularization for linear and nonlinear Volterra equations of the first kind (Russian), *Issled. Integro-Diff. Urav* 14 (1979), 249–53.

H. M. Srivastava, R. G. Buschman
1. *Convolution Integral Equations*, John Wiley, New York, 1977.

D. F. St. Mary
1. Riccati integral equations and nonoscillation of self-adjoint linear systems, *J. Math. An. Appl.* 121 (1987), 109–18.

I. Stakgold
1. *Green's Function and Boundary Value Problem*, Interscience Publ. (Wiley), New York, 1979.

O. J. Staffans
1. Nonlinear Volterra integral equations with positive definite kernels, *Proc. AMS* 51 (1975), 103–8.
2. Tauberian theorems for a positive definite form, with applications to Volterra equations, *Trans. AMS* 218 (1976), 239–59.
3. On the asymptotic spectrum of a convolution, *SIAM J. Math. An.* 10 (1979), 1138–43.
4. Convergence theorems for semilinear Volterra equations with infinite delay, *J. Int. Equations* 8 (1985), 261–74.
5. On a nonconvolution Volterra resolvent, *J. Math. An. Appl.* 108 (1985), 15–30.
6. Extended initial and forcing function semigroups generated by a functional equations, *SIAM J. Math. An.* 16 (1985), 1034–48.
7. On the almost periodicity of the solutions of an integrodifferential equation, *J. Int. Equations* 8 (1985), 249–60.
8. A direct Liapunov approach to Volterra integrodifferential equations, Helsinki University of Technology Institute of Mathematics, Research Report A244 (1987), 25 pp.

V. S. Sunder
1. Unitary equivalence to integral operators, *Pacific J. Math.* 92 (1981), 211–15.

M. Svec
1. Equivalence of Volterra integral equations (Russian. & Czech.) *Casopis Pest. Mat.* 11 (1986) 185–202.

S. Szeptycki
1. Domain of integral transformations on general measure spaces, *Math. Annalen* 242 (1979), 267–71.

S. Szufla
1. On Volterra integral equations in Banach spaces, *Funk. Ekv.* 20 (1977), 247–58.

2. On the Hammerstein integral equation in Banach spaces, *Math. Nachr.* **124** (1985), 7–14.

J. D. Tamarkin
1. On integrable solutions of Abel's integral equation, *Ann. Math.* **31** (2) (1930), 219–29.

H. Tanabe
1. *Equations of Evolution*, Pitman, London, 1979.

H. R. Thieme
1. Density-dependent regulation of spatially distributed populations and their asymptotic speed, *J. Math. Biol.* **8** (1979), 173–87.
2. Asymptotic estimates of the solutions of nonlinear integral equations and asymptotic speeds for the spread of populations, *J. Reine Angew. Math.* **306** (1979), 94–121.
3. On a class of Hammerstein integral equations, *Manuscr. Math.* **29** (1979), 49–84.
4. On the boundedness and the asymptotic behavior of the nonnegative solutions of Volterra-Hammerstein integral equations, *Manuscr. Math.* **31** (1980), 379–412.

L. Tonelli
1. Su un problema di Abel, *Math. Ann.* **99** (1928), 183–99.
2. Sulle equazioni funzionali di Volterra, *Bull. Calcula. Math. Soc.* **20** (1929), 31–48 (Opere scelte 4, 198–212).

F. Tricomi
1. *Integral Equations*, Interscience Publ. (Wiley), New York, 1957.

C. P. Tsokos, W. J. Padgett
1. *Random Integral Equations with Applications to Life Sciences and Engineering*, Academic Press, New York, 1974.

C. Tudor
1. On Volterra stochastic equations, *Boll. UMI* **5-A** (6) (1986), 335–44.
2. On weak solutions of Volterra stochastic equations, *Boll. UMI* **1-B** (7) (1987), 1033–54.

M. Turinici
1. Coincidence points and applications to Volterra functional equations, *An. St. Univ. 'Al. I. Cuza', Iasi*, Ia **23** (1977), 51–7.
2. Abstract monotone mappings and applications to functional differential equations, *Rend. Accad. Naz. Lincei, Sc. fis., mat. nat.* **66** (1979), 189–93.
3. Volterra functional equations via projective techniques, *J. Math. An. Appl.* **103** (1984), 211–29.
4. Abstract comparison principles and multivariable Gronwall-Bellman inequalities, *J. Math. An. Appl.* **117** (1986), 100–27.

A. N. Tychonoff
1. Sur les équations functionnelles de Volterra et leurs applications a certains problemes de la Physique Mathématique, *Bull. Univ. Moscow (Serie Intern.)* Al No. 8 (1938), 25 pp.

V. A. Tyshkevich
1. *Some Problems of the Stability Theory of Functional Differential Equations* (Russian), Naukowa Dumka, Kiev, 1981.

M. M. Vainberg
1. *Variational Method and Method of Monotone Operators in the Theory of Nonlinear Equations*, John Wiley, 1973.

A. F. Verlan', V. C. Sizikov
1. *Integral Equations: Methods, Algorithms, Programs* (Russian), Naukova Dumka, Kiev, 1986.

H. D. Victory, Jr
1. On linear integral operators with nonnegative kernels, *J. Math. An. Appl.* **89** (1982), 420–41.
V. R. Vinokurov
1. Optimal control of processes described by integral equations I, II, III, *SIAM J. Control* **7** (1969), 324–55.
V. R. Vinokurov, Yu. N. Smolin
1. On the asymptotics of Volterra equations with almost periodic kernels and lags, *Soviet Math. Dok.* **12** (1971), 1704–17.
2. Some properties of the fundamental matrix of a system of Volterra integrodifferential equations, *Yzv. Vysch. Ucheb. Zaved. Mat.*, No. 9 (1982), 77–80.
A. Visintin
1. A model for hysteresis of distributed systems, *Annali. Mat. Pura. Appl.* **131** (1982), 203–31.
R. Vittal Rao, N. Sukavanam
1. Spectral analysis of finite section of normal integral operators, *J. Math. An. Appl.* **115** (1986), 23–45.
G. Vivanti
1. *Elementi della teoria delle equazioni integrali lineari*, Milano, U. Hoepli, 1916.
V. Volterra
1. *Leçons sur les équations intégrales et les équations integro-différentielles*, Gauthier-Villars, Paris, 1913.
2. *Opere Matematiche*, Vols 1, 2, 3, Accad. Naz. Lincei, Cons. Naz. Ricerche, Roma, 1954–1955.
V. Volterra, U. d'Ancona
1. *Les Associations Biologiques au Point de Vue Mathématique*, Hermann, Paris, 1935.
I. I. Vrabie
1. *Compactness Methods for Nonlinear Evolutions*, Pitman Publ. Co., London, 1987.
W. Walter
1. *Differential and Integral Inequalities*, Spring-Verlag, 1970.
J. Warga
1. *Optimal Control of Differential and Functional Equations*, Academic Press, New York, 1972.
2. Controllability, Externality, and Abnormality in Nonsmooth Optimal Control, *J. Optimization Theory Appl.* **A1** (1983), 239–60.
G. F. Webb
1. Volterra integral equations and nonlinear semigroups, *Nonlinear Analysis, TMA* **1** (1977), 415–27.
2. Volterra integral equations as functional differential equations on infinite intervals, *Hiroshima Math. J.* **7** (1977), 61–70.
3. An abstract second order semilinear Volterra integrodifferential equation, *Proc. AMS* **69** (1978), 255–60.
4. *Theory of Nonlinear Age-dependent Population Dynamics*, M. Dekker, New York, 1985.
L. Weis
1. Integral operators and changes of density, *Indiana Univ. Math. J.* **31** (1982), 83–96.
H. Weyl
1. Singuläre Integralgleichungen, *Math. Annalen* **66** (1909), 273–324.
G. Wiarda
1. *Integralgleichungen unter besonderer Berücksichtigung der Anwendungen*, Teubner, Leipziz-Berlin, 1930.

H. Widom
1. Asymptotic behavior of the eigenvalues of certain integral equations, *Trans. AMS* **109** (1963), 278–95.
2. Asymptotic behavior of the eigenvalues of certain integral equations II, *Arch. Rat. Mech. Analysis* **17** (1964), 215–29.
3. *Lectures on Integral Equations*, Van Nostrand Reinhold, New York, 1969.

N. Wiener
1. *Nonlinear Problems in Random Theory*, MIT Press, Cambridge, MA, 1958.

J. C. Williams
1. *The Analysis of Feedback Systems*, Research Monograph No. 62, MIT Press, Cambridge, MA, 1971.

C. C. Yeh
1. Bellman–Bihari integral inequalities in several independent variables, *J. Math. An. Appl.* **87** (1982), 311–21.

K. Yosida
1. *Functional Analysis* (5th edition), Springer-Verlag, Berlin, 1978.

A. C. Zaanen
1. *Integration*, North Holland Publ. Co., Amsterdam, 1967.

P. P. Zabreyko
1. Nonlinear integral operators (Russian), *Trudy Sem. Funk. Analizu Voronez SU*, **8** (1966), 3–148.
2. On the spectral radius of Volterra operators (Russian), *Usp. Mat. Nauk.* **22** (1967), 167–8.

P. P. Zabreyko, A. Z. Koshelev, M. A. Krasnoselskii, S. G. Mikhlin, L. S. Rakovshchik, V. Ya. Stetsenko
1. *Integral Equations–A Reference Tet*, Noordhoff Int. Publ., Leyden, 1975.

P. P. Zabreyko, N. L. Mayorova
1. On the solvability of the nonlinear integral equation of Urysohn (Russian), Qualit. and Approx. Methods for the Invest. of Operator Equations, Yaroslav. Gos. Univ., 1978, pp. 61–73.

E. Zeidler
1. *Nonlinear Functional Analysis and its Applications*, Vol. 1 (*Fixed-Point Theorem*), Springer-Verlag, Berlin, 1986.

L. A. Zhivotovskii
1. *Existence Theorems and Classes of Uniqueness for Solutions of Integro-Functional Equations of Volterra*, Differential Equations (Transl.), Vol. 7, 1971, pp. 1043–9.

INDEX